S0-AGF-862

Fatigue and Fracture in Steel Bridges

Fatigue and Fracture in Steel Bridges

CASE STUDIES

JOHN W. FISHER

Professor of Civil Engineering
Fritz Engineering Laboratory
Lehigh University

A WILEY-INTERSCIENCE PUBLICATION

JOHN WILEY & SONS

New York · Chichester · Brisbane · Toronto · Singapore

Library of Congress Cataloging in Publication Data:

Fisher, John W., 1931–
Fatigue and fracture in steel bridges.
"A Wiley-Interscience publication."
Includes index.
1. Bridges, Iron and steel. 2. Steel, Structural—Fatigue. 3. Steel, Structural—Fracture. I. Title.
TG380.F57 1984 624′.252 83-23495
ISBN 0-471-80469-X

Printed in the United States of America

10 9 8 7 6 5 4 3 2 1

To Nelda, Tim, Chris, Beth and Nevan

Preface

Since 1967 a number of highway and railroad bridge structures in the United States and Canada have experienced fatigue cracking which sometimes resulted in brittle fracture as a result of service loading. This book provides a detailed review and summary of 22 case studies of bridges that have experienced crack growth.

My objective in preparing this book is to make the reader aware of the many types of cracks that have developed in bridge structures under service loads, examine the reasons for their occurrence, and provide a fracture mechanics evaluation in order to relate the parameters of crack size, stress, detail geometry, crack propagation, and material toughness. The individual cases provide valuable insight into the causes of cracking, the significance of details, and the significance of defects on the performance of cyclic loaded structures. All of the causes of cracking discussed in this book can be prevented with the engineering knowledge and other tools available today. The lessons learned from the past should assist with an understanding of the behavior of structures and the importance of detail and execution. They assist toward the objective of enhanced reliability for structural systems subjected to cyclic loads.

The book is divided into two parts. The first deals with cracks that have formed as a result of low fatigue resistant details or large initial discontinuities. The large discontinuities often resulted because attachments were not considered as important as a groove welded tension flange. The welded joints that were used did not have adequate quality control requirements imposed on the welded joints. In several cases the structural detail created a cracklike geometric condition that was not recognized.

The second part of the book deals with fatigue cracks that form as a result of unanticipated secondary or displacement-induced stresses. These cases of cracking have developed in many types of structures with great frequency. They have generally formed in small web gaps between attachments and the girder flanges. The interaction between the main longitudinal members and the transverse framing such as cross-frames, diaphragms, and floor beams has resulted in out-of-plane distortion in the web gap that was not anticipated. Out-of-plane distortion has developed from handling and shipping of individual members, as well as the interaction between intersecting elements under everyday traffic. A large number of fatigue cracks generally form in a given structure because many small gaps normally exist.

Acknowledgment is due the U.S. Department of Transportation Federal Highway Administration for sponsoring a survey of localized failures due to cracking. A number of the examples given in this book resulted from information acquired during that study. Other cases were acquired over the years as a result of studies and evaluations that were carried out to determine the causes of cracking and to provide recommendations for the repair and retrofit of the damaged structure.

I am also indebted to the following individuals and organizations for their support and assistance on a variety of cases: C. F. Scheffey, C. F. Galambos, J. Nishanian, B. Brakke, and F. D. Sears—Federal Highway Administration; R. E. Cassano—California Department of Transportation; L. Koncza—Chicago Bridge Department; R. A. Norton and J. Cavanaugh—Connecticut Department of Transportation; C. E. Thunman—Illinois Department of Transportation; C. Pestotnic—Iowa Department of Transportation; L. Garrido—Louisiana Department of Transportation; D. Carpenter—Maryland Department of Transportation; K. V. Benthin—Minnesota Department of Highways; W. Sunderland—New Jersey Department of Transportation; R. B. Pfeifer—Ohio Department of Transportation; W. S. Hart—Oregon Department of Transportation; B. F. Kotalik—Pennsylvania Department of Transportation; K. C. Wilson—South Dakota Department of Transportation; W. Henneberger—Texas Department of Transportation; S. Gloyd—Washington Department of Transportation; W. A. Kline and S. Wood—Wisconsin Department of Transportation; P. F. Csagoly and A. Radkowski—Ontario Ministry of Transportation and Communications; Z. L. Szeliski and R. A. P. Sweeney—Canadian National Railroad; Delaware River Port Authority; Ammann and Whitney; DeLeuw-Cather; H. W. Lockner Inc.; Modjeski and Masters; Richardson Gordon and Associates and Wiss, Janney, Elstner and Associates.

Special thanks are due my colleagues Professors G. R. Irwin, A. W. Pense, R. Roberts, and B. T. Yen for their suggestions and assistance on a number of the case studies. Thanks are also due K. H. Frank, H. Hausammann, H. T. Sutherland, and D. R. Mertz for their assistance and help with a number of these studies.

The final manuscript was prepared while I was a Visiting Professor at the Institut de la Construction Métallique (ICOM), École Polytechnique Fédérale de Lausanne, Switzerland. I am particularly indebted to Professor J. C. Badoux for his support and assistance. The manuscript was in part typed at ICOM and in part by my secretary, Ruth Grimes, at Lehigh University. I appreciate the care and exactness that was taken. Thanks are also due R. N. Sopko, photographer at Fritz Engineering Laboratory, for the many photographs that are included in this book. I am also indebted to those who provided original prints of previously published photographs.

I am also grateful for the understanding and support provided by my wife, Nelda, and my children, J. Timothy, Christopher, Elizabeth, and Nevan, while I worked on this project.

JOHN W. FISHER

Bethlehem, Pennsylvania
March 1984

Contents

PART II SECONDARY STRESSES AND DISTORTION-INDUCED STRESS

Abbreviations and Symbols

ABBREVIATIONS

AASHO	American Association of State Highway Officials
AASHTO	American Association of State Highway and Transportation Officials
ADT	Average daily vehicle traffic
ADTT	Average daily truck traffic
AISC	American Institute of Steel Construction
AREA	American Railway Engineering Association
ASTM	American Society for Testing and Materials
CVN	Charpy V-notch impact test, or pendulum energy loss measured in that test

SYMBOLS

A	Proportionality factor in relationship between cycles, N, and stress range
a	Half-length of a central (through-the-thickness) crack, length of an edge (through-the-thickness) crack, depth of a surface crack
$\frac{a}{c}$	Depth to surface-half-length ratio for a surface crack with the crack front shape approximated as half of an ellipse
C	Proportionality factor in relationship between fatigue crack growth rate and ΔK^n
c	Surface half-length of a surface crack
D	Weld leg size
$\frac{da}{dN}$	Fatigue crack growth rate

$E(k)$	Elliptical integral $= \int_0^{\pi/2} (1 - k^2 \cos u)^{1/2}\, du$ where $k = 1 - (a/c)^2$
$F_{(i)}$	Correction factors on $\sigma\sqrt{\pi a}$ to assist K value estimates
F_e	Related to a/c ratio for half elliptical shape
F_g	Related to stress gradient across the plane of the crack
F_s	Related to the free surface influence for a surface crack
F_w	Related to finite plate width relative to crack depth or crack length
H	Weld leg size
K	Stress intensity factor
K_c	Generic symbol for estimate of the opening mode K near the onset of rapid crack extension. For bridge steels, unless otherwise stated, the loading time is approximately one second.
K_{Ic}	Value of K_c for crack front stress state corresponding nearly to plane-strain
K_{Id}	Value of K_{Ic} for impact loading (a loading time less than 2 msec)
K_Q	An estimate of K_c or K_{Ic} from a fracture test in which allowance must be made for plastic yielding of the net section
ΔK	Maximum K value minus the minimum K value during a stress fluctuation
K_{tm} (or SCF)	Stress concentration factor, for example, at a notch root due to weldment contour
L	Length of web gap
l	Distance between point load and point on crack tip
M	Bending moment
N (or N_T)	Total number of fatigue cycles
N_i	Number of cycles corresponding to fatigue strength at stress range level S_{ri}
n_i	Number of cycles experienced at stress range level S_{ri}
n	Exponent for the crack growth relationship and for the stress range–cycle life relationship
P_{rs}	Point load residual stress force equal to product $\sigma_{rs} \cdot dA$
r_y	Plastic zone size radius
S_r	Stress range equal to the maximum stress minus the minimum stress during a stress fluctuation

S_r^D	Design stress range
S_{ri}	Stress range level i
$S_{r\text{Miner}}$	Effective stress range, calculated using Miner's rule and used to estimate effective ΔK values for variable amplitude loading
T	Temperature
T_s (or T-shift)	Estimated temperature shift of K_{Ic} versus T due to change of loading time
t	Thickness
t_{cp}	Cover plate thickness
t_f	Flange thickness
t_w	Web plate thickness
t'_w	Effective web plate thickness
t_p	Plate thickness for load-carrying cruciform joint
α	Reduction factor to correct for differences in calculated and measured stress
α_i	Frequency of occurrence of cyclic stress at stress range level S_{ri}, equal to n_i/N_T
Δ	Relative out-of-plane deflection in web gap
δ	Relative deflection between adjacent beams
θ	Rotation
ϕ	Angle between the major axis of an elliptical-shaped crack and a point on the crack front
ϕ_i	Ratio of gross vehicle weight GVW_i to design vehicle weight
σ	The nominal tensile fiber (or section) stress as commonly used for design purposes
σ_{rs}	Residual stress from welding
σ_Y	Tensile yield strength
$\Delta\sigma$	Cyclic stress range equal to S_r

CHAPTER 1

Introduction

1.1 GENERAL OVERVIEW

Over the past two decades a number of localized failures developed in components of steel bridges, due to fatigue and brittle fracture. Where there were rapid cleavage separations, these were with few exceptions preceded by fatigue crack propagation on the crack surface. A survey [1.1] carried out between 1978 and 1981 including about 20 states and Ontario, Canada, amassed information on cracking that had developed at 142 bridge sites. Often several types of cracking were found at a single site, occurring in different details of a structure.

Table 1.1 provides a summary of the types of design details, identified by member or type of connection, at which cracks have developed in steel bridges. Listed in Column 2 are descriptions of the cracking. Twenty-eight general categories of cracking appear. This column indicates whether or not a large defect initiated the cracking or identifies the critical design condition. Often more than one factor contributed to the cracking, and each is identified. Column 3 gives the number of bridge sites at which the identified condition and cracking had developed. Usually there was more than one bridge at a given site. The last column provides the specification detail classification if no other unusual condition was observed. The classification system referred to here is one in general use in the various specifications of U.S. engineering associations such as the AASHTO Bridge Specification, the AREA Specifications for Steel Railway Bridges, and the AISC Specification for Steel Buildings. Where applicable a preexisting crack or an out-of-plane distortion that led to fatigue is indicated instead. Large cracks and distortion-induced cracking provide conditions that are not accounted for by the design provisions.

Table 1.1 shows that at least 60 bridge sites developed fatigue cracks as a result of out-of-plane distortion in a small gap. Most often this in-

Table 1.1 Summary of Types of Details Experiencing Cracking

Detail	Initial Defect or Condition	Number of Bridges	Fatigue Category
1. Eyebars	Stress corrosion	1	
	Forge laps, unknown defects	12	Initial crack
2. Pin plates	Frozen pins	1	Out of plane
	Other	1	D
3. Cover-plated beams	Normal weld toe	3	E'
	Fabrication cracks	1	$<E'$
4. Flange gussets	Welds toe	5	E or E'
5. Flange or web groove	Lack of fusion	6	Large initial crack
6. Coverplate groove	Lack of fusion	4	Large initial crack
7. Electroslag welds	Various flaws	6	
8. Longitudinal stiffeners	Lack of fusion, poor weld	4	Large initial crack
9. Web gusset	Intersecting welds	5	$<E'$
	Gap between stiffener and gusset	2	Out of plane
10. Flanges and brackets through web	Flange tip crack	3	$<E'$
11. Welded holes	Lack of fusion	3	Large initial cracks
12. Cantilever brackets	Tack welds	1	Out of plane
	Riveted connection	2	Out of plane
13. Lamellar tearing	Restraint	1	
14. Transverse stiffeners	Shipping and handling	4	Out of plane
15. Floor-beam connection plates	Web gaps	26	Out of plane
16. Diaphragm connection plates	Web gaps	9	Out of plane
17. Diaphragm and floor-beam connection plates at piers	Web gaps	4	Restraint
18. Tied arch floor beams	Web gaps	8	Out of plane
19. Tied arch floor-beam connection	Weld root	2	Restraint
20. Coped members	Flame-cut notch	13	Restraint
21. Welded web inserts	Lack of fusion	1	Large initial crack
22. Plug welds	Crack	1	Large initial crack
23. Gusset plates	Lateral bracing vibration	3	Out of plane

Table 1.1 (*Continued*)

Detail	Initial Defect or Condition	Number of Bridges	Fatigue Category
24. Box Girder corner welds	Transverse weld cold cracks	4	Large initial crack
25. Stringer–floor-beam brackets	Web gaps	4	Out of plane
26. Stringer end connections	Restraint	3	Weld termination
27. Hangers (truss and arches)	Vibration—wind	4	Aeroelastic instability
28. Welded repair	Lack of fusion, weld termination	2	$<E'$

volved a segment of a girder web. When distortion-induced cracking develops in a bridge, large numbers of cracks usually form before corrective action is taken. The cyclic stress amplitudes in small gaps caused by distortion are often very high, and therefore many cracks form simultaneously in the structural system. However, a low fatigue strength detail or a large built-in defect may produce only a single significant crack. Other potential crack locations can then be identified and retrofitted before significant damage develops elsewhere.

Displacement-induced fatigue cracking has developed in a wide variety of bridge structures. Among these are suspension bridges, two-girder floor-beam bridges, multiple beam bridges, tied-arch bridges, and box girder bridges. Cracks form initially in planes parallel to the tensile stresses considered in the structure's design. These cracks being parallel to the tensile stresses may not be detrimental to the performance of the structure provided they were discovered and retrofitted before turning perpendicular to the tensile stresses expected from loads. In some structures the cracks arrested in low stress areas served to relieve the restraint condition.

Large initial defects or cracks make up the next largest category of cracked members and components summarized in Table 1.1. In several cases the defect was due to poor quality welds which were produced before nondestructive test methods were well developed. A larger number of these cracks occurred because the groove welded component was considered a secondary member or attachment, so no weld quality criteria were established nor were nondestructive test requirements imposed on the affected weldment. Splices in continuous longitudinal stiffeners are a common condition that falls into this category. A related condition has been found where backup bars were used to make a groove weld between transverse stiffeners and a lateral gusset plate. Often lack of fusion adjacent to the girder web in the transverse groove welds led to the cracking. The transverse welds that intersect with the longitudinal welds provide a path for a crack to enter the girder web.

A related lack-of-fusion type of defect (and crack) that has been among the most severe encountered occurred at details where a plate component was inserted through an opening cut into girder webs. The resulting joint was usually welded into place with either fillet or groove welds. In either case there were large cracks at the edge of the flange plate where short vertical weld lengths left large unfused areas. Cracklike defects also resulted from weld-filled holes and the use of plug-welded slots or holes. Most of the remaining conditions summarized in Table 1.1 resulted from low fatigue strength details that were not anticipated to have such a low fatigue resistance at the time of the original design.

A representative sample of these structures were selected for a detailed study of the causes of cracking. This book provides a description of each structure, a summary of the cracking and known characteristics of the material and crack surface, and an analysis of the cracking, using fracture mechanics models of the cracks to evaluate and assess the fatigue and fracture behavior of the details. The study also provides a discussion and review of the repair and retrofit scheme that was used to restore the cracked section and prevent its reoccurrence elsewhere.

1.2 FRACTURE MECHANICS OF CRACK GROWTH

The fracture mechanics of both stable and unstable crack growth provides a useful means of analyzing the crack condition. The analysis of cracking in steel bridges can usually be accomplished by using linear-elastic fracture mechanics. With this method the stresses very close to the crack front, which cause crack extension, are treated as proportional in a fixed known way to a single parameter, K, termed the stress intensity factor. The size, shape, and orientation of the crack plays a major role in determining the applicable K value. Knowledge from experiments of K values, termed K_c, which are large enough to cause onset of rapid fracturing, and knowledge of the expected fatigue crack growth rate in terms of fluctuations of K, termed ΔK, assist our understanding of cracking in steel bridges.

Several books provide a good introductory discussion of fracture mechanics concepts and their development; see, for example, [1.7], [1.9], and [1.10]. The reader is urged to refer to one or more of these for general background on the subject and its application to cracks.

For complex details such as those in common use in welded structures, the stress intensity factor, K, for a surface crack of depth, a (see Figure

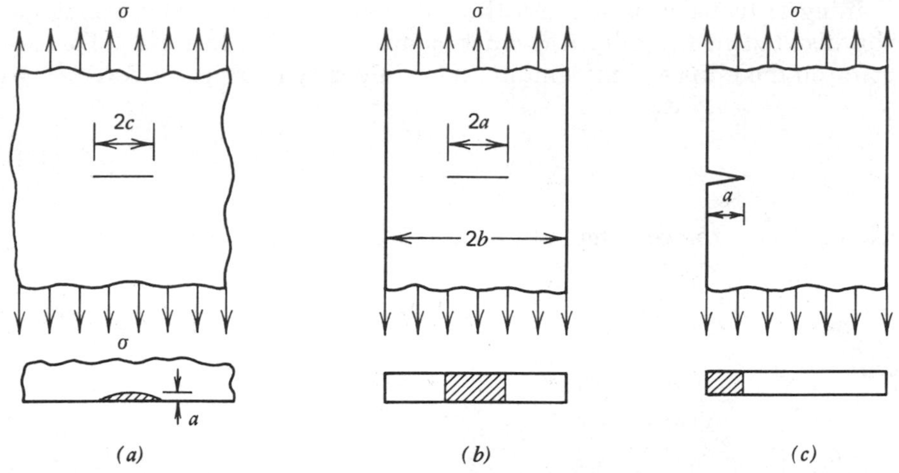

Figure 1.1 Idealized crack conditions. (*a*), Surface crack; (*b*), through crack; and (*c*), edge crack.

1.1a), can be conveniently related to the well-known expression for a central through crack in an infinite plate by use of correction factors [1.2, 1.3, 1.4]. The resulting generalized stress intensity factor is expressed

$$K = F_e \cdot F_s \cdot F_w \cdot F_g \cdot \sigma\sqrt{\pi a} \tag{1.1}$$

These correction factors modify $\sigma\sqrt{\pi a}$ (for the idealized case) to account for effects of free surface F_s, the finite width (or thickness) F_w, nonuniform stresses acting on the crack F_g, and the crack shape F_e. To evaluate fracture instability, the total sum of stresses due to residual welding or rolling stresses, dead load, and live loads must be considered. For cyclical fatigue loading due to traffic, $\Delta\sigma$ is the live load variation in stress which results in a ΔK stress intensity value range.

Numerous solutions for the correction factors F_s, F_w, F_g, and F_e, both empirical and exact, are to be found in the literature [1.2]. A few of these have received frequent usage. A free surface correction of

$$F_s = 1.211 - 0.186\sqrt{\frac{a}{c}} \tag{1.2}$$

is employed for a semicircular crack in a semi-infinite plate subjected to uniform stress [1.2]. For a central crack in a plate of finite width shown in Figure 1.1b, the function

$$F_w = \sqrt{\sec\frac{\pi a}{2b}} \tag{1.3}$$

has an accuracy of 0.3 percent for an a/b ratio less than 0.7 [1.2].

Integral transformation of a three-dimensional elliptical crack shape has resulted in the elliptical crack shape correction factor F_e. For the point on the ellipse of maximum stress intensity [1.5] its value is

$$F_e = \frac{1}{E(k)} \tag{1.4}$$

where $E(k)$ is the complete elliptical integral of the second kind:

$$E(k) = \int_0^{\pi/2} [1 - k^2 \sin^2 \theta]^{1/2} \, d\theta \tag{1.5}$$

where

$$k^2 = \frac{c^2 - a^2}{c^2}$$

Equation 1.5 is dependent only on the minor to major axis semidiameter ratio a/c.

Relationships between the minor axis semidiameter and major axis semidiameter have been empirically determined for different structural detail geometries and are presented in Figure 1.2. Also shown are experimental observations on fatigue-cracked stiffener and cover-plated beam details. Use of these relationships with Eq. 1.5 results in a crack shape correction factor, F_e, as a function of the crack size a.

Expressions for the stress gradient correction factor F_g can be very complex. Numerical results often require a procedure involving first determination of the stress field with finite elements in the uncracked structure and then removal of these stresses from the crack surface by integration. An outline for this procedure is given in [1.4].

Approximate equations for the stress gradient correction factor, F_g, have been derived for several details. One approximation that appears applicable to a number of structural details such as stiffeners, attachments, cover plates, and gusset plates has the form [3.9]:

$$F_g = \frac{K_{tm}}{1 + G\alpha^{\beta}} \tag{1.6}$$

where G and β are dimensionless constants, α = ratio of crack size to plate thickness, a/t, and K_{tm} is the maximum stress concentration factor at the weld toe.

An approximate equation for the stress gradient correction factor, F_g, at the toe of an end-welded cover-plated beam has the value [3.9]:

$$F_g = \frac{K_{tm}}{1 + 6.789(a/t_f)^{0.4348}} \tag{1.7}$$

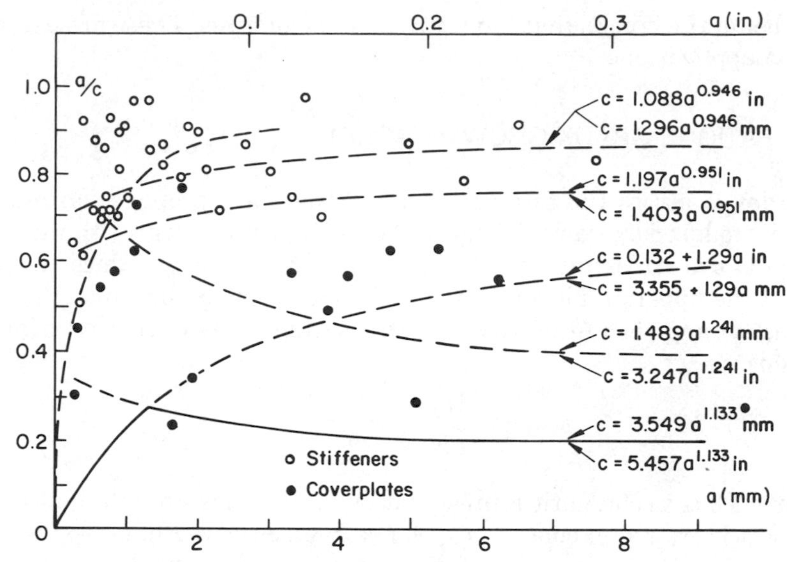

Figure 1.2 Crack shape measurements and empirical relationships.

where K_{tm} = stress concentration factor
a = crack depth
t_f = flange thickness

The stress concentration factor at the toe of the weld for the uncracked section is approximated by the following equation [3.9]:

$$K_{tm} = -3.539 \ln\left(\frac{Z}{t_f}\right) + 1.981 \ln\left(\frac{t_{cp}}{t_f}\right) + 5.798 \tag{1.8}$$

where Z = weld leg size
t_{cp} = cover plate thickness
t_f = flange thickness

Cracks growing at weld toes tend to form at numerous sites along the transverse width perpendicular to the applied stresses. These cracks start to coalesce and form a common crack front very early in the crack growth process. Figure 1.2 shows crack shape measurements and several empirical relationships that have been used to account for the crack geometry. For cover-plated beams a lower bound is provided by

$$c = 5.457a^{1.133} \quad \text{(in.)} \tag{1.9}$$

Other relationships for the stress gradient correction and crack shape are used for other welded details when evaluating the fatigue behavior of a number of the cracks that have formed in structures. These are described in the applications.

1.3 FATIGUE CRACK GROWTH MODEL

In order to assess the fatigue behavior, the crack propagation relationship introduced by Paris [1.6] is used to relate the crack growth rate to the range of the stress intensity factor, $\Delta K = K_{\max} - K_{\min}$. Since the crack size at the upper and lower limits of the load cycle is the same, the stress intensity range is a function of the stress range. The Paris power law has the form

$$\frac{da}{dN} = C\Delta K^n \tag{1.10}$$

Figure 1.3 is a schematic representation of the crack growth relationship. The crack growth exponent, $n = 3$, has been observed to be applicable to basic crack growth rate data for structural steels as well as test data on welded members [1.7, 1.8]. The corresponding average crack growth constant was found to be [1.8] $C = 2 \times 10^{-10}$, using units of inches for crack size and ksi $\sqrt{\text{in.}}$ for ΔK. An upper bound value of $C = 3.6 \times 10^{-10}$ has been suggested by Rolfe and Barsom [1.7]. In this book the relationship

$$\frac{da}{dN} = 3.6 \times 10^{-10}\Delta K^3 \tag{1.11}$$

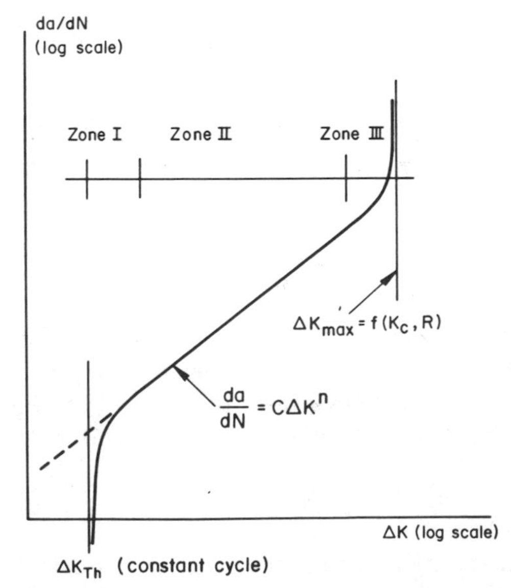

Figure 1.3 General crack propagation relationship.

is used to relate the crack growth rate and stress intensity range for all of the examples evaluated (units of inches for crack size and ksi $\sqrt{\text{in.}}$ for ΔK).

The corresponding upper bound crack growth constant for SI units is $C = 2.18 \times 10^{-13}$, using units of mm for crack size and MPa $\sqrt{\text{m}}$ for ΔK.

Other variations of the crack growth equation have been proposed by Foreman [1.11], McEvily [1.12], and Pearson [1.13]. These generally attempt to fit the crack growth threshold where crack propagation approaches zero and the region of accelerated crack growth as the stress intensity range approaches the material toughness, K_c. Since welded structures have high residual tensile stresses, the mean stress and the R ratio ($R = K_{\text{min}}/K_{\text{max}}$) tend to be high, and the crack growth threshold, ΔK_{th}, for zero growth rate is itself nearly zero. For these reasons and others noted later, Paris's simple law has been found to provide as good an estimate of crack propagation as do the other crack propagation relationships when applied to complex structural details.

Since randomly variable loading was involved in every case of fatigue crack propagation, an effective stress intensity range was used based on Miner's rule and the corresponding Miner's effective stress range $S_{r\text{Miner}}$. The threshold ΔK_{th} was ignored when some of the stress cycles in the random variable spectrum exceeded the constant cycle limit. Recent crack growth studies under random variable loading have demonstrated that the Paris power law can be extended below the constant cycle crack growth threshold when some of the cycles in the spectrum loading exceed ΔK_{th}; see [1.14]. Hence ΔK in Eq. 1.11 can be defined as

$$\Delta K_e = S_{r\text{Miner}} F_i \sqrt{\pi a} \tag{1.12}$$

where

$$S_{r\text{Miner}} = [\Sigma \alpha_i S_{ri}^3]^{1/3}$$

1.4 DAMAGE ACCUMULATION

Since bridge structures are subjected to loads at variable times and with different characteristics and magnitude, a random variable-amplitude type of loading results. For the most part in bridges a single dominant stress range excursion results, although smaller amplitude vibrations may be experienced as well when the structure is excited. Additional stress cycles may result under some conditions.

The effects of variable loading are normally accounted for by applying a cumulative damage rule. Although many rules have been proposed, the linear damage rule suggested by Miner [1.15] seems to provide a reasonable means of accounting for random variable loading.

Miner's rule suggests that damage from variable loading is given by

$$\sum \frac{n_i}{N_i} = 1 \tag{1.13}$$

where the ratio n_i/N_i is the incremental damage that results from the block of stress range cycles S_{ri} that occurs n_i times. Failure is defined when the sum of the increments of damage equals or exceeds unity.

Criticism of Miner's rule has been expressed because it fails to agree well with some of the experimental data [1.16]. However, the AASHO Road Test bridges [1.17] and random-variable tests of welded details have demonstrated that the variation in the cumulative damage ratio and variable load test data was no greater than the variation in the constant cycle test data. One should not expect a reduction in variability under random loading as compared to the variability experienced by constant cycle loading. Both Barsom's crack growth studies [1.7] and the tests on welded details [1.14, 1.18] have shown good correlation with constant cycle test data when Miner's rule is used to provide the correlation.

Schilling and Klippstein have shown that Miner's rule can be used to develop an equivalent stress range of constant amplitude that equals the variable amplitude damage for the total number of applied stress cycles [1.18]. This derivation is based on the use of an exponential model of the stress range life relationship which has the form [1.19]

$$N = AS_r^{-n} \tag{1.14}$$

where A is a proportionality coefficient dependent on detail.

Substitution of Eq. 1.14 into Eq. 1.13 provides

$$\sum \frac{n_i}{AS_{ri}^{-n}} = \sum \frac{\alpha_i N_T}{AS_{ri}^{-n}} = \sum \frac{\alpha_i S_{re}^{-n}}{S_{ri}^{-n}} = 1 \tag{1.15}$$

or

$$S_{re}^n = \Sigma \alpha_i S_{ri}^n$$

and

$$S_{re} = S_{r\text{Miner}} = [\Sigma \alpha_i S_{ri}^n]^{1/n} \tag{1.16}$$

where $n_i = \alpha_i N_T$ = number of occurrences at stress range level S_{ri}. The exponent, n, for most structural steel details is about 3 which is compatible with the crack growth rate constant given in Eq. 1.12.

Generally, stress range values, S_{ri}, greater than 25 to 30% of the constant cycle fatigue limit can be expected to contribute to crack growth.

The constant cycle fatigue limit can be taken as the allowable stress range for a given detail class that corresponds to the over 2,000,000 cycle design case for redundant load path structures. These values are provided in the AASHTO [1.20], AREA [1.21], and AISC [1.21] specifications.

The effective stress range, $S_{r\text{Miner}}$, can be utilized in evaluating fatigue crack growth using Eqs. 1.11 and 1.12 or comparing the accumulated stress cycles with the appropriate stress range–cycle life relationship.

1.5 CRACK EXTENSION BEHAVIOR

The crack extension behaviors of main interest in bridge structures are those associated with fatigue and with fracture toughness. Laboratory studies of the cycles of fatigue life for structural details as a function of stress range [1.19] and measurements of fatigue crack growth rate as a function of stress intensity range, ΔK, have been conducted for the commonly used bridge steels [1.7]. The results are essentially independent of yield strength, temperature, and cyclic frequency for the range of parameters that are of interest. Applicable test results of fatigue life and growth rate are described in later chapters where their use is illustrated.

The measurement point for fracture toughness was initially thought of as "onset of rapid fracture." In the development of standard ASTM test methods, for example, method E399, a more conservative choice better described as "initiation of crack extension" was used. However, for bridge steels, at temperatures and toughness levels such that low toughness may represent a hazard, crack extension begins essentially in the dominant cleavage manner, which is always fast, and the original measurement point idea "onset of rapid fracturing" is applicable.

The K values at this measurement point are generally termed K_c and K_{Ic} (for a degree of constraint across the crack front corresponding nearly to plane strain). These fracture toughness values decrease with decrease of test temperature and with increase of loading speed.

For a crack in a bridge structure the loading differs considerably from that used in fracture toughness testing and has to be given special attention. With the passage of truck traffic on a bridge, the region of the structure containing a crack is subjected to live load stress elevations somewhat like that illustrated in Figure 1.4. The increase in nominal stress due to live load, BC, is usually small in comparison with the dead load, OA. From measurements on bridges with truck or railroad traffic, the live load increases occur in times on the order of 0.1 sec or longer. In Figure 1.4 the loading speed for this increment of time is shown to be equivalent to a zero-to-full-load path, OBC, across a time period of at least 1 sec. The loading path ABC is less severe than the loading path OBC with regard to the influence of loading speed on the decrease of fracture toughness, K_c. Fracture toughness measurements using a load

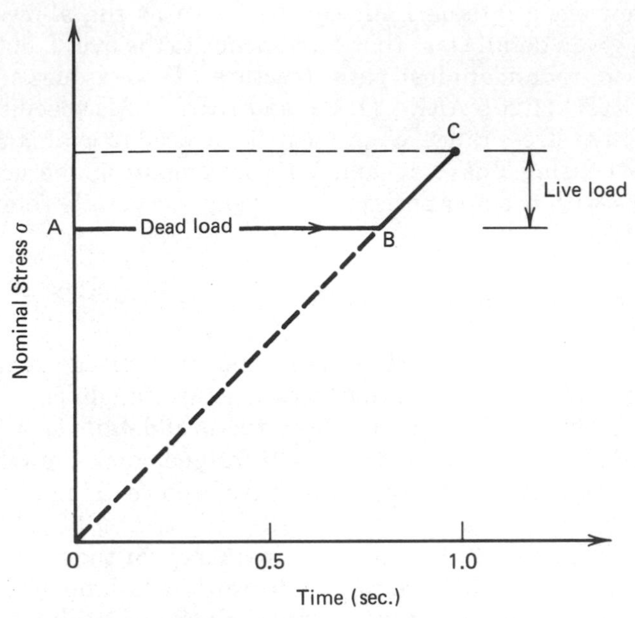

Figure 1.4 Schematic representation of live load stress elevation across a region of the bridge structure containing a crack.

rise time of about 1 sec have been used frequently to provide a moderately conservative match for the loading speed influences in bridge structures. On the other hand, when the ratio of live load to dead load is quite small, the appropriate K_c or K_{Ic} value may be the toughness value measured at customary "static" loading rates with a load rise time on the order of 1 min. An illustration of this is provided in Chapter 2.

Charpy V-notch tests and impact-K_c tests provide load rise times in the range of 0.1 to 1.0 msec. These are regarded as "dynamic" tests. Fracture tests at the three load rise times of approximately 1 min, 1 sec, and 1 msec are referred to as static, intermediate, and dynamic fracture tests and loading rates.

1.6 SUMMARY

Estimates of the stress intensity factor, K, are basic to the use of linear-elastic fracture mechanics in applications. Sections 1.2 to 1.6 have presented those K analysis techniques that have been most helpful for cracks in steel bridges. Similar analysis techniques are followed for fatigue cracking and the more rapid fracture extension. Section 1.5 provided a brief review of the crack extension behaviors observed in bridge steels. All of these analysis and testing methods are given specific illustration in the following chapters.

REFERENCES

1.1 Fisher, J. W., and Yuceoglu, U., A Survey of Localized Cracking in Steel Bridges, Interim Report, Federal Highway Administration, DOT-FH-11-9506, December 1981.

1.2 Tada, H., Paris, P. C., and Irwin, G. R., The Stress Analysis of Cracks Handbook, Del Research Corp., Hellertown, Pa., 1973.

1.3 Maddox, S. J., Assessing the Significance of Flaws in Welds Subject to Fatigue, *Welding J.* 53 (September 1974).

1.4 Albrecht, P. A., and Yamada, K., Rapid Calculation of Stress Intensity Factors, *J. Struct. Div.*, ASCE, 103 (February 1977).

1.5 Irwin, G. R., Crack Extension Force for a Part Through Crack in a Plate, *Trans. ASME* E 29 (December 1962).

1.6 Paris, P. C., and Erdogan, F., A Critical Analysis of Crack Propagation Laws, *Trans. ASME* D 85 (December 1963):528–534.

1.7 Rolfe, S. T., and Barsom, J. M., Fracture and Fatigue Control in Structures, *Applications of Fracture Mechanics* (Englewood Cliffs, N.J.: Prentice-Hall, 1977).

1.8 Hirt, M. A., and Fisher, J. W., Fatigue Crack Growth in Welded Beams, *Engrng. Fracture Mech.* 5 (1973).

1.9 Broek, D., *Elementary Engineering Fracture Mechanics,* 3rd ed. (Hingham, Mass.: Martinus Nijhoff, 1982).

1.10 Hertzberg, R. W., *Deformation and Fracture Mechanics of Engineering Materials,* 2nd ed. (New York: Wiley, 1983).

1.11 Foreman, R. G., Numerical Analysis of Crack Propagation in Cyclic Loaded Structures, *J. Basic Engr., Trans. ASME* D 89 (September 1967).

1.12 McEvily, A. J., The Fracture Mechanics Approach to Fatigue, Significance of Defects in Welded Structures, Proc. U.S.–Japan Seminar, University of Tokyo, 1974.

1.13 Pearson, S., The Effect of Mean Stress on Fatigue Crack Propagation in ½ in. Thick Specimens of Aluminum Alloys of High and Low Fracture Toughness. *Engrng. Fracture Mech.* 4 (1972).

1.14 Fisher, J. W., Mertz, D. R., and Zhong, A., Steel Bridge Members under Variable Amplitude, Long Life Fatigue Loading, NCHRP Report 267, Transportation Research Board, 1983.

1.15 Miner, M. A., Cumulative Damage in Fatigue, *J. Appl. Mech.* 12 (September 1945).

1.16 Committee on Fatigue and Fracture Reliability, P. Wirsching, Chairman, Fatigue and Fracture Reliability: A State of the Art Review, *J. Struct. Div.,* ASCE, 108 (January 1982).

1.17 Fisher, J. W., and Viest, I. M., Fatigue Life of Bridge Beams Subjected to Controlled Truck Traffic, Preliminary Publication, 7th Cong., IABSE, 1964.

1.18 Schilling, C. G., Klippstein, K. H., Barsom, J. M., and Blake, G. T., Fatigue of Welded Steel Bridge Members under Variable Amplitude Loadings, NCHRP Report 188, Transportation Research Board, 1978.

1.19 Fisher, J. W., *Bridge Fatigue Guide—Design and Details* (Chicago: American Institute of Steel Construction, 1977).

1.20 AASHTO, Standard Specifications for Highway Bridges, 12th ed., The American Association of State Highway and Transportation Officials, Washington, D.C., 1977.

1.21 AREA, Manual for Railway Engineering, Chapter 15: Steel Structures, American Railway Engineering Association, Washington, D.C., 1983.

1.22 AISC, Specification for the Design, Fabrication and Erection of Structural Steel for Buildings, American Institute of Steel Construction, Chicago, 1978.

PART I

Fatigue and Fracture of Structural Members at Details or Large Defects

Fatigue crack growth and eventual brittle fracture have developed in a number of structures because details were used that were not anticipated to have such low fatigue resistance when originally designed. Either specification provision was overly optimistic as a result of lack of test data, or the detail classification was not known. An example is the cover-plated beam with flange thickness greater than 0.8 in. (20 mm). The actual fatigue resistance was much lower than test data indicated at the time of many original designs. The actual lower bound resistance was not established until 1978.

More often large initial defects and cracks have been fabricated into the structure because of lack of quality in certain welds, which were not considered important, or because quality control was not able to disclose the crack and defect condition. This has resulted in fatigue crack enlargement, as the applied cyclic stresses exceeded the crack growth threshold. The subsequent enlargement of the crack resulted, in some cases, in brittle fracture of the cross section.

The initial cases of cracking that are reviewed and analyzed in this book generally are concerned with low fatigue strength details or the consequence of fabricating a large defect into a given structural member or attachment. Often the defect was not recognized as being a cracklike condition due to the geometry and complexity of the connection. This appeared to be the case for plate penetrations of members where a girder flange was passed through a web, cut out, and then connected with fillet welds or partial penetration groove welds. Another common cracklike condition has resulted from weld-filled holes and short groove-welded inserts. These have created severe cracklike discontinuities which were further subjected to high residual tensile stresses from the welding.

In a few instances brittle fractures have occurred as a result of a low fracture toughness material before any significant fatigue crack extension developed. This may happen during construction or after a short period of service. However, this is not the usual condition. More often a large initial defect has led to significant fatigue crack growth in a short period of time, as nearly all of the fatigue resistance was exhausted at the time the structure was placed in service. Although the final mode of crack extension was due to crack instability (brittle fracture), not much additional life would have resulted with higher levels of fracture toughness. A standard Charpy V-notch test requirement will generally screen out very low toughness material and ensure the fatigue resistance of the detail.

Eleven different bridge structures are examined in this section, and these provide examples of a wide variety of conditions that have led to fatigue crack growth and/or fracture. The table that follows gives a summary of the various types of details or defects that are examined and the locations of the bridges involved. The structural details include cover-plated beams and lateral gusset plates; details with defects include groove-welded connections in attachments, such as longitudinal stiffeners, flange cover plates and in web insert plates; plate penetrations; welded bolt holes; transverse hydrogen-induced cracking in longitudinal welds; lamellar tearing; and stress corrosion cracking in eyebars.

SUMMARY OF CASES AT DETAILS OR LARGE DEFECTS

Case	Type Detail or Defect	Bridge
2.1	Stress corrosion crack in eyebar head	Silver Bridge, Point Pleasant, W.V.
2.2	Fatigue cracking from bending of hangers	Illinois Route 157 over St. Clair Avenue
3.1	Cover plate termination	Yellow Mill Pond Bridge, Bridgeport, Conn.
4.1	Lateral connection plate	Lafayette Street Bridge, St. Paul, Minn.

Case	Type Detail or Defect	Bridge
5.1	Weld defect in web groove weld	Aquasabon River Bridge, Ontario
5.2	Weld defect in longitudinal stiffener groove weld splice	Quinnipiac River Bridge, New Haven, Conn.
5.3	Weld defect in cover plate groove weld splice	U.S. 51 Illinois River Bridge, Peru, Ill.
6.1	Web plate penetration by girder flange	Dan Ryan Elevated, Chicago, Ill.
7.1	Welded holes	County Highway 28 Bridge over I-57 at Farina, Ill.
8.1	Box girder corner welds with transverse cold cracks	Gulf Outlet Bridge New Orleans, La.
9.1	Rigid frame joints, lamellar tearing	Ft. Duquesne Approaches, Pittsburgh, Pa.

CHAPTER 2

Eyebars and Pin Plates

The failure of the Point Pleasant or Silver Bridge, in West Virginia, in 1967 [2.1] which was caused by the fracture of an eyebar is well known, so only a summary of this failure is provided here. Of course eyebars have cracked in several other bridges although none have resulted in a collapse. However, at least two railroad bridges in the United States developed fatigue cracks in eyebar heads that resulted in fracture [1.1]. One such failure that occurred in a Santa Fe bridge in Oklahoma was found to be caused by a forge lap in the eyebar head that happened to be oriented normal to the largest tension from the pin, as can be seen in Figure 2.1. This served as an initial crack that led to fatigue crack growth and eventual fracture.

Five Japanese railway bridges built between 1900 and 1913 have developed fatigue cracks at the eyebar head normal to the tension from the pin loading. A description of the Fujigawa Bridge is given in [2.2]. About 820,000 trains have crossed over most of these structures.

Two bridge structures showed cracking in pin plates that supported suspended spans [1.1]. One such crack formed on the net section at the pin hole, much like an eyebar failure. However, in the second structure several pin plates failed in the gross section as a result of corrosion product that "welded" the pin plate heads to the girders. This produced bending stresses in the pin plates, and since these ends were fixed because of the corrosion, the area cracked. Bending stresses were measured in the pin plates of two bridges that were instrumented during the summer of 1981. Friction and corrosion product provided fixity at the pins and caused rigid link behavior and bending.

A summary of structures with cracked pin plates is provided in Section 2.2 in the discussion of the Illinois Route 157 Bridge.

Figure 2.1*a* Cracked eyebar showing forge laps on the eyebar surface (courtesy of Santa Fe Railroad).

Figure 2.1*b* Fracture surface showing initial forge lap crack, fatigue crack, and final fracture surface.

2.1 FRACTURE ANALYSIS OF POINT PLEASANT ("SILVER") BRIDGE OVER THE OHIO RIVER

2.1.1 Description and History of the Bridge

Description of Structure

The Point Pleasant Bridge which carried U.S. 35 highway over the Ohio River was located between Point Pleasant, West Virginia, and Kanauga, Ohio. The bridge was also known as the "Silver Bridge" because it was one of the major structures to be painted with aluminum paint [2.1]. It was one of two nearly identical and unique eyebar chain suspension bridges in the United States. The other bridge, also spanning the Ohio River, was at St. Mary's, West Virginia, until it was dismantled in 1969.

This bridge was an important link between major cities on both sides of the Ohio River. It was also an important local artery for industrial plants in Point Pleasant and the Gallipolis areas of Ohio.

The Point Pleasant Bridge was an eyebar chain suspension bridge with its axis in an east–west direction over the Ohio River. It had a 700 ft (213.4 m) center or main span and two 380 ft (115.8 m) side spans, as shown in Figure 2.2. In addition there were two approach spans on each side of the bridge. They were plate girder spans 75.25 ft (22.94 m) and 71.50 ft (21.8 m) in length supported on concrete piers. The two suspension bridge towers extended 103 ft 10.25 in. (39.88 m) above the tops of the two main piers. The total length of the bridge was 1753 ft (534.3 m).

The roadway of the suspended span, as originally built in 1928, consisted of a timber deck and sidewalks. In 1941 the timber deck was replaced with a 3 in. (76 mm) deep steel grid floor filled with concrete. The new roadway was 21 ft (6.4 m) wide for two lanes of traffic with a 5 ft 8 in. (1.725 m) sidewalk. The new deck caused no significant change in the dead load of the structure.

The bridge superstructure was unique in that the stiffening trusses of both the center span and the two side spans were framed into the eyebar chain to make up the top chord of about half of the length of the stiffening trusses (see Figure 2.2). Most of the eyebars were between 45 and 55 ft (13.72 to 16.76 m) in length of varying thickness (1.5 to 2.2 in., or 38 to 56 mm) depending on the truss panel in which they were located. The shank width of eyebars was 12.0 in. (305 mm), and the head width was 8.5 in. (216 mm). They were made of heat treated rolled carbon 1060 steel bars with forged heads designed to break at ultimate loading in the shank. Each joint was composed of four eyebar heads and a connecting pin or pin rod, and was secured by double nuts on each end of the pin rod, as shown in Figures 2.3 and 2.4. Some typical stiffening truss and joint details are given in Figures 2.4 and 2.5.

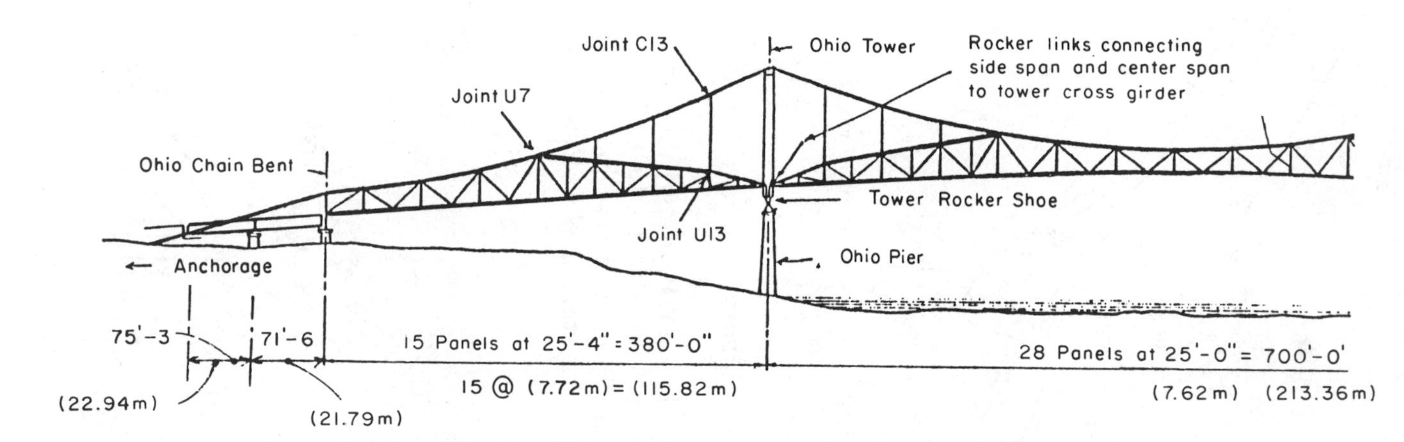

Figure 2.2 Elevation of Point Pleasant Bridge showing location of joints C13, U7, and U13.

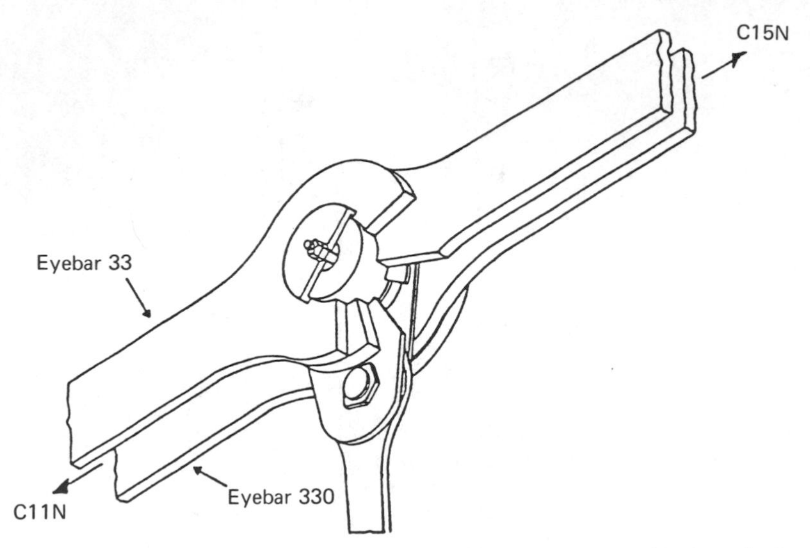

Figure 2.3 Eyebar chain joint at C13 where the chain was connected to truss by hanger and hanger strap plates.

History of Structure and Cracking

Construction began on the Point Pleasant Bridge on May 1, 1926, and the bridge was opened to traffic on May 19, 1928. The bridge originally was built and owned by the West Virginia–Ohio River Bridge Corporation. On December 24, 1941, the state of West Virginia purchased the Point

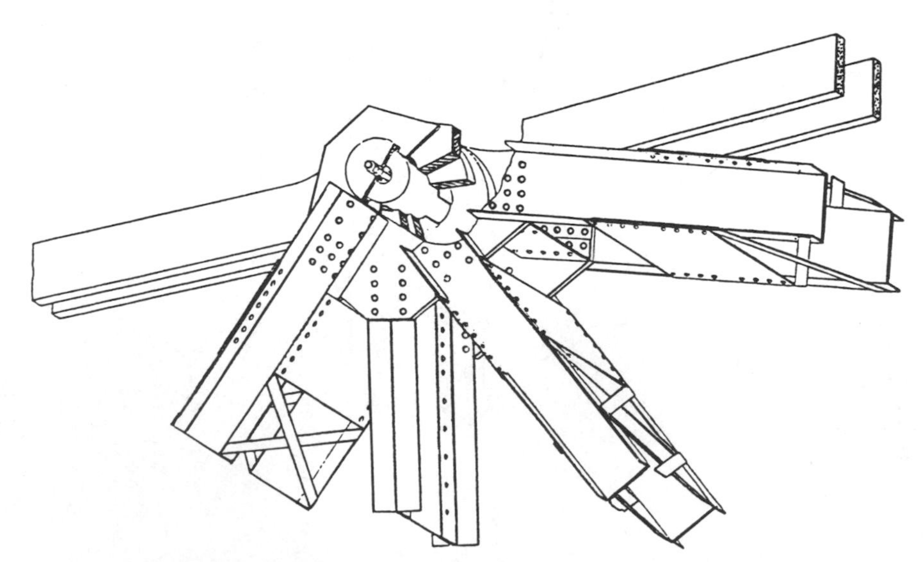

Figure 2.4 Details of joint U7 where eyebars connected to truss.

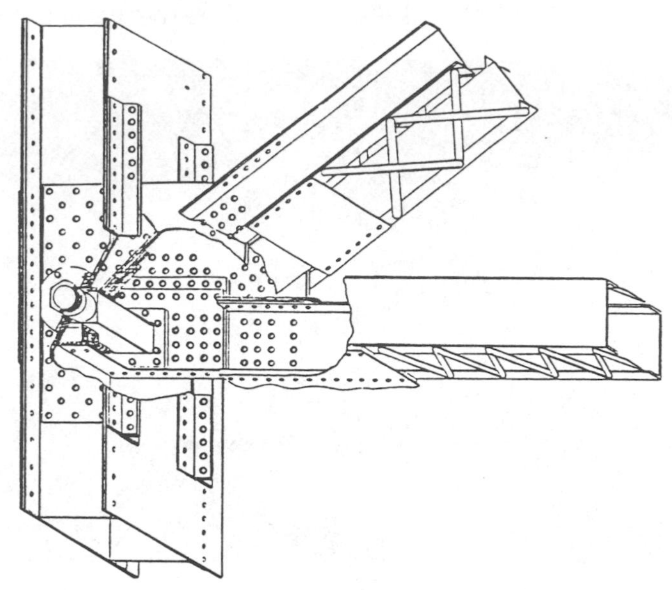

Figure 2.5 Details at Ohio end of the Point Pleasant Bridge where lower chords of the side spans were connected to vertical chain bent post at the tower.

Figure 2.6 Point Pleasant Bridge after collapse (courtesy of Federal Highway Administration).

Figure 2.7 One of the eyebar joints after collapse (courtesy of Federal Highway Administration).

Pleasant Bridge and continued to operate the bridge under the authority of the State Highway Commission as a toll bridge until January 1952, when it became toll free.

For almost 40 years it carried an increasing number of vehicles across the Ohio River. On December 15, 1967, the Point Pleasant Bridge collapsed suddenly without any warning about 5:00 P.M. with the loss of 46 lives and 37 vehicles of all types [2.1]. The three suspended sections fell within 60 sec. The collapse was immediately preceded by several loud "cracking" sounds on the Ohio side span superstructure. According to eyewitness accounts, the bridge started to fall, hesitated, and then collapsed completely with a "roaring" sound that lasted during the collapse. The temperature at the time of collapse was recorded as 30°F (−1°C).

It has been clearly established by the investigation following the failure that the nominal static stresses in all eyebar elements at the time of collapse did not exceed those provided for in the original design [2.1]. The main cause of collapse was traced back to the unstable extension or brittle fracture of two corrosion cracks located at the pinhole of one of the four eyebars (eyebar 330) of joint C13 north shown schematically in Figure 2.3 [2.3, 2.4, 2.5]. The design of the chain was unique in that only two eyebars were used in each chain segment, and the chain also formed a portion of the top chord of the stiffening trusses. Thus failure of any one eyebar in the bridge would cause the complete collapse of the structure (see Figures 2.6 and 2.7).

2.1.2 Failure Modes and Analysis

Cyclic Loads and Stresses

A traffic survey conducted by Ohio-DOT in 1964 indicated that 6640 vehicles (average daily traffic) crossed the bridge which was designed for two-way traffic. Of the vehicles 5040 were passenger cars, and 1600 trucks.

Although the bridge was not posted for weight limits, an "overload" permit was required for all vehicles over 70,000 lb (313.6 kN). Thus it was recorded that on December 7, 1967, a tractor trailer weighing 98,000 lb. (439 kN) crossed the bridge without any unusual occurrences.

A computer analysis of the bridge was conducted after the collapse, and it was found that the unit stress at collapse was less than the 1927 design unit stress for all essential structural members in the bridge [2.1]. In the eyebar where failure occurred, the dead load stress was 36.6 ksi (252.5 MPa), and the estimated live load was 5 ksi (86 MPa).

Model test results on an eyebar and joint assembly made up from the structural elements recovered from the wreckage similar to those of joint C13N indicate that the separation of joint C13N was the key event in the sequence of collapse of Point Pleasant Bridge [2.4]. Several possibilities for separation of C13N were considered, and tests were made of cyclic and static load conditions imitating those at the time of failure. In the test model the components of the joint at which brittle fracture was induced through the lower limb of eyebar 330, as shown in Figure 2.8, separated in essentially the same way as the physical evidence. Also the opening of the eye of the eyebar due to plastic strain as well as the appearance of the second fracture in the upper limb of this same eye were essentially similar. However, in the model the branching of the opposite limb of the eyebar introduced a slightly different appearance after midwidth of the section [2.4]. And it was observed during the test that there was an unexpectedly long time interval between the instant of brittle fracture in the lower limb of eyebar 330 of the model and the subsequent complete separation of the joint. This indicated that some time interval would likely be present before collapse of the bridge structure.

Temperature and Environmental Effects

Other evidence showed that the atmosphere provided the necessary elements for the development of stress corrosion cracking in the heat-treated 1060 eyebar steel, and that the size of the corrosion-induced cracks had been increasing with passage of time. In the heat treatment of the steel eyebar, the outer layer was decarburized, but a hard layer of martensite remained beyond a depth of 0.1 to 0.4 in. (2.5 to 10 mm) which

Figure 2.8 Fractured C13N eyebar (courtesy of Federal Highway Administration).

was the region where the cracks were located (see Figure 2.9). In addition at the time of collapse the recorded temperature was 30°F (−1°C) [2.1]. The fracture resistance was then low enough in comparison with crack size (~40 ksi $\sqrt{\text{in.}}$ or 44 MPa $\sqrt{\text{m}}$) so that onset of rapid brittle fracturing developed.

The measured static fracture toughness, K_{Ic}, of the specimens from the eyebar material and the estimated dynamic toughness, K_{Id}, are plotted in Figure 2.10. Approximate values of K_{Ic} and K_{Id} corresponding to the temperature of 30 to 32°F (0°C) at the time of collapse are indicated in Figure 2.10.

Mechanical, Chemical, and Fracture Characteristics of Material

Several specimens were also tested from the shank portion of the eyebar members of the collapsed Point Pleasant Bridge. The chemical and metallurgical characteristics of the material were reported and evaluated in [2.3]. The tensile and fatigue characteristics of the eyebar steel from the collapsed bridge were also evaluated, and the results are reported in [2.5].

The tensile strength values obtained from standard and special tension specimens at room temperatures met the minimum strength requirements of the original specifications. The average yield strength of the eyebar shank was found to be 79 ksi (545 MPa). The elongation in 2 in. (51 mm) was 21.6%. The specification required a yield point of 75 ksi (518 MPa) and a tensile strength of 100 to 105 ksi (690 to 725 MPa). No major variation in the tensile properties of the eyebar steel was apparent.

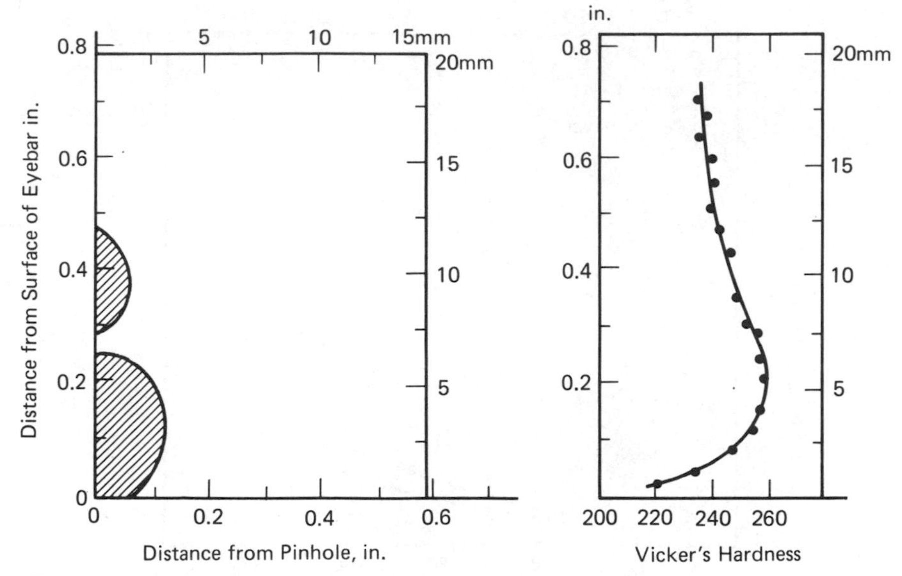

Figure 2.9 Location of initial eyebar cracks and the local hardness of the material.

Charpy V-notch tests were carried out on the eyebar material and are summarized in Figure 2.11. The 15 ft-lb (20 J) transition temperature was about 212°F (100°C).

The measured static toughness, K_{Ic}, and the estimated dynamic toughness, K_{Id}, of the eyebar material are shown in Figure 2.10 [2.8]. A temperature shift of 100°F (55°C) was used to estimate the dynamic frac-

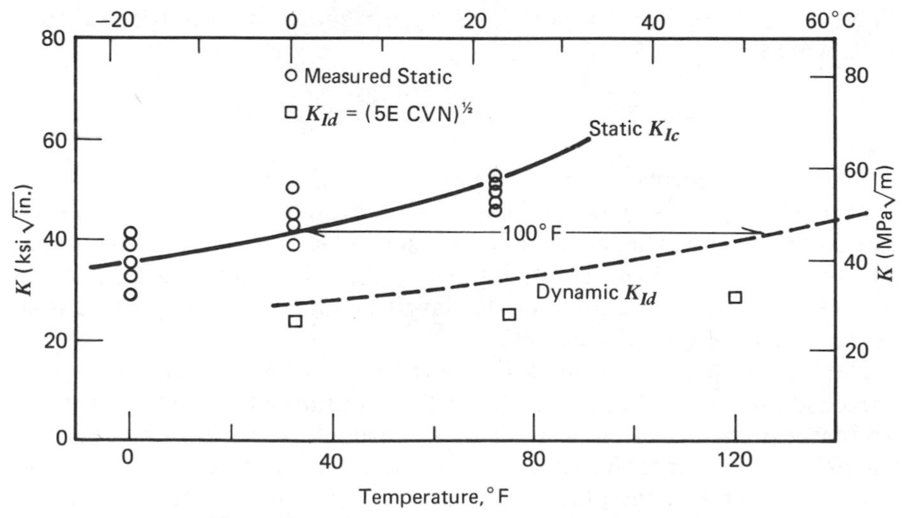

Figure 2.10 Fracture toughness of eyebar steel.

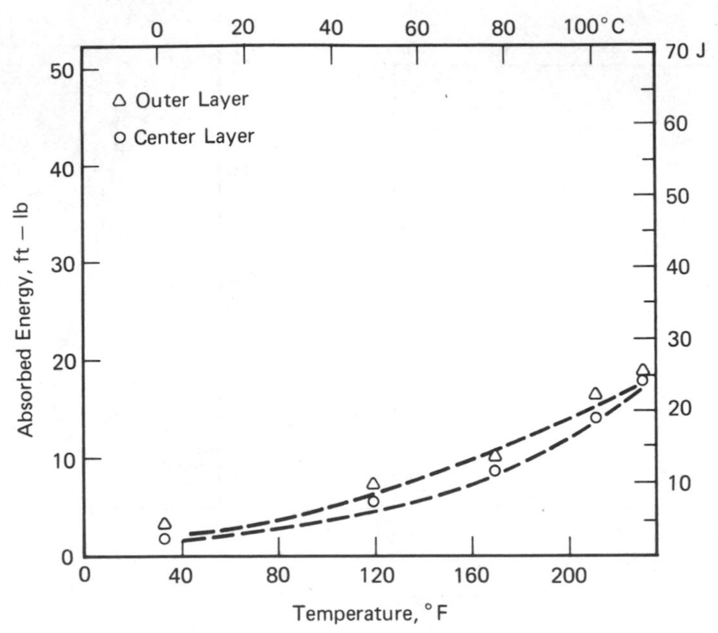

Figure 2.11 Average CVN test data from eyebars.

ture toughness. At the failure temperature of 30 to 32°F (0°C), the measured value of K_{Ic} varied between 41 ksi $\sqrt{\text{in.}}$ and 50 ksi $\sqrt{\text{in.}}$ (45 to 55 MPa $\sqrt{\text{m}}$). The dynamic fracture toughness K_{Id} was estimated to be 26 ksi $\sqrt{\text{in.}}$ (28.6 MPa $\sqrt{\text{m}}$).

Also plotted in Figure 2.10 are the dynamic fracture toughness estimates from the Charpy V-notch data suggested by Barsom, $K_{Id} = [5E\ \text{CVN}]^{1/2}$. They confirm the applicability of the strain rate temperature shift, where

$$T_s(°\text{F}) = 215 - \tfrac{3}{2}\sigma_Y \simeq 95°\text{F}\ [T_s(°\text{C}) = 120 - 0.12\sigma_Y]$$

Other studies corroborated the finding that the mechanism of stable crack extension was stress corrosion. This eventually caused the crack that triggered the collapse [2.5]. It was concluded that fatigue was probably not the cause of the crack extension [2.5]. The low live load stress range in the eyebar would not be large enough to cause appreciable fatigue crack propagation.

The initial flaws or cracks from which the fracture of the eyebar started are shown in Figures 2.12 and 2.13. Figure 2.9 also shows results of a test to determine the hardness of the through thickness of the eyebar. The heat treatment of the eyebar during manufacturing caused the material in the center of the plate to become harder than the material on the surface. The flaws are obviously located in the area of maximum hard-

Figure 2.12 Fracture surface of C13N eyebar (courtesy of Federal Highway Administration).

Figure 2.13 Initial cracks in pinhole of eyebar C13N at arrows (see Figure 2.9 for dimensions of cracks) (courtesy of Federal Highway Administration).

ness which produces lower fracture toughness. The larger crack was 0.12 in. (3 mm) deep. It seems probable that the two corrosion cracks would have eventually coalesced and increased K substantially. However, such fracture would have also eventually occurred with a higher toughness material that was stress corrosion sensitive.

Visual and Fractographic Analysis of Failure Surfaces

Examination of the flaw surfaces on the C13N eyebar revealed considerable quantities of corrosive products. Stress corrosion tests on the material showed that it was corrosion sensitive. The fatigue tests on specimens with machined surfaces indicated that eyebars of the structure should have adequate fatigue resistance [2.5].

The two corrosion-fatigue cracks from which the fracture of the eyebar started are shown in Figure 2.13. Figure 2.9 shows a sketch of the flaws and their dimensions [2.3]. Due to heat treatment of the eyebar a harder layer existed at a depth of 0.1 to 0.4 in. (2.5 to 10 mm) from the surface. This also decreased the toughness of the material. Since the two flaws lie in the area of highest hardness, the local fracture toughness was less than the average value shown in Figure 2.10 [2.3].

Examination of the flaws and the crack surface revealed considerable corrosion and corrosion product. At the location of the flaws contaminated water could be trapped in a crevice area of high stress and high local hardness. Hence stress corrosion appears to have been the dominant mechanism for the creation of the two initial flaws or cracks. These flaws coupled with the low toughness at the time of collapse caused the sudden brittle fracture phase in the lower limb of the eye of the eyebar and consequently the catastrophic failure of the Point Pleasant Bridge.

Failure Analysis

An analytical and experimental study of the fatigue characteristics of the small specimens fabricated from the eyebars of the collapsed bridge showed that the small stress range levels probable at eyebar C13N would be too low to cause the initiation and extension of the flaws and cause failure by fatigue [2.5, 2.6]. The design live load stress in the shank of the eyebar was 12.5 ksi (86 MPa). The dead load design stress was 36 ksi (248 MPa). At the time of failure the total shank stress was estimated to be 41 ksi (283 MPa).

The elastic stress concentration at the pinhole as a function of the stress in the eyebar shank was estimated to be between 2.3 and 3.1 for full-size tests of the eyebar and from a finite element analysis (see Figure 2.14) [2.7]. The full-size eyebar tension tests also demonstrated that significant yielding occurred at the pinhole at the working load level.

The stress intensity factor for the eyebar head can be estimated from the relationship

$$K = F_e F_s F_w F_g \sigma \sqrt{\pi a} \tag{2.1}$$

The free surface correction was taken as

$$F_s = 1.12 \tag{2.1a}$$

since the crack was small and there was an adjacent small crack. The crack shape correction was calculated as

$$F_e = \frac{1}{E(k)} \tag{2.1b}$$

where

$$\begin{aligned} E(k) &= \int_0^{\pi/2} [1 - k^2 \sin^2 \theta]^{1/2} \, d\theta \qquad k^2 = \frac{c^2 - a^2}{c^2} \\ &\simeq \frac{3\pi}{8} + \frac{\pi}{8}\left(\frac{a}{c}\right)^2, \quad \text{for } c < 2a \\ &= \frac{\pi}{2}, \quad \text{since } c \sim a \end{aligned}$$

The back surface correction for small cracks is negligible and $F_w = 1.0$.

The stress gradient correction, F_g, was estimated from the stress field gradient obtained by Poletto [2.7]. A higher-order polynomial was fitted to the stress gradient shown in Figure 2.14. This resulted in the following relationship for F_g (3.9) with SCF = 2.65 and with $w = 8.5$ in. (216 mm), the eyebar head halfwidth:

$$F_g \cong \text{SCF}\left[1 - \frac{2}{\pi}(2.6196)\frac{a}{w} + \frac{1}{2}(2.5994)\left(\frac{a}{w}\right)^2 - \frac{4}{3\pi}(1.1064)\left(\frac{a}{w}\right)^3\right] \tag{2.1c}$$

Under dead load alone, the stress intensity estimate is 40.8 ksi $\sqrt{\text{in.}}$ (44.9 MPa $\sqrt{\text{m}}$). Under the estimated live load, the stress intensity increases to 46.4 ksi $\sqrt{\text{in.}}$ (51 MPa $\sqrt{\text{m}}$). Allowance for plastic yielding would decrease this computed value of K to no less than 40 ksi $\sqrt{\text{in.}}$ (44 MPa $\sqrt{\text{m}}$). A comparison with the fracture toughness shown in Figure 2.10 indicates that the estimated stress intensity factor is bracketed by the measured fracture toughness of the critical eyebar at 30 to 32°F (0°C). Hence crack instability would be expected and did occur on December 15, 1967.

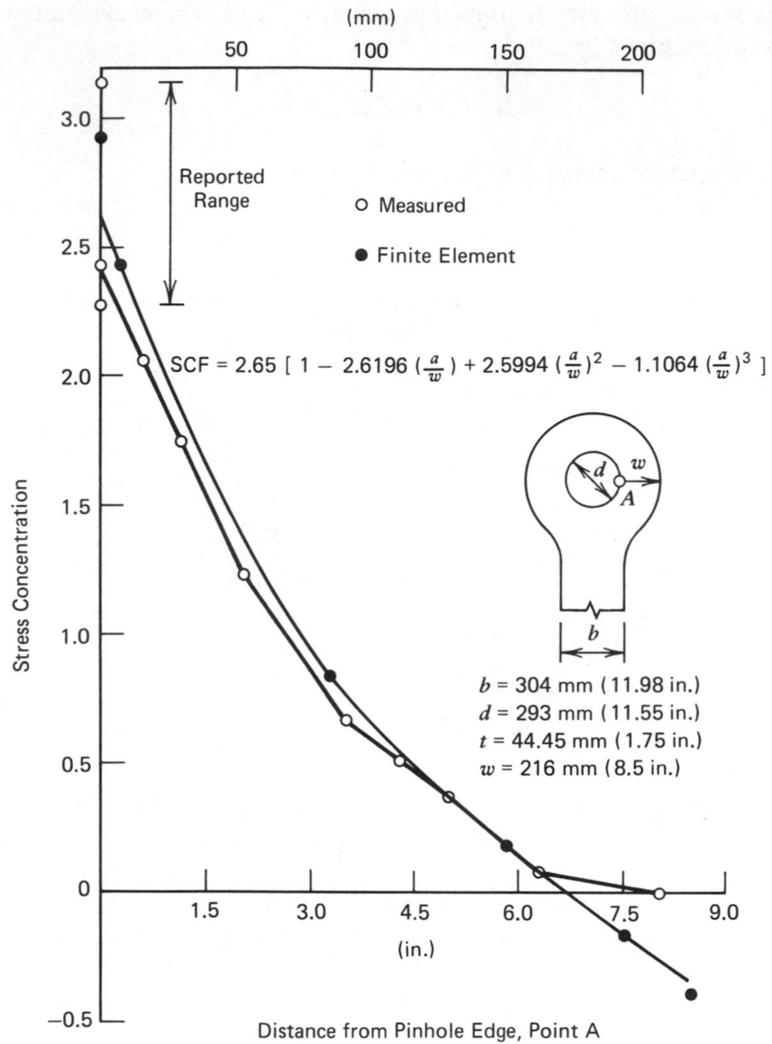

Figure 2.14 Comparison of finite element analysis and measured values.

2.1.3 Conclusions

The collapse of the Point Pleasant Bridge was extensively evaluated and reviewed in [2.5, 2.6]. A summary of the findings follows:

1. The tensile properties of the eyebars in the structure met the minimum strength requirements of the original design.
2. The eyebar material proved to be stress corrosion sensitive. The mechanism of crack development and extension in eyebar C13N head was likely caused by stress corrosion. These cracks also devel-

oped in an area of high local hardness. The combination of high hardness and the stress corrosion sensitivity of the eyebar steel were the primary causes of the failure.

3. With the cracks in the eyebar head, the structure was predicted to become unstable at the time of failure because of its material characteristics, the crack size, and the stress conditions. The fracture mechanics model of the flaw provided a stress intensity factor estimate that exceeded the fracture toughness of the eyebar material.

2.2 ILLINOIS ROUTE 157 OVER ST. CLAIR AVENUE

2.2.1 Description and History of the Bridge

Description of Structure

The Route 157 bridge located in St. Clair County, Illinois, is a skewed seven-span continuous structure 474 ft 6 in. (144.7 m) long over St. Clair Avenue. It is composed of 13 rolled WF beams spaced at 5 ft 6 in. (1.67 m) intervals and varying spans. Spans 2 and 6 are each 100 ft (30.5 m) long with 60-ft (18.3 m) suspended segments supported by adjacent cantilever spans. The suspended spans are supported by a rocker bearing on one end and by pin plate links on the other end, as illustrated in Figure 2.15. Figure 2.16 shows a view of span 2. Spans 1 and 7 are 54 ft 9 in. (16.7 m) long, and spans 3, 4, and 5 are 55 ft (16.78 m) long.

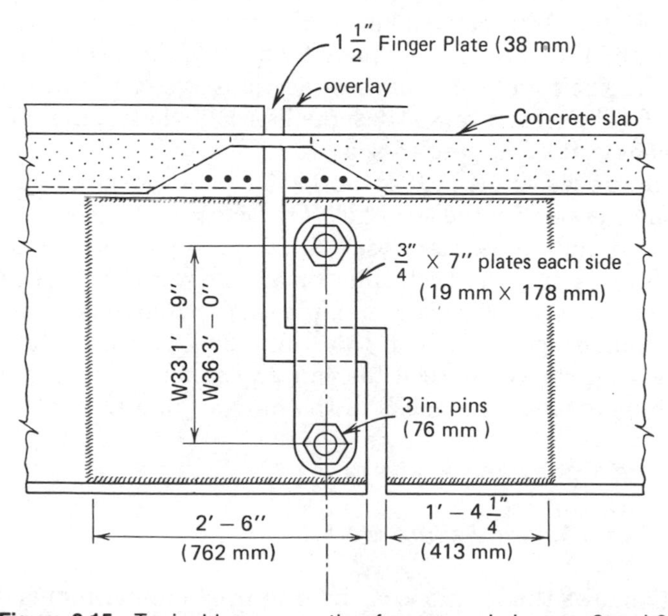

Figure 2.15 Typical hanger section for suspended spans 2 and 6.

Figure 2.16 View showing suspended span 2 over St. Clair Avenue (courtesy of Illinois Department of Transportation).

The original A7 steel rolled sections were W36 × 230 in spans 1, 3, 5, and 7. The suspended spans were W36 × 150 sections as was span 4. Two additional A36 steel beams were added in 1966 to the west side of the structure. They were W33 × 200 sections between pier 3 and the north abutment, W36 × 160 sections between pier 4 and the south abutment, and W33 × 160 sections between piers 3 and 4. Welded cover plates were attached to the bottom flanges of these sections over the supports at piers 1, 2, 5, and 6. The structure was designed for H20-44 loading. Figure 2.17 shows a schematic of suspended span 2.

The structure was built in 1945. In 1966 it was widened by adding two beams to the west side of the structure. On October 23, 1978, the pin links supporting the ends of beams (Nos. 8, 9, 10) in the passing lane of span 2 were found to be fractured, as illustrated in Figure 2.18. The cracked plates were discovered by a district employee driving under the structure. The beams had dropped between $\frac{1}{2}$ to $\frac{3}{4}$ in. (12 to 19 mm) below the deck slab. Span 6, which is identical to span 2, was found to be structurally sound with no evidence of cracks in the hanger plates.

2.2.2 Failure Modes and Analysis

Location of Cracks and Failure Modes

Figure 2.17 shows the locations of the fractured hanger plates in span 2. In addition to the hangers supporting beams 8, 9, and 10, seven addi-

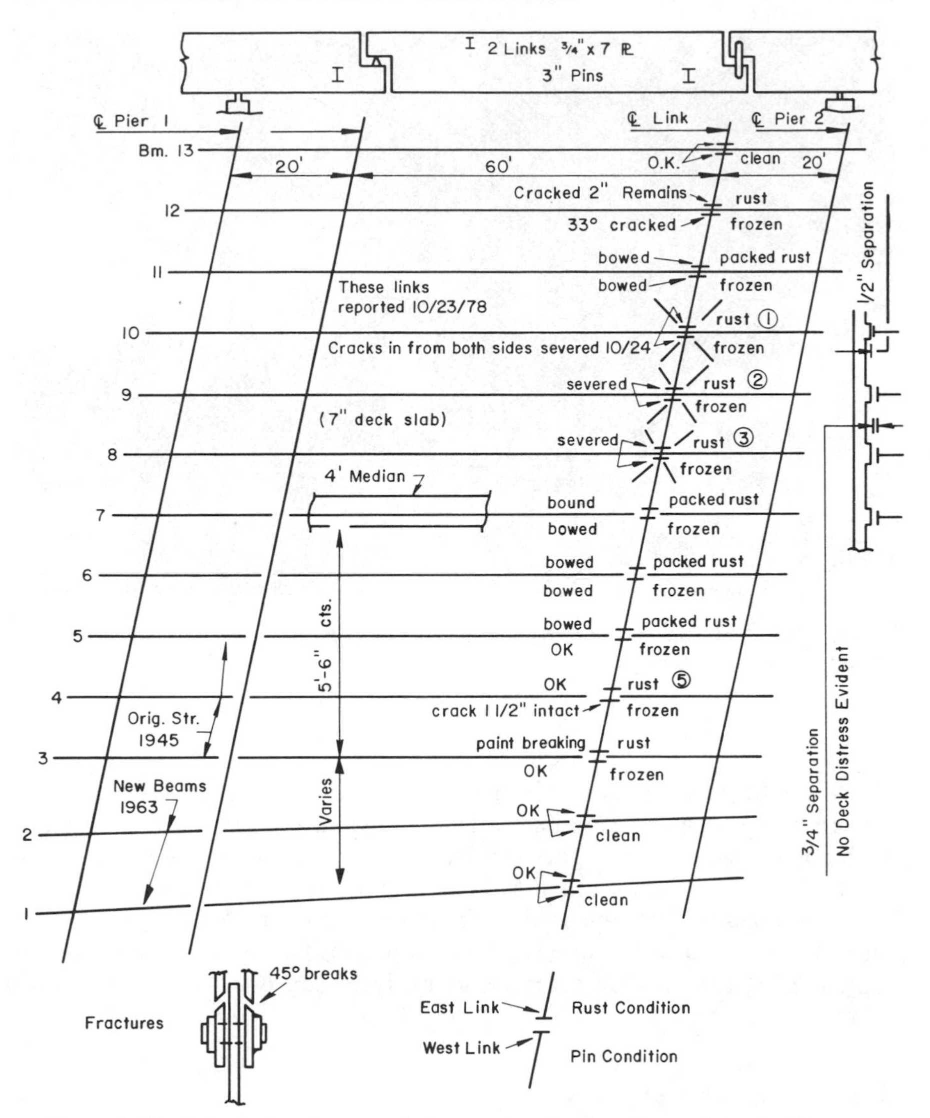

Figure 2.17 Schematic of suspended span showing locations of cracked hangers.

tional beams had plate hangers that were either bowed or partially cracked, and these are also identified in Figure 2.17.

The failures appear to be due to rust and other corrosion products which had built up between the beam web reinforcement and the hanger plates. Since the hangers were located beneath the expansion joint finger plates, water, salt and other debris came in contact with the pinned connections of the hanger plates. This caused the lower pins to freeze. Hence, over time the joints became rigid.

With the joints frozen, the hanger plates were subjected to a large in-

Figure 2.18 View showing one of the cracked hangers in span 2 (courtesy of Illinois Department of Transportation).

plane bending stress as traffic passed over the bridge and temperature change brought on thermal expansion. Measurements on other bridges with suspended span hangers confirmed that even with the relative end rotation that can be assumed to occur, frozen hanger plates cause large bending stresses. The repeated loading resulted in cracking and the eventual failure of the $\frac{3}{4} \times 7$ in. (19×178 mm) pin plates. As can be seen in Figure 2.18, the cracking formed away from the net section of the pin plate.

Cyclic Loads and Stresses

The average daily truck traffic (ADTT) crossing the structure in 1980 was 300. The total daily traffic in 1979 was 18,600 vehicles, and 15 to 18% were observed to be trucks. If it is assumed that 110,000 vehicles cross the bridge each year, then an estimated 3.5 million have crossed the structure since 1935 when it was opened.

No stress measurements were available for the Route 157 bridge. However, measurements on the U.S. 309 bridge near Philadelphia provide some insight into the cracking that developed [5.8]. The U.S. 309 bridge had a 100 ft (30.5 m) suspended span, 18 ft (5.5 m) cantilevers, and 77 ft (23.5 m) side spans. Two 1×10 in. (25×250 mm) pin plates supported

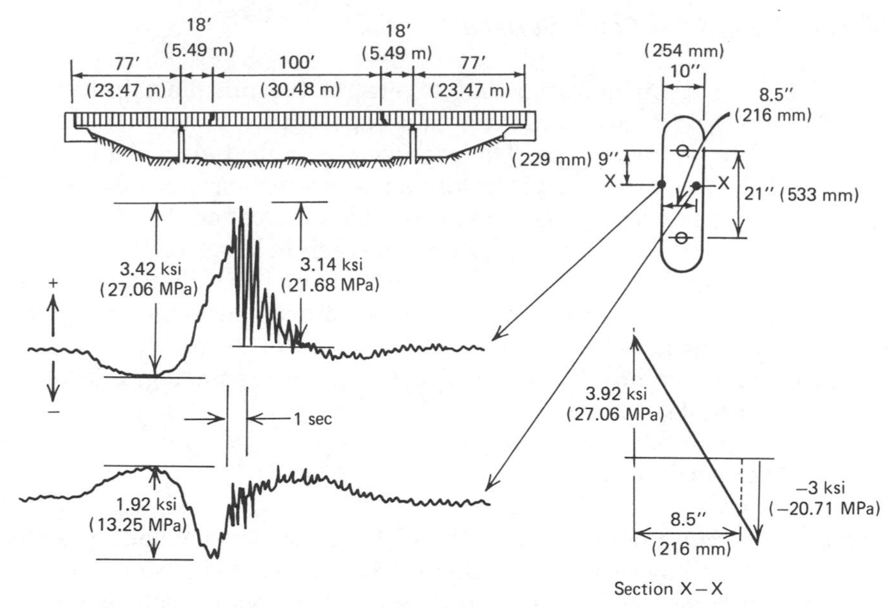

Figure 2.19 Strain-time response of pin plates of U.S. 309 Bridge.

each of the six girders. The center-to-center spacing of the pins was 21 in. (533 mm) which was identical to the Route 157 bridge. Although different in geometry, the strain measurements demonstrated that the pin plates acted as rigid links between the suspended span and the adjacent cantilever section. Figure 2.19 shows the strain-time response at a section 9 in. (229 mm) from the center of the top pin. The measurements indicated that the primary stresses on the pin plate cross section under live load was bending. Figure 2.19 also shows the stress gradient across the pin width. Strain measurements on the cross section at the pin were very small, as the corrosion product "welded" the plates and beam webs together. Extrapolation to the fixity point nearest the pin indicated that the cyclic stress could be as high as 15 ksi (103.5 MPa) for the 45 k (201 kN) test truck. These measurements suggest that high bending stresses were introduced into the pin plates of the Route 157 bridge. Also apparent in Figure 2.19 is the fact that a single passing vehicle produces several large stress cycles as the structure vibrates and deforms under loading. At least 5 stress cycles per vehicle passage occurred at the U.S. 309 bridge. Assuming similar behavior, the traffic on the Route 157 bridge would produce 15 to 20 million random variable stress cycles. Therefore higher loads would seem to cause crack initiation at the edge of the cutedge of the hanger plate, which is augmented by thermal expansion and contraction which introduces a daily thermal stress cycle into the pin plates.

Examination of the Crack Surfaces

Visual examination indicated that several of the pin plate links were completely cracked in two. Others had edge cracks between 1½ in. (38 mm) and 2½ in. (64 mm) long. Their location is indicated on Figure 2.17. Several of the cracks had a 45° failure surface which seems to be related to the bowed condition of the pin plates which introduces bending about the weak axis of the plate as well. As can be seen in Figure 2.18, the crack propagated straight across the plate width, about 1 in. (25 mm) from the edge of the nut and near the edge of the reinforcement plates that were welded to the surface of the web at the pinholes. No fractographic studies were carried out on any of the crack surfaces, nor are photographs of the surfaces available.

Failure Analysis

The primary cause of failure is believed to be the development of frozen pin joints caused by corrosion products. The hanger and girder web reinforcement became rigidly attached because of the water, salt, and other debris to which the hangers were exposed. This bonded the elements together and prevented the links from rotating and accommodating the repeated loading of traffic and thermal expansion of axially loaded members. The cracks developed in the cross section at the edge of the frozen joints, as illustrated in Figure 2.18.

The corrosion product effectively welded the hanger to the web and prevented the net section from becoming critical. The fracture surface of the hanger from beams 8 and 9 appeared to be "very old." The hangers on beams 5, 6, 7, and 11 were all bowed.

Although the cyclic stresses were not measured, it is likely that the stress range at the edge of the hanger plates exceeded the yield point when the heaviest truck traffic and high thermal stress cycles combined. Such circumstances may be expected to induce cracking and the subsequent growth of fatigue cracks from the plate edge.

2.2.3 Conclusions

In sum, the cracked hangers of the Route 157 bridge in Illinois appear to have been induced by repeated loads after the pin connection froze. The suspended span end displacements subjected the hanger to high cyclic in-plane bending stresses from traffic and thermal conditions. The resulting cumulative damage caused cracks to form and propagate at the edge of the corrosion "welded" region.

The highest thermal expansion and contraction stresses were probably induced during June to August and December to February. This corresponds to the highest and lowest temperature periods. Final failure may

Figure 2.20 Steel shapes shoring up cracked beams in span 2 (courtesy of Illinois Department of Transportation).

have occurred during the winter when the material toughness would be at its lowest level.

2.2.4 Repair and Retrofit

Until the broken hangers were replaced, the northbound passing lane was closed, and beams 8, 9, and 10 were shored, as illustrated in Figure 2.20. Fourteen additional hangers were replaced during subsequent maintenance operations. All work was carried out by Illinois DOT. The hangers on the beams which were added in 1966 were inspected and found to be satisfactory. No evidence of cracking was observed. All corrosion product was removed, and all the pins were freed for movement.

REFERENCES

2.1 National Transportation Safety Board, Collapse of US 35 Highway Bridge Point Pleasant, West Virginia, December 15, 1967, U.S. Department of Transportation, Washington, D.C., October 4, 1968.

2.2 Nishimura T., and Miki, C., Fracture of Steel Bridges Caused by Tensile Stress, *J. Japanese SCE* (November 1975, in Japanese).

2.3 Ballard, D. B., and Yakowitz, H., Mechanisms Leading to the Failure of the Point Pleasant, West Virginia Bridge—Part 3, National Bureau of Standards Report No. 9981 to U.S. Bureau of Public Roads, September 1969.

2.4 Scheffey, C. F., and Cayes, L. R., Model Tests of Modes of Failure of Joint C13N of Eyebar Chain—Point Pleasant Bridge Investigation, Federal Highway Administration Report No. FHWA-RD-74-19, U.S. Department of Transportation, Washington, D.C., January 1974.

2.5 Nishanian, J., and Frank, K. H., Fatigue Characteristics of Steel Used in the Eyebars of the Point Pleasant Bridge, Final Report, Federal Highway Administration Report No. FHWA-RD-7, No. FHWA-RD-73-18, U.S. Department of Transportation, Washington, D.C., June 1972.

2.6 Frank, K. H., and Galambos, C. F., Application of Fracture Mechanics to Analysis of Bridge Failures, Specialty Conference Safety and Reliability of Metal Structures of ASCE, Pittsburgh, Pa., November 2–3, 1972, pp. 279–306.

2.7 Poletto, R. J., Stress Distribution in Eyebars, M. S. Thesis, Lehigh University, Bethlehem, Pa., 1970.

2.8 Barsom, J. M., Investigation of Toughness Criteria for Bridge Steels, ARLR 97.018.001 (5), U.S. Steel Corp., Monroeville, Pa., February 8, 1973.

CHAPTER 3

Cover-Plated Beams and Flange Gussets

The possibility of fatigue cracks forming at the ends of welded cover plates was demonstrated at the AASHO Road Test in the 1960s [3.1]. Multiple beam bridges subjected to relatively high stress range cycles (~12 ksi, or 83 MPa) under controlled truck traffic experienced cracking after 500,000 vehicle crossings. In general, not many cracked details were known to exist until a cracked beam was discovered in span 11 of the Yellow Mill Pond Bridge in 1970. Between 1970 and 1981 the Yellow Mill Pond multibeam structures located at Bridgeport, Connecticut, have developed extensive numbers of fatigue cracks at the ends of cover plates. These cracks resulted from the large volume of truck traffic and the unanticipated low fatigue resistance of the large-sized cover-plated beam members (category E').

The King's Bridge in Australia is also well known for its failure due to fatigue cracking and fracture from very large weld toe cracks that developed during fabrication in all four girders [3.2, 3.3].

A third structure experienced cracking at the cover plate end of one beam at Route 21, Clifton City, New Jersey [1.1]. However, no detailed information is available on the crack or material.

In addition to welded cover plates, several other category E or E' details have experienced fatigue cracking at weld terminations. These include gusset plates welded to the flange and the termination of longitudinal stiffener welds at transverse stiffeners or connection plates. Three structures have exhibited cracking at the ends of flange gusset plates [1.1].

Typical of these structures and cracking is the crack formed in the Vermilion River Bridge on Illinois 1 at Danville originated at the end of a

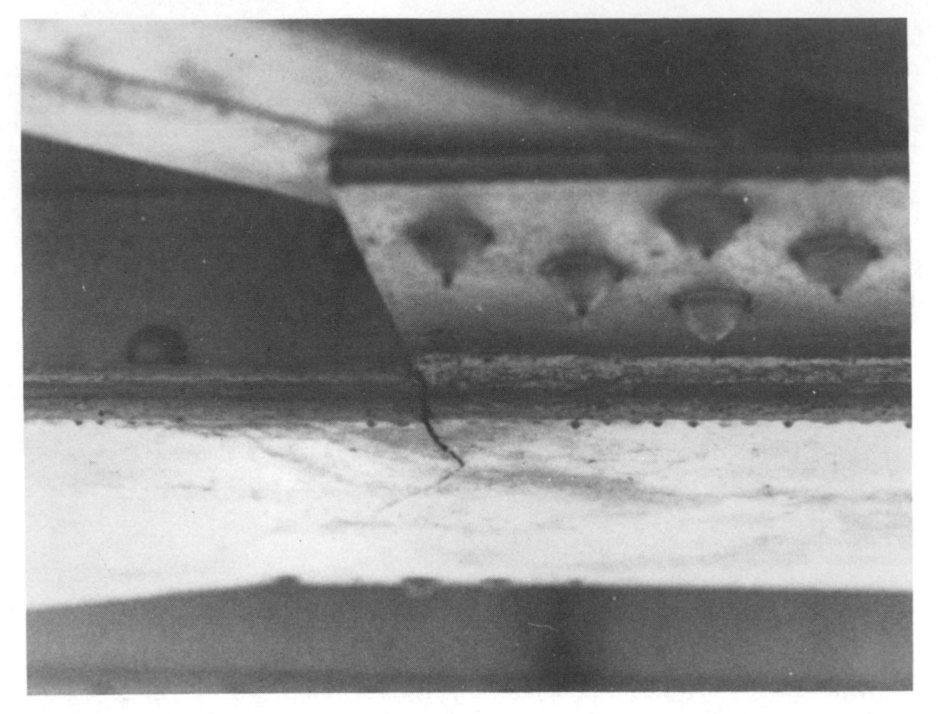

Figure 3.1 Crack at end of lateral connection plate (courtesy of Illinois Department of Transportation).

longitudinal fillet weld that was used to attach a 3 ft (0.9 m) long lateral connection plate to the edge of a 8 × 8 × 1 in. (200 × 200 × 25 mm) flange angle, as shown in Figure 3.1. The gusset plate lapped over the flange angle surface and was welded along both edges. An 18 in. (46 cm) wide cover plate was also welded to the flange angles.

3.1 FATIGUE-FRACTURE ANALYSIS OF YELLOW MILL POND BRIDGE AT BRIDGEPORT, CONNECTICUT

3.1.1 Description and History of the Bridge

Description of Structure

The Yellow Mill Pond Bridge carries Interstate I-95 (Connecticut Turnpike) over the Yellow Mill Channel (an extension of Bridgeport Harbor on Long Island Sound). The simple spans of the bridge complex were designed with consideration to the composite action between the rolled cover-plated steel beams and the 7.25 in. (184 mm) reinforced concrete

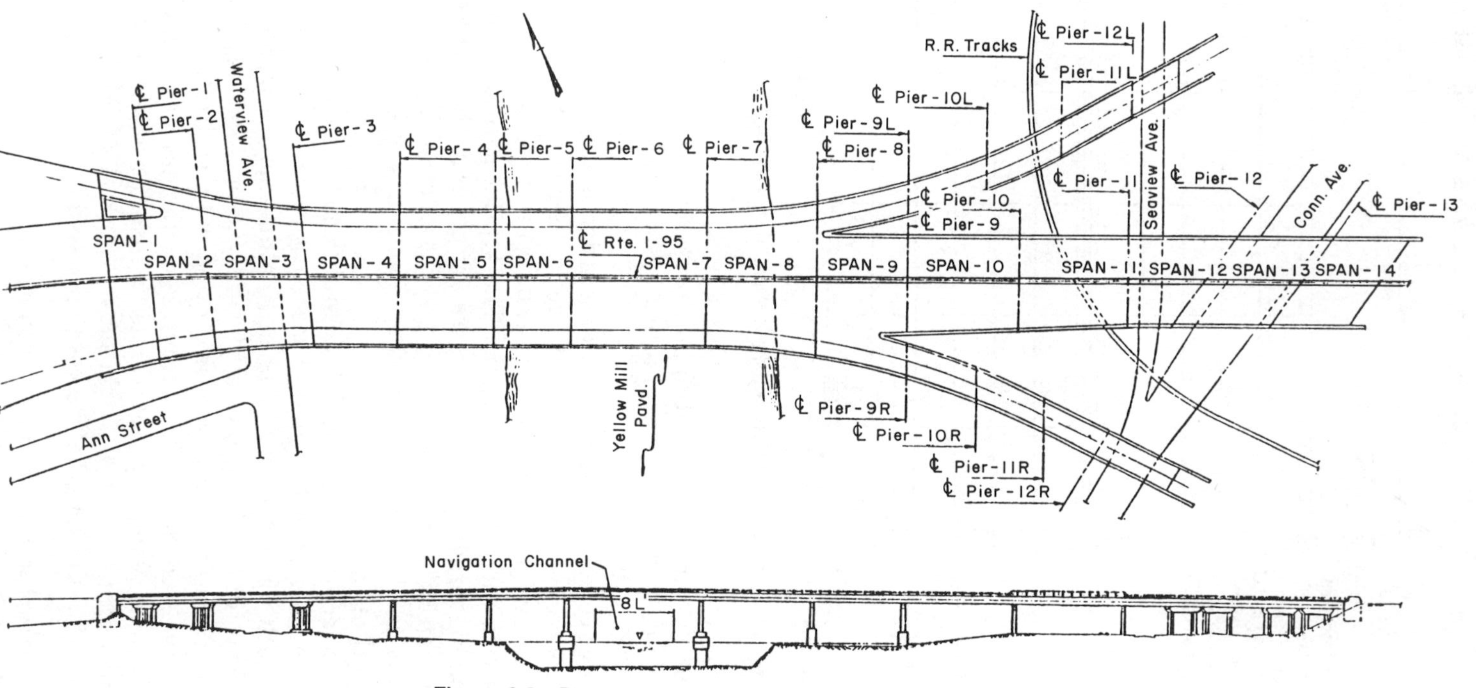

Figure 3.2 Plan and elevation of Yellow Mill Pond Bridge.

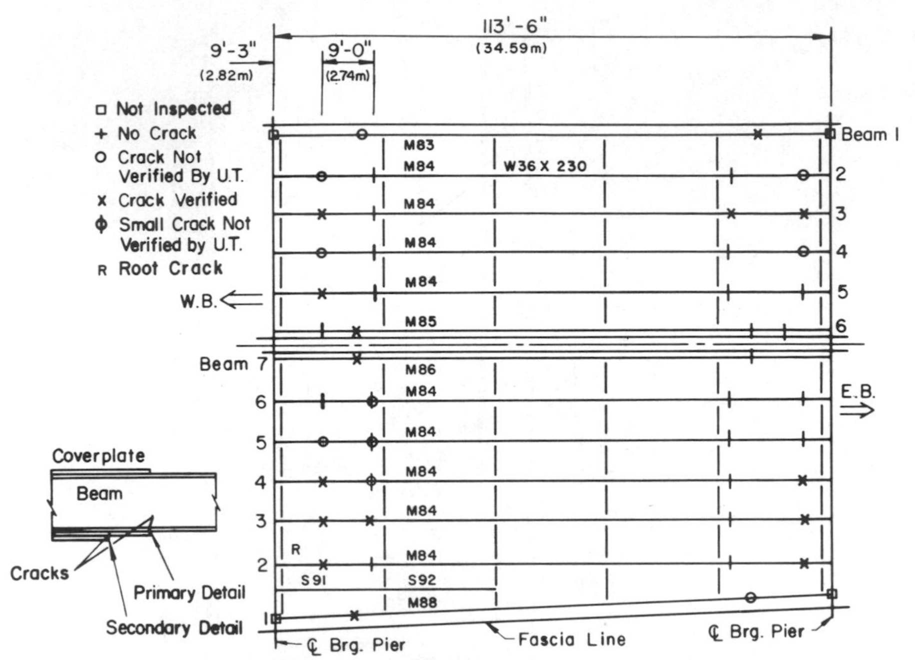

Figure 3.3 Plans of inspected details in span 10, Yellow Mill Pond Bridge.

slab. In 1969 a 2 in. (51 mm) bituminous concrete overlay was placed on the concrete deck. The A242 steel-rolled beams are W36 × 230, W36 × 280, and W36 × 300 sections.

The bridge complex consists of 14 consecutive simple span cover-plated steel and concrete composite beam bridges crossing the Yellow Mill Pond Channel (14 bridges in each direction of traffic); see Figure 3.2. Each bridge carries three lanes of traffic. The typical position of the main beams and diaphragms of the structure are given in the plan and cross section of span 10* in Figures 3.3 and 3.4. The external fascia beam (M88) of the eastbound bridge is skewed where four lanes of through traffic are reduced to three lanes.

In order to accommodate the Yellow Mill Pond Bridge to tidal shipping channel clearances, the designers chose to use cover-plated beams in lieu of deeper built-up girders. In the span 10 all beams are W36 A242 steel sections. The tension flanges of all beams except (M86) interior fascia beam of the eastbound roadway are fitted with multiple cover plates (primary and secondary). The compression flanges have single cover plates. All cover plates are partial length with the exception of the primary plates of the external fascia tension flanges of (M88) and (M83) which extend to full length. The cover plate ends are square, and their

* This is the typical span at which field strain measurements were recorded and stress range histograms developed; see [3.4].

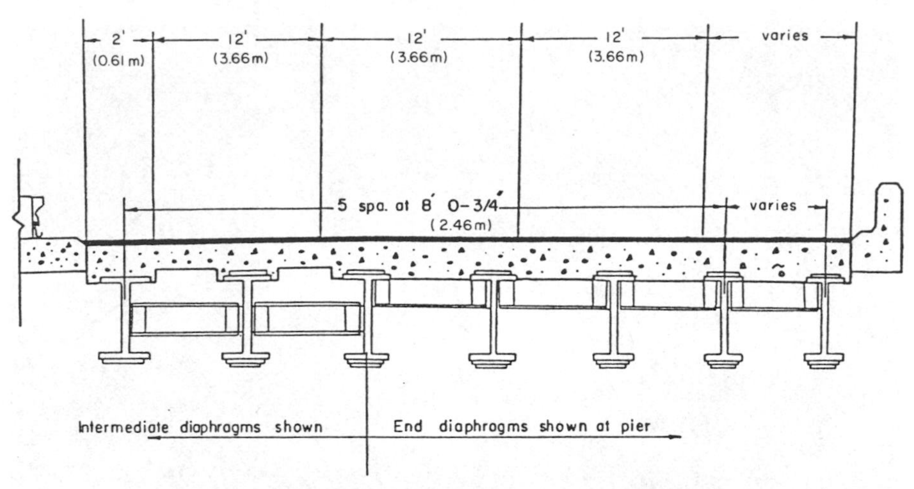

Figure 3.4 Typical expressway cross section (eastbound).

corners are rounded to a 3 in. (76 mm) radius of curvature (for the cover plate details, see Figure 3.6*a*).

History of the Structure and Cracking

The Yellow Mill Pond Bridge was constructed in 1956 to 1957. It was opened to traffic in January 1958. During October and November 1970 the steel superstructure of the Yellow Mill Pond Bridge was inspected, following the cleaning and repainting done by a contractor prior to that time. On November 2, 1970, during the inspection of the repainting workmanship, a crack was discovered in span 11 of the eastbound roadway. The crack extended from the west end of the cover-plated beam through the flange up to 16 in. (400 mm) into the web of one of the main girders (girder 4, span 11). A photograph of the fracture in the beam appears in Figure 3.5. The crack originated at the toe of the transverse fillet weld connecting the cover plate to the tension flange of the beam, as shown in Figures 3.6*a* and 3.6*b*.

Inspection of 15 similar locations on different girders of the bridge revealed incipient cracks of varying magnitude along eight of the cover plate end welds. The two beams adjacent to the fractured beam had fatigue cracks that extended halfway through the tension flange at the ends of cover plates. These were beams 3 and 5 in span 11 of the eastbound roadway. The crack depths in beams 3 and 5 were measured by ultrasonic testing, revealing a depth of 0.625 in. (16 mm) in each beam.

In December 1970 a section of the fractured girder was removed, and all three damaged girders were repaired with bolted web and flange splices. In November 1973, the east ends of beams 2 and 3 of the east-

Figure 3.5 Cracked girder G4 in eastbound span 11 of Yellow Mill Pond Bridge (courtesy of Connecticut Department of Transportation).

Figure 3.6a Typical large crack at weld toe at end of cover plate.

Figure 3.6*b* Smaller crack at weld toe of cover plate (crack $2\frac{3}{4}$ in. long).

bound roadway of span 10 were reinspected. Fatigue cracks were discovered in both girders at the toes of the primary cover plate transverse fillet welds. A crack depth of 0.375 in. (10 mm) in beam 2 was verified by a magnetic crack definer.

In June 1976 the cover plate details in the eastbound and westbound lanes of span 10 were inspected for fatigue cracking by a visual dye-penetrant and ultrasonic techniques. Twenty-two out of 40 details were found to have cracks. The smallest crack visible to the naked eye was 0.25 in. (6 mm) long. Ultrasonic inspection provided confirmation of cracks about 0.5 in. (13 mm) long at the west end of the eastbound lane in beams 3 and 7 of span 10. The results of the inspection in terms of crack depth and locations are given in [3.4] and [3.5].

In November 1976 during a brief inspection of span 13 with field glasses from the ground level four large cracks were observed. These cracks were 6 to 10 in. (150 to 250 mm) long and about 0.5 in. (13 mm) deep. They had in fact broken through the paint film at the weld toe (see Figure 3.6*a*).

In September 1977 span 10 was inspected again. The secondary details at beam 5 of the westbound bridge and the secondary details of beams 5 and 6 of the eastbound bridge at the east end were inspected for the first time. No cracks were observed. Small cracks were found at the west end of the secondary details of beams 4, 5, and 6 of the eastbound bridge. (These details had been previously inspected in 1976, but no cracks were found at that time.) In addition several cracks were found in details in span 12. The cracks were up to 15 in. (381 mm) long.

In November 1979 several details in span 10 were inspected again by sight and by ultrasonic means. The ultrasonic inspection verified the crack found during the 1977 inspection. The secondary detail on beam 4 of the eastbound bridge near the west end was cracked. The retrofitted details were also inspected, and only one crack was detected, which was growing from the root of the primary detail on the west end of beam 2 of the eastbound structure. The crack was 1.25 in. (32 mm) on the surface, and ultrasonic inspection indicated a 6 in. (152 mm) length at the root. Small cracks were found for the first time at the west end of the secondary detail on beam 2 of the eastbound bridge.

The history of inspection of span 10 and the location of the cracks detected or found each time are summarized in [3.4]. Figure 3.3 shows the location of these details.

3.1.2 Failure Modes and Analysis

Crack Locations and Failure Modes

In the case of the Yellow Mill Pond Bridge all cracks were observed to have formed at the ends of welded cover plates. They in fact propagated into the beam flange at the transverse weld toe in the same way that fatigue cracks do in laboratory tests.

Except for a small length of web at the first fracture location, none of the cracks exhibited a brittle fracture mode of crack propagation. The estimated minimum service temperature at the bridge site is about −10°F (−23°C). However, it is not known when the cleavage fracture occurred in the beam web. No other detail appeared to exhibit unstable crack extension even when the flange nearly cracked in two.

Cyclic Loads and Stresses

The eastbound and westbound bridges of span 10 were selected for two major stress history studies. Both of these studies were conducted by the state of Connecticut (CONN-DOT) and the Federal Highway Administration (FHWA). The first study was done in July 1971 [3.6], and the second between April 1973 and April 1974 [3.7].

In the July 1971 study two electrical resistance strain gauges were placed on the interior beams. One gauge was placed midspan and the other 4 in. (102 mm) from the end of the primary cover plate (the gauges were either on the bottom flange or on primary cover plates). A typical stress histogram is given in Figure 3.7 for the westbound lanes. The distribution and position of truck traffic in the lanes were simultaneously recorded, and truck weights were sampled [3.6].

The stress range histograms of the westbound bridges indicated that 94 to 99% of the stress ranges occurring at the cover plate ends fell within

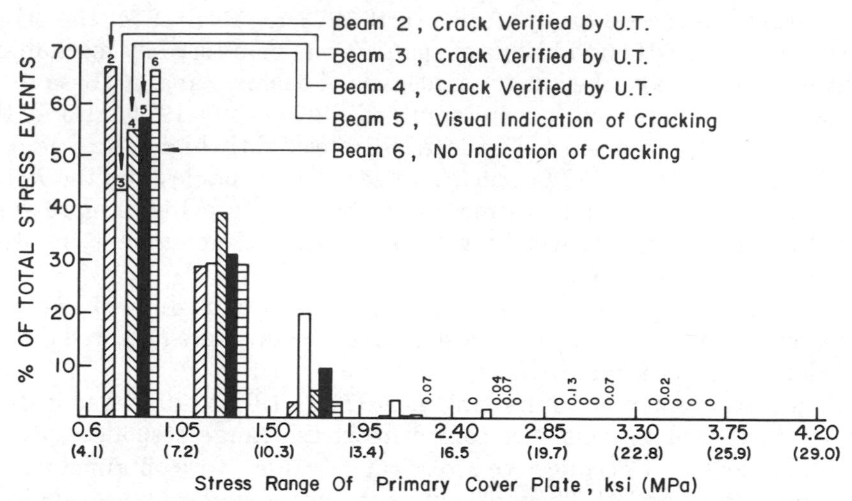

Figure 3.7 Stress range histogram, 1971 westbound span 10.

the limits 0.60 to 2.85 ksi (4.14 to 19.65 MPa). Thirty-five events exceeded 3.0 ksi (20.68 MPa). The overwhelming majority of stress ranges were below 3.6 ksi (24.82 MPa). A large event equal to 7.2 ksi (49.64 MPa) occurred in only one case. The gross truck weights were fairly evenly distributed between 10,000 and 70,000 lb, with the maximum weights recorded between 90,000 and 100,000 lb. The truck distribution reading indicated that approximately 55% of the trucks used the outer lane and 45% in the middle lane. Less than 1% were in the inner lane. The composition of the average daily truck traffic (ADTT) at the Yellow Mill Pond Bridge was about the same as the gross vehicle weight distribution developed from the 1970 FHWA nationwide Loadometer survey. The one-way average daily traffic (ADT) and the average daily truck traffic (ADTT) on span 10 was available for 1958 through 1975 [3.4]. From January 1958 to June 1976 approximately 259 million vehicles crossed span 10. Since approximately 13.5% of this traffic flow was truck traffic, approximately 35 million trucks crossed the eastbound and westbound bridges of span 10 between 1958 and 1976.

The reading of a gauge on the diaphragm provided some of the most surprising information. Within 63 hours, the bottom flange at midspan of the diaphragm was stressed to 3.0 ksi (20.7 MPa) or more 2400 times. There were 27 stress cycles greater than 5.4 ksi (37.3 MPa). The only explanation for this phenomenon has to do with a possible rotation and differential deflection of the main girders. Movement of the internal beams is restrained by the diaphragm at their points of intersection with the girders unless some lateral movement of the girders is permitted. This unexpected development in the Yellow Mill Pond structure may explain the high incidence of bolt failures in these connections.

During the second study, from April 1973 to April 1974, the stress history of the eastbound bridge of span 10 was monitored by mechanical strain recorders attached at the midspans of beams 3 and 4. These readings were compared with some obtained during July 1971, also at the midspans of beams 3 and 4. The data indicate slightly higher stress range events in the 1973 to 1974 period. This seems reasonable since the 1973–74 studies involved a longer time frame and likely provided a more unbiased survey, since overweight vehicles tended to circumvent the July 1971 measurement sites.

A more limited strain history record was acquired in June 1976 on the eastbound bridge at span 10. These measurements were acquired during a pilot retrofit program [3.5].

The measurements acquired in July 1971, in 1973 to 1974, and in June 1976 were used to construct composite stress range response spectra. Both Miner and RMS effective stress range values were obtained for the gauged locations [3.4]. The measurements for the highest stressed girder were

$$S_{r\,\mathrm{Miner}} = [\Sigma \alpha_i S_{ri}^3]^{1/3} = 1.98 \text{ ksi } (13.1 \text{ MPa})$$

These measurements indicated that the effective stress range varied from 1.1 ksi (7.6 MPa) up to the 1.98 ksi (13.1 MPa) value. Those beams located under the outside lane tended to provide the higher stress range values.

An examination of the stress response to truck passage also indicated that more than one stress cycle occurred. The variable stress spectrum appears to correspond to about 1.8 events per truck. This would correspond to 62,800,000 cycles if it were applied to each vehicle passage.

Mechanical, Chemical, and Fracture Properties of Materials

In December 1970, following the discovery of cracks, a portion of the flange and web adjacent to one of the cracks was removed for testing.

The chemical analysis of the material from the web and flange of the W36 × 300 rolled steel beam indicated that excessive manganese was present. The chemistry check based on ASTM A242 specifications yielded 1.69% manganese which is above the limit value of 1.45%. All other chemical elements appeared to be satisfactory and within tolerance limits [3.4].

Mechanical tests of the flange and web material provided a 57.8 ksi (398.5 MPa) yield point for the flange and a 95.1 ksi (656 MPa) tensile strength. The web material had a yield point of 57.2 ksi (394.4 MPa) and a tensile strength of 87.5 ksi (603 MPa). Both flange and web exhibited satisfactory elongation characteristics (i.e., 28%).

The fracture characteristics of the web and flange material were assessed using Charpy V-notch impact tests. These tests revealed an interesting difference in transition temperature of the flange and web mate-

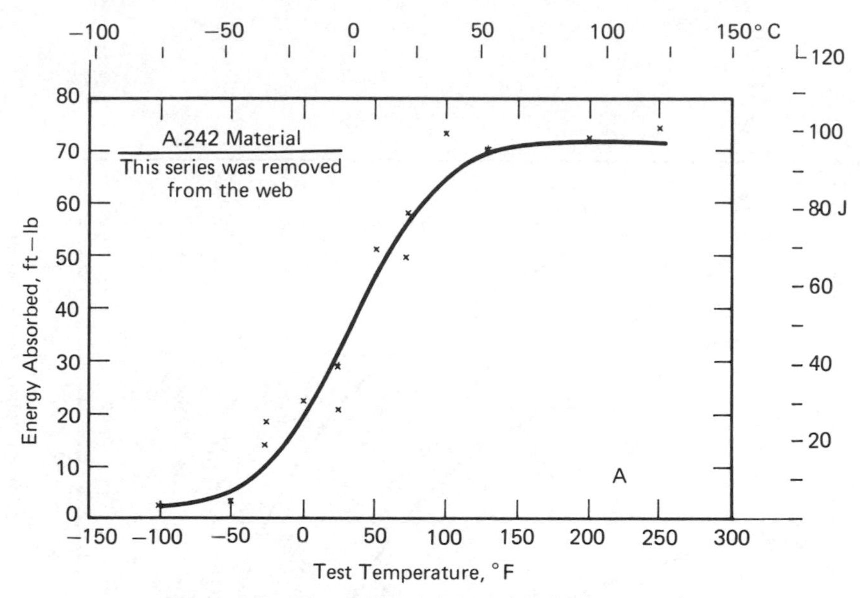

Figure 3.8 Charpy V-notch test data for beam web.

rial. The flange material provided a +55°F (13°C) transition temperature at 15 ft-lb (20 J), whereas the web provided a −10°F (−23°C) transition temperature. This is most likely due to the excess manganese and increased thickness of the flange material. The Charpy V-notch curves for both flange and web material are given in Figures 3.8 and 3.9. The transition temperature at 15 ft-lb (20 J) for the beam flange was 55°F (13°C).

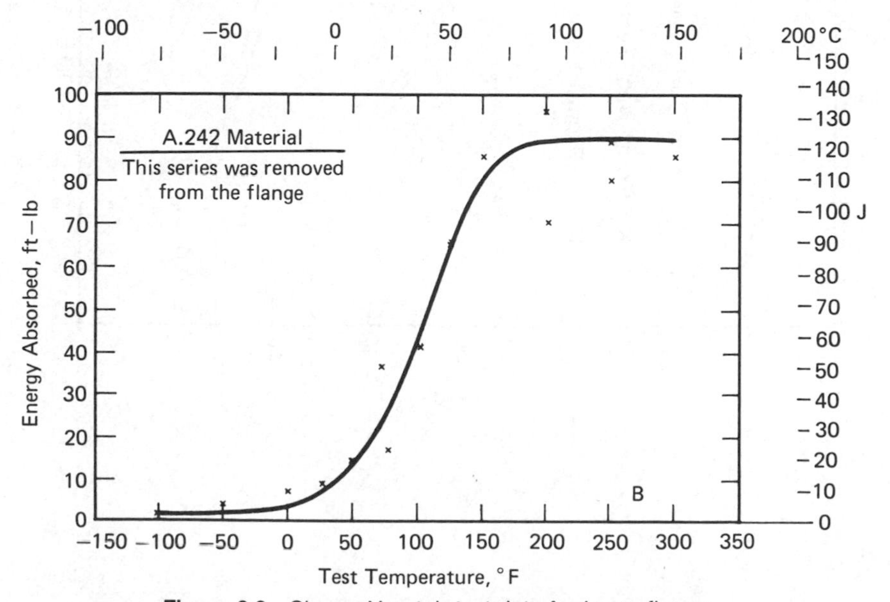

Figure 3.9 Charpy V-notch test data for beam flange.

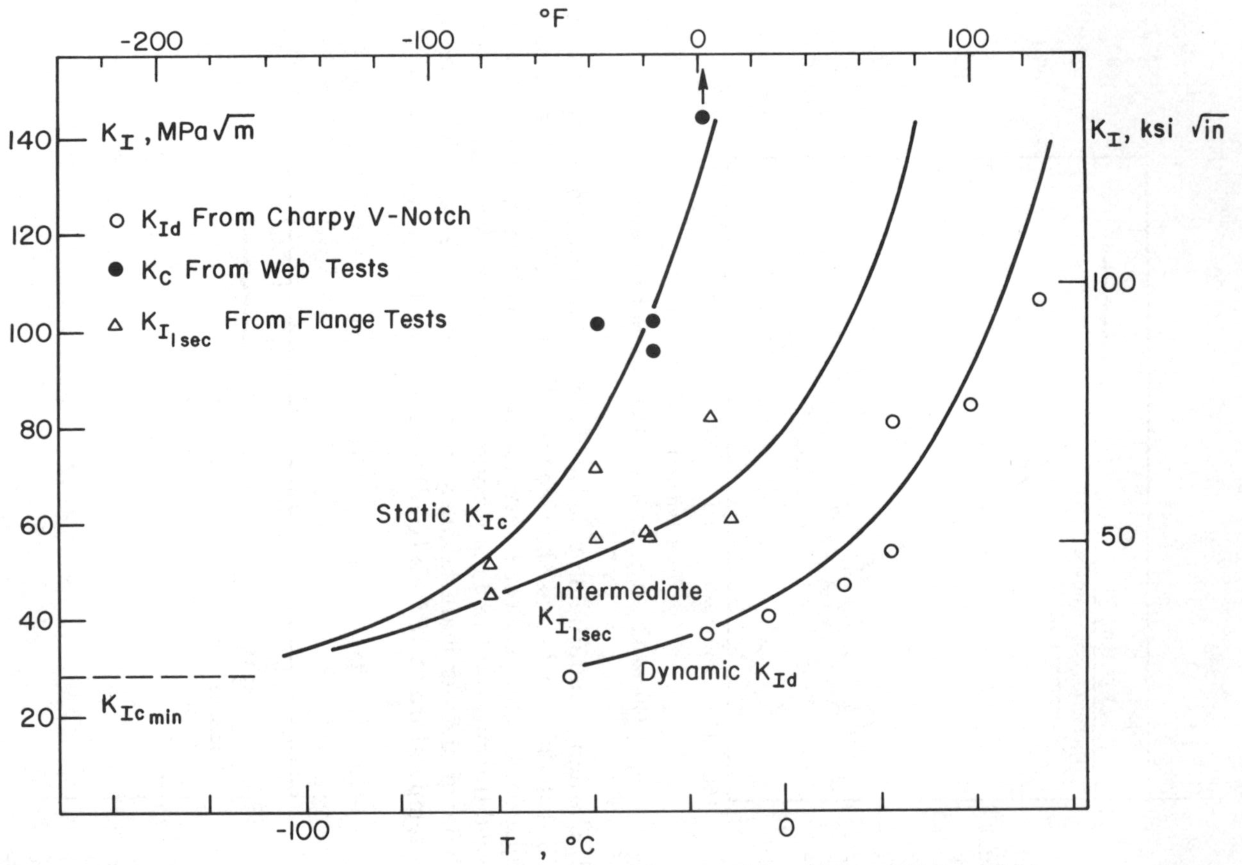

Figure 3.10 Fracture toughness for material removed from Yellow Mill Pond Bridge.

The beam section satisfied the material requirements of the AASHTO specification for temperature zone 1. Compact tension tests were also carried out, and these provided 1-sec fracture toughness values of about 80 ksi $\sqrt{\text{in.}}$ (88 MPa $\sqrt{\text{m}}$) at -10°F (-23°C), as illustrated in Figure 3.10.

Visual and Fractographic Examination of Crack Surfaces

The crack surfaces in the web and flange were subjected to visual examination in order to evaluate crack growth regions. Figure 3.11 shows the crack surface in its heavily oxidized and corroded condition. Three distinct areas or regions were noted. The fracture surface examinations indicated that fatigue crack growth started at the weld toe and grew completely through the flange and into the web about 2 in. (51 mm). The crack path changed when it encountered a lamination condition near the center of the flange. The lamination retarded the advancement of the path.

Near the edges of the beam flange the crack path moved into a second crack growth region away from the weld toe. At this point, tensile fracture partially occurred, as evidenced by some apparent necking. The fracture surface in the web characteristically indicates that rapid crack growth developed after a 2 in. (51 mm) fatigue crack had penetrated into the web. The crack extended about 7 in. (178 mm) up the web and was

Figure 3.11 Corroded and paint-covered crack surfaces. (*a*) Crack surface of beam flange.

Figure 3.11 Corroded and paint-covered crack surfaces. (*b*) Crack surface adjacent to cover plate weld.

finally arrested near middepth. The fracture surface suggested a "brittle fracture" mode in the web. There was also some evidence of shear lips at the surface of the web plate. It is highly probable that the rapid crack extension occurred in September 1970 when a large overload crossed the bridge. A third region of apparent slow crack growth, which likely occurred after the large overload, was also observed.

The fracture surface was coated with oxide and paint. This suggested that the crack existed for several months prior to its discovery on Novem-

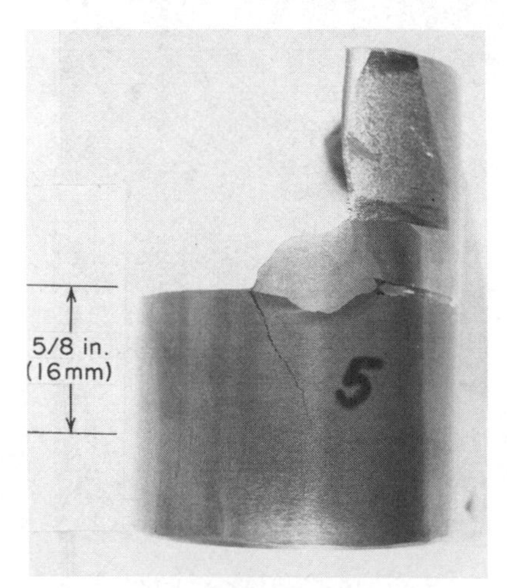

Figure 3.12 Core showing part through flange crack.

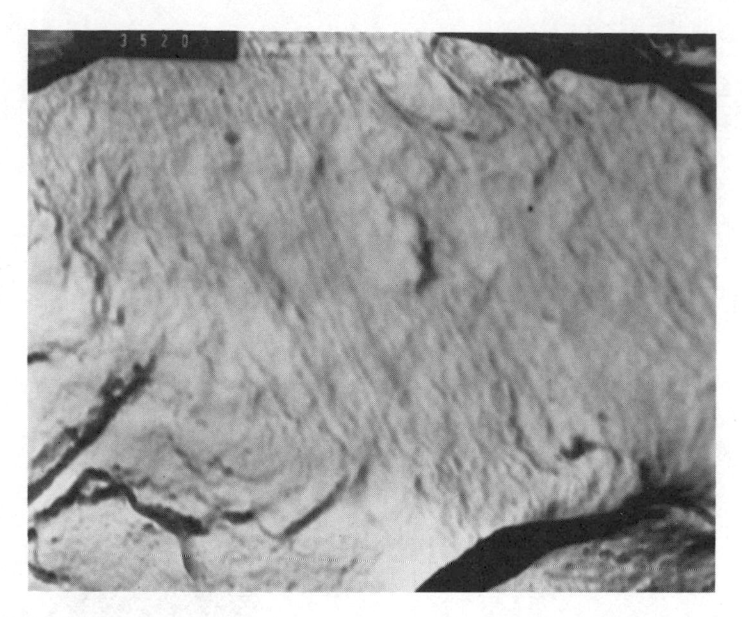

Figure 3.13 Fatigue crack striations near crack tip from transmission electron microscope fractograph, 56,000×.

ber 2, 1970, after the structure was painted during the summer and early fall of 1970.

During 1981 cores were removed from several of the beams in order to examine the crack surfaces and the retrofit conditions carried out in 1976. Figure 3.12 shows the edge of a polished and etched core with the crack at the weld toe which can be seen to extend more than halfway through the beam flange. The crack surface of the core was exposed and examined with an electron microscope. Since the surface was relatively clean and not extensively corroded, the crack surface features were not destroyed. Figure 3.13 shows a transmission electron microscope fractograph at a magnification of 56,000×. Striationlike features can be seen on the crack surface at various locations. The random variable loading produces striations of variable height and spacing. The measured spacing varied from 10^{-6} to 7×10^{-7} in./cycle (4 to 3×10^{-8} mm/cycle).

Failure Analysis

The cracks forming at the cover plate weld toes were modeled as semielliptical surface cracks in the flange [3.8, 3.9]. The stress intensity was defined as

$$K = F_e F_s F_w F_g \sigma \sqrt{\pi a} \tag{3.1}$$

where

$$F_e = \frac{1}{E(k)} \qquad E(k) = \int_0^{\pi/2} [1 - k^2 \sin^2 \theta]^{1/2}\, d\theta \qquad k^2 = \frac{c^2 - a^2}{c^2} \tag{3.1a}$$

$$F_s = 1.211 - 0.186\sqrt{\frac{a}{c}} \tag{3.1b}$$

$$F_g = \frac{K_{tm}}{(1 + 1/0.1473)[(a/t_f)^{0.4348}]} \tag{3.1c}$$

$$K_{tm} = 3.539 \ln\left(\frac{Z}{t_f}\right) + 1.981 \ln\left(\frac{t_{cp}}{t_f}\right) + 5.798$$

$$F_w = 1.0 \quad \text{and} \quad c = 5.46a^{1.133}\ \text{(in.)}$$

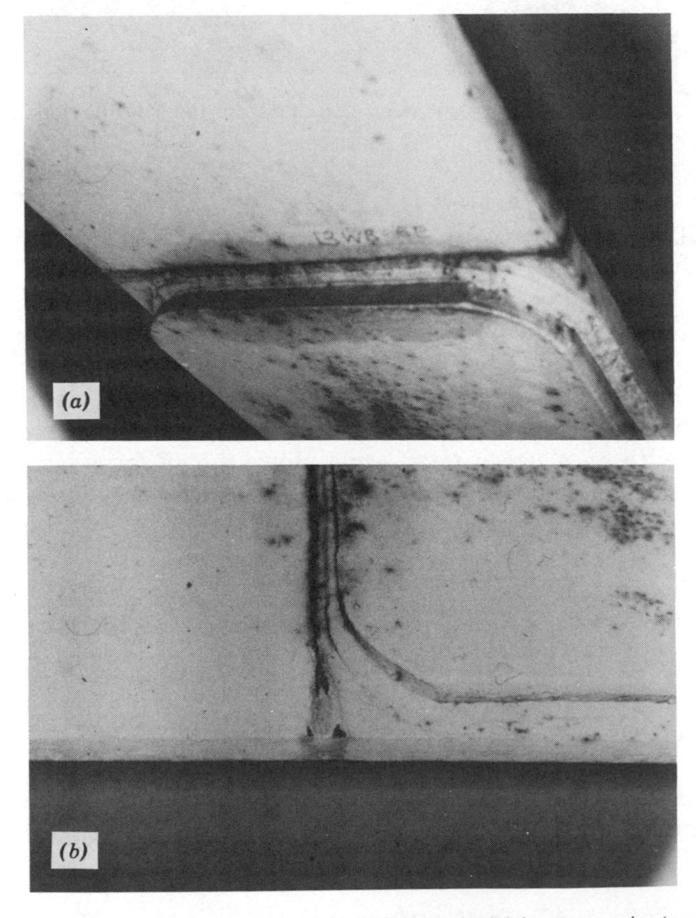

Figure 3.14 Beam flange nearly fatigue crack through. (*a*) Large crack at end of cover plate; (*b*) segment of beam flange showing ductile yielding.

With $Z = 0.63$ in. (16 mm), $t_f = 1.26$ in. (32 mm), $t_{cp} = 1\frac{1}{4}$ in. (32 mm), $a_i =$ 0.03 in. (0.75 mm), and $S_{r\text{Miner}} = 1.9$ ksi (13 MPa), the cycles required to grow a fatigue crack 1 in. (25 mm) deep was estimated as

$$N = \int_{0.03}^{1.0} \frac{da}{3.6 \times 10^{10} \Delta K^3} = 36 \times 10^6 \text{ cycles} \tag{3.2}$$

This is in reasonable agreement with the larger fatigue cracks that formed between 1958 and 1976 when an estimated 35 million trucks crossed the bridges.

The minimum fracture toughness of the flanges of the A242 steel beams was estimated to be 80 to 100 ksi $\sqrt{\text{in.}}$ (88 to 110 MPa $\sqrt{\text{m}}$) at the minimum service temperature of −10°F (−18°C) on the basis of the test results plotted in Figure 3.10. Since the crack tip is far removed from the local stress concentration and surface residual tensile stress due to the fillet welded cover plate, the stress gradient correction factor, $F_g \simeq 1.0$. Furthermore the residual stresses are relatively small near the crack tip. An estimated value of $\sigma_{rs} \simeq 10$ ksi (69 MPa) was used for the residual stress field in the flange-web core. The dead load stress at the cover plate termination was about 4.5 ksi (31 MPa), and the maximum live load stress was about 6 ksi (55 MPa). Hence under the extreme load condition the maximum stress intensity was

$$K = F_e F_s F_w \sigma \sqrt{\pi a} = 91 \text{ ksi} \sqrt{\text{in.}} < K_c \tag{3.3}$$

where:

$$F_e = 0.96, \quad F_s = 1.211 - 0.186\sqrt{\frac{1}{5.5}} = 1.132$$

$$F_w = \sqrt{\sec \frac{\pi}{2 \times 1.4}} = 2.3, \qquad \text{and} \qquad \sigma = 21.5 \text{ ksi (148 MPa)}$$

Hence it is readily apparent that the entire beam flange can be cracked through without initiating crack instability. This phenomenon may be observed in Figure 3.14 where the flange is nearly cracked in two. It occurs because the outer surface layers are generally higher in toughness than the full thickness estimates, so the residual stresses may have less effect than estimated.

3.1.3 Conclusions

The field observations on cracking of the Yellow Mill Pond Bridge yielded the following conclusions:

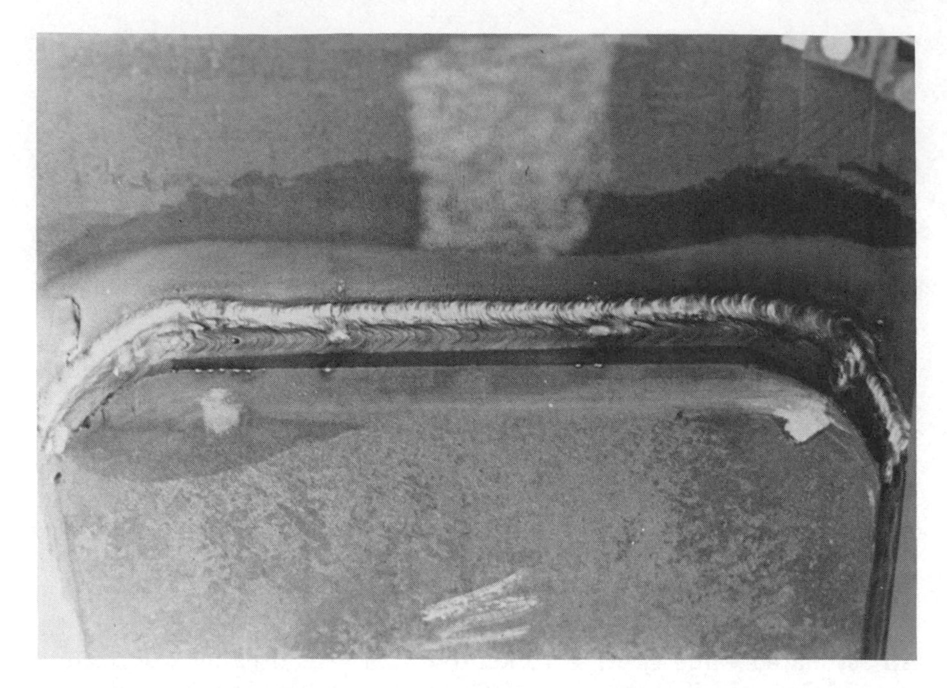

Figure 3.15 Gas tungsten arc remelted weld toe.

Figure 3.16 Peened weld toe.

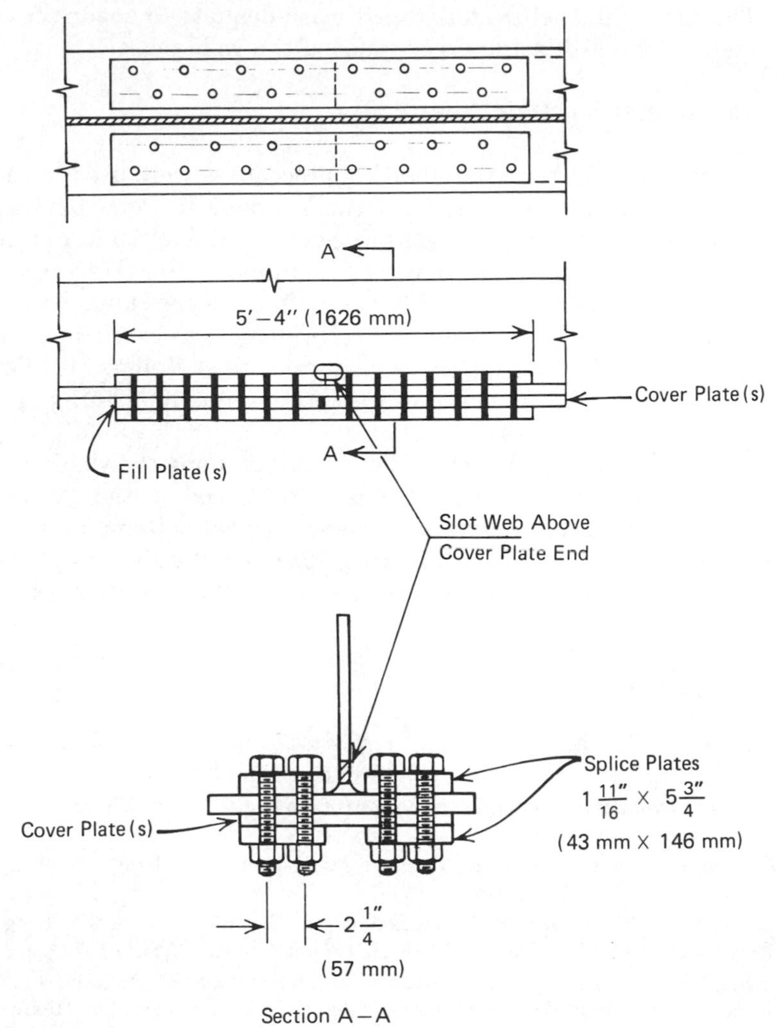

Figure 3.17 Retrofit details of bolted splices.

1. Significant fatigue crack growth can occur at welded cover plate ends when a few stress cycles in the variable stress spectrum exceed the constant cycle fatigue limit and a large number of stress cycles accumulate ($\geqq 10^7$).
2. The observed field behavior at the Yellow Mill Pond Bridge is compatible and corroborated by the results of laboratory tests on full-scale cover-plated beams. The results on small-size specimens underestimate the actual field behavior of full-sized cover-plated beams.

3. The material fracture toughness was adequate to ensure development of the full fatigue resistance of the weld detail.

3.1.4 Repair and Fracture Control

Peening and gas tungsten arc remelting procedures were used to retrofit the cover-plated beams in span 10 of the Yellow Mill Pond Bridge as a pilot study. Twenty-five of the cover plate details in span 10 were treated. Of these, 14 were peened, and 11 were gas tungsten arc (GTA) remelted. Figure 3.15 shows a transverse fillet weld after the gas tungsten retrofit at each original cover plate weld. All retrofitting was carried out under normal traffic without shoring. A peened weld toe at Yellow Mill Pond is shown in Figure 3.16. The depth of indentation due to peening was approximately 0.03 in. (0.8 mm).

All cracks that exceeded $1\frac{1}{2}$ in. (38 mm) length along the weld toe were spliced with bolted butt splices. Holes were placed in the girder web directly above the crack. Figure 3.17 shows typical bolted joints used to splice across the crack locations. During 1981 the remaining 427 details at the Yellow Mill Pond structures were retrofitted by peening or bolting.

REFERENCES

3.1 Fisher J. W., and Viest, I. M., Fatigue Life of Bridge Beams Subjected to Controlled Truck Traffic, Preliminary Publication, 7th Cong. IABSE, 1964.

3.2 Report of Royal Commission into the Failure of the King's Bridge, Victoria, Australia, 1963.

3.3 Madison, R. B., and Irwin, G. R., Fracture Analysis of King's Bridge, Melbourne, *J. Struct. Div.,* ASCE, 97 (September 1971).

3.4 Fisher, J. W., Slockbower, R. E., Hausammann, H., and Pense, A. W., Long Time Observation of a Fatigue Damaged Bridge, *J. Tech. Councils,* ASCE, 107 (April 1981).

3.5 Fisher, J. W., Hausammann, H., Sullivan, M. D., and Pense, A. W., Detection and Repair of Fatigue Damage in Welded Highway Bridges, National Cooperative Highway Research Program Report 206, June 1979.

3.6 Bowers, D. G., Loading History Span No. 10, Yellow Mill Pond Bridge I-95, Bridgeport, Connecticut, Bureau of Highways Report, Department of Transportation, State of Connecticut, May 1972.

3.7 Dickey, R. L., and Severga, T. P., Mechanical Strain Recorder on a Connecticut Bridge, Report FCP 45 G1-222, 1974.

3.8 Fisher, J. W., Frank, K. H., Hirt, M. A., and McNamee, B. M., Effect of Weldments on the Fatigue Strength of Steel Beams, NCHRP Report 102, 1970.

3.9 Zettlemoyer, N., and Fisher, J. W., Stress Gradient Correction Factor for Stress Intensity at Welded Stiffeners and Cover Plates, *Welding J.* 56 (December 1977): 393s–398s.

CHAPTER 4

Web Connection Plates

Cracks that have developed in the web at lateral connection plates have generally occurred because of intersecting welds. The lateral connection plate is often framed around a transverse stiffener. It was used to connect diaphragms and lateral bracing members to the longitudinal girders of bridges. One of the first bridge structures to exhibit this cracking was the Lafayette Street Bridge over the Mississippi River at St. Paul, Minnesota. Figure 4.1 shows the cracked girder at the connection plate to the web. The primary problem was the large defect in the weld attaching the lateral connection plate to the transverse stiffener. Since this weld was perpendicular to the cyclic stresses, and intersected with the vertical welds attaching the stiffener to the web and the longitudinal welds of the connection plate, a path was provided into the girder web.

Similar cracks have formed in at least five other bridges with similar intersecting welds (1.1). In two of these cases the cracks were discovered before the girder flange cracked in two.

Cracking was also observed in two other bridges where the lateral connection plates were installed on each side of the transverse stiffeners but only welded to the girder web. This often provided an intersecting weld or short gap condition between the transverse stiffener welds and the longitudinal welds attaching the connection plates to the girder web. Since no connection was provided between the lateral connection plate and the transverse stiffeners, out-of-plane movement occurred in the web gap or at the intersecting welds. The resulting high cyclic out-of-plane bending stress conditions in the web caused vertical cracks to form in the web. Figure 4.2 shows a web crack that resulted from such distortions.

Figure 4.1 Cracked girder of the Lafayette Street Bridge.

4.1 FATIGUE-FRACTURE ANALYSIS OF LAFAYETTE STREET BRIDGE

4.1.1 Description and History of the Bridge

Description of Structure

The Lafayette Street Bridge spans the Mississippi River at St. Paul, Minnesota. The main channel crossing consists of two parallel structures composed of two main girders extending three spans over the Mississippi River with the central main span (span 10) of 362 ft (110 m) and side spans of 270 ft (82 m) and 250 ft 6 in. (76 m) corresponding to spans 9 and 11, respectively (see Figure 4.3). These continuous main girders extend beyond piers 8 and 11 as 40 ft (12 m) cantilevers. The transverse cross section consists of two main girders connected by transverse floor beams and by transverse bracing, as shown in Figure 4.3. The transverse floor beams support two WF stringers.

Figure 4.2 Crack in girder web at lateral gusset plate–transverse stiffener web gap.

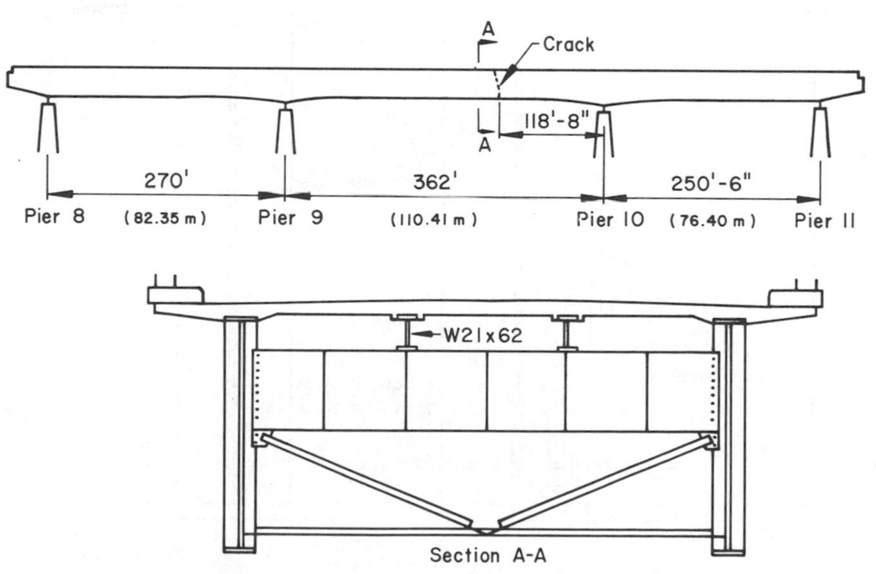

Figure 4.3 Schematic of span and cross section.

The web and flanges of the main girders were fabricated from ASTM A441 steel.

History of Structure and Cracking

The Lafayette Street Bridge was opened to traffic on November 13, 1968. The southbound lanes were closed between May 20 and October 25, 1974, for repairs to the deck and overlays.

On May 7, 1975, a crack was discovered in the east main girder of the southbound structure in Span 10 (see Figure 4.1). Figure 4.4 is a sche-

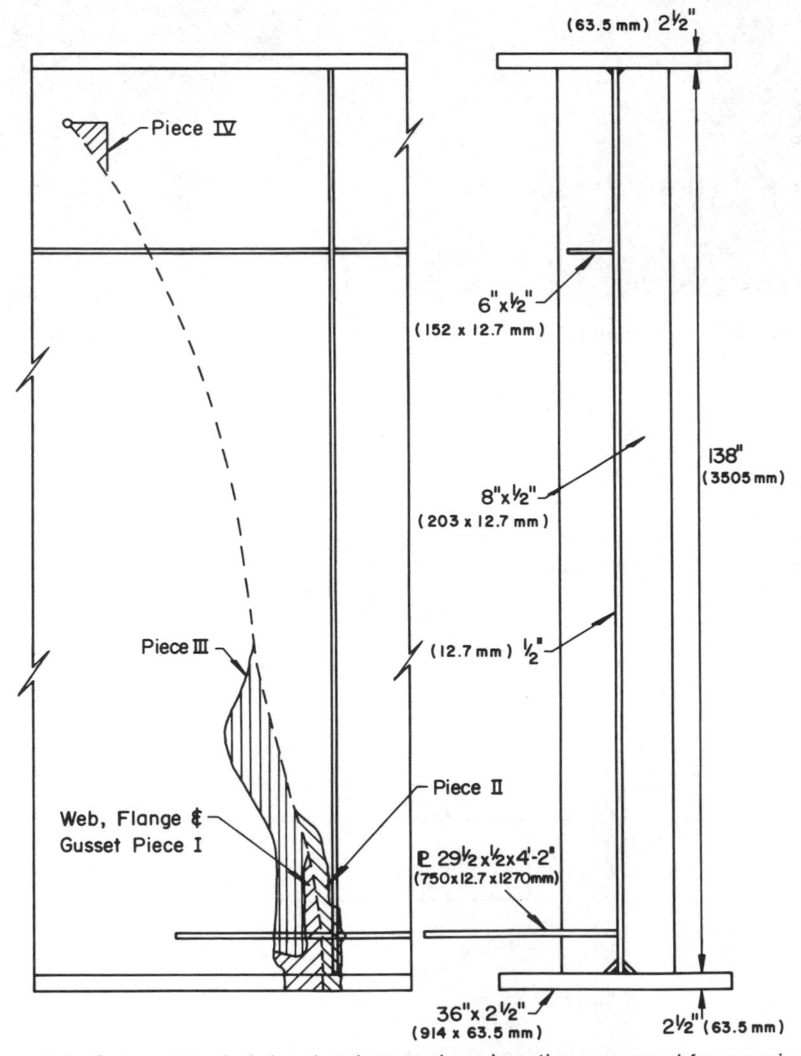

Figure 4.4 Schematic of girder showing crack and sections removed for examination.

matic of the girder section containing the crack and also shows the location of three segments that were removed for investigation. Figures 4.5, 4.6, and 4.7 show portions of the crack surface in the main girder web and flange and the lateral bracing gusset plate–stiffener connection.

The crack occurred in the "east" girder of the southbound lane in span 10, 118 ft 8 in. (36 m) from pier 10 (see Figure 4.3). The web crack had propagated to within 7.5 in. (19 cm) of the top flange when it was discovered that day. The entire bottom flange was fractured (see Figure 4.7).

A detailed study [4.1, 4.2] and visual examination of fractured portions of the girder web, flange, and the gusset plate for the lateral bracing indicated that fatigue crack growth originated in the weld between the gusset plate and the transverse stiffener as a consequence of a large lack of fusion discontinuity in this location (see Figures 4.6 and 4.10). The fracture surface of the web shows that a brittle or cleavage fracture occurred after the fatigue crack propagated into the web through the gusset plate–stiffener weld. As can be seen in Figure 4.7, the cleavage

Figure 4.5 Fracture surface of gusset to web showing lack of fusion. Piece II.

Figure 4.6 Fracture surface of gusset to stiffener weld and adjacent web. Piece II.

Figure 4.7 Fracture surface of web-flange boundary.

fracture of web continued and also extended down into the bottom flange, and consequently the entire bottom flange was broken. The cleavage fracture in the web was arrested 4 to 6 in. (100 to 150 mm) above the gusset. The web fracture surface was reported as shiny metal without a significant oxide coating from a point 4 to 6 in. (100 mm to 150 mm) above the gusset to the end of the crack near the top flange [4.1]. The balance of the cracked section in the web and the flange was coated with corrosion product.

4.1.2 Failure Modes and Analysis

Cyclic Loads and Stresses

The estimated average daily truck traffic (ADTT) crossing the bridge was 1500 vehicles during the period November 1968 to May 1975. Thus approximately 3,300,000 trucks crossed the bridge prior to the time the crack was discovered.

Stress range histograms for the main girders and other structural members are not available. However, the Miner effective stress range was approximated from the nationwide gross vehicle weight distribution [4.3]. A standard HS20 truck was used to generate a moment range in the main girder which varied from +3814 ft-kips (+5.17 kNm) to −1622 ft-kips (−2.20 kNm). This results in a stress range of 4.68 ksi (32 MPa) in the girder flange and 4.13 ksi (28 MPa) at the gusset-web connection. A few strain measurements were acquired on the main girders near the fracture during passage of an HS20 type truck and resulted in a stress range of about 2 ksi (14 MPa). Based on the measured strains and the gross vehicle weight distribution, an effective stress range $S_{r\text{Miner}} = 2$ ksi (13.8 MPa) was used to assess the fatigue resistance.

During the winter of 1975 the temperature at St. Paul, Minnesota, was recorded at −8°F (−22°C) on March 13, 1975, and on February 9, 1975, the temperature in St. Paul reached −22°F (−30°C).

Mechanical, Chemical, and Fracture Properties of Material

Four pieces (I to IV, as shown in Figure 4.4) of material from the flange, web, and gusset plate were removed for testing. Charpy V-notch tests, tensile tests, and chemical analysis were performed.

The $2\frac{1}{2}$ in. (64 mm) thick flange plate was found to have a yield strength of 46 ksi (317 MPa) and a tensile strength of 76.2 ksi (525 MPa). The $\frac{1}{2}$ in. (13 mm) web plate had a yield strength of 53.7 ksi (370 MPa) and a tensile strength of 81.8 ksi (564 MPa). The Charpy V-notch test results are summarized in Figure 4.8. The A36 steel gusset plate had a yield strength of 37.9 ksi (260 MPa) and a tensile strength of 67 ksi (462 MPa). In addition a few compact tension tests (K_Q) were conducted on the web

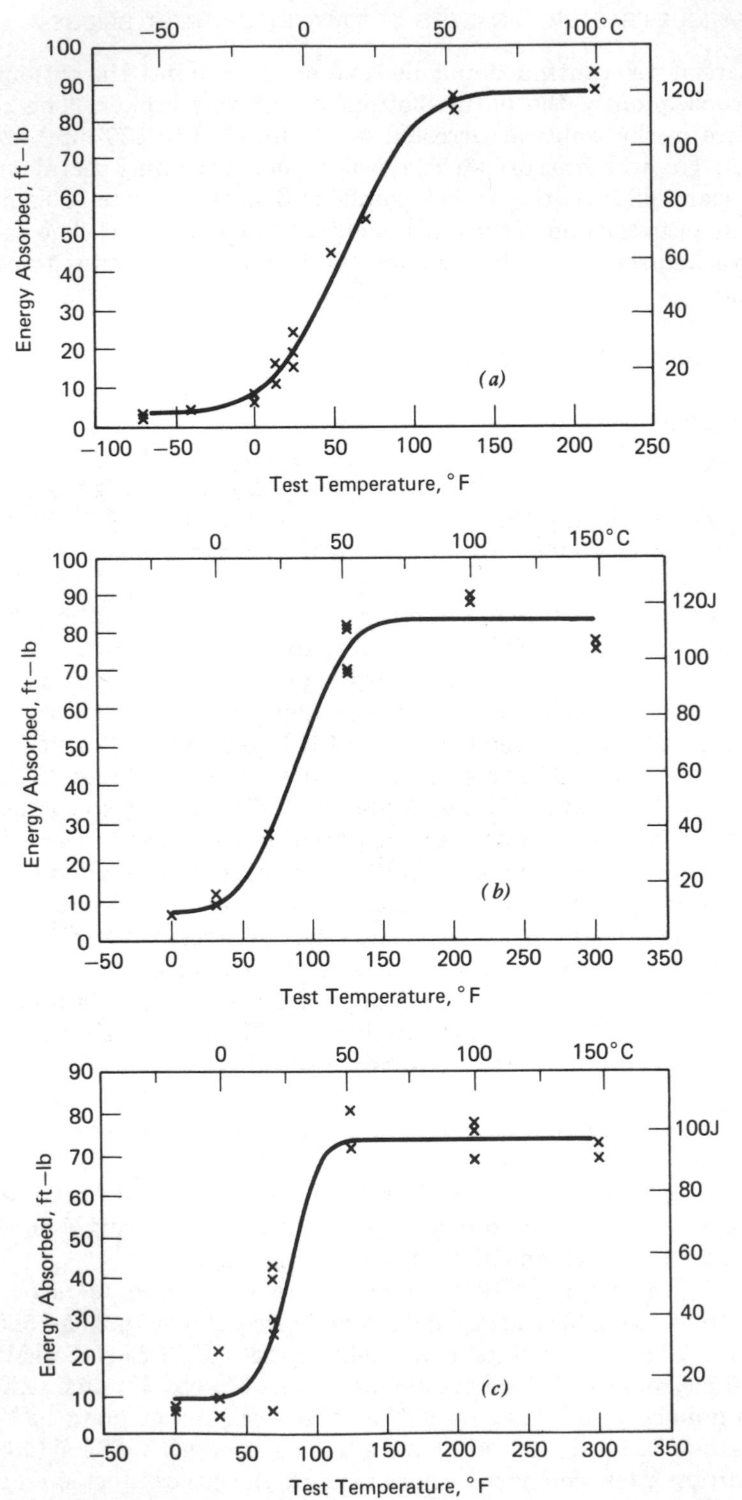

Figure 4.8 Charpy V-notch test results. (*a*) Web; (*b*) flange; and (*c*) gusset.

plate. The Charpy V-notch tests were transformed into dynamic fracture toughness values by using the Barsom correlation equation [2.8]:

$$K_{Id} = (5E\,\text{CVN})^{1/2} \quad (\text{ksi}\ \sqrt{\text{in.}}) \tag{4.1}$$

The measured K_Q values and the dynamic fracture toughness values from the Charpy V-notch tests are plotted in Figure 4.9. A static fracture toughness was also estimated by shifting the dynamic fracture toughness curve by

$$T(°\text{F}) = 215 - \frac{3}{2}\sigma_Y \tag{4.2}$$

The test results suggest a fracture toughness at -22°F between 70 ksi $\sqrt{\text{in.}}$ and 90 ksi $\sqrt{\text{in.}}$ (88 MPa $\sqrt{\text{m}}$ and 99 MPa $\sqrt{\text{m}}$).

The web plate was observed to satisfy the fracture toughness requirements of the 1974 AASHTO Specifications for Temperature Zone 2 (15 ft-lb., or 20 J, reached at 15°F or -9°C, for the web material). The flange and the gusset plates, however, only met the requirements of Temperature Zone 1.

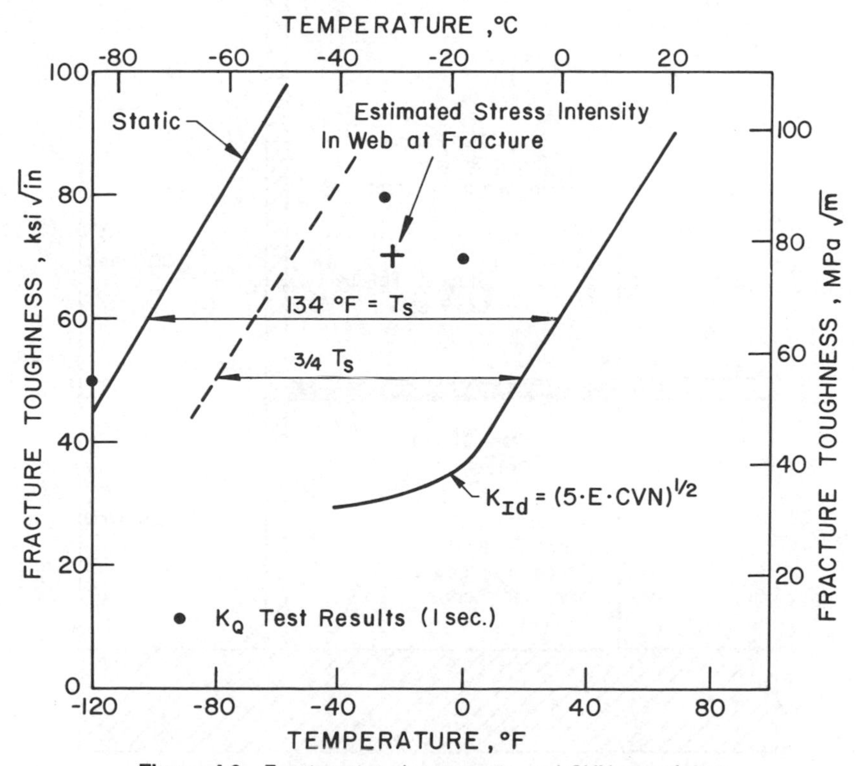

Figure 4.9 Fracture toughness tests and CVN correlation.

The flange and web plate show correspondence with the requirements of A441 and the gusset plate with A36 steel.

Visual and Fractographic Examination of Crack Surfaces

The fracture surfaces of part of the web, flange, and gusset plate are given in Figures 4.5, 4.6, and 4.7. From these photographs it can be seen that the transverse and longitudinal single bevel groove welds that connect the gusset plate to the transverse stiffener and web did not penetrate to the backup bar at the root of the weld near the web. This resulted in a significant lack of fusion in a plane perpendicular to the stress field.

Further visual examination of the fracture surface indicated that fatigue crack growth initiated in the weld between the gusset plate and the transverse stiffener as a consequence of the fusion discontinuity mentioned earlier. The fracture surface of the web indicates that mainly a

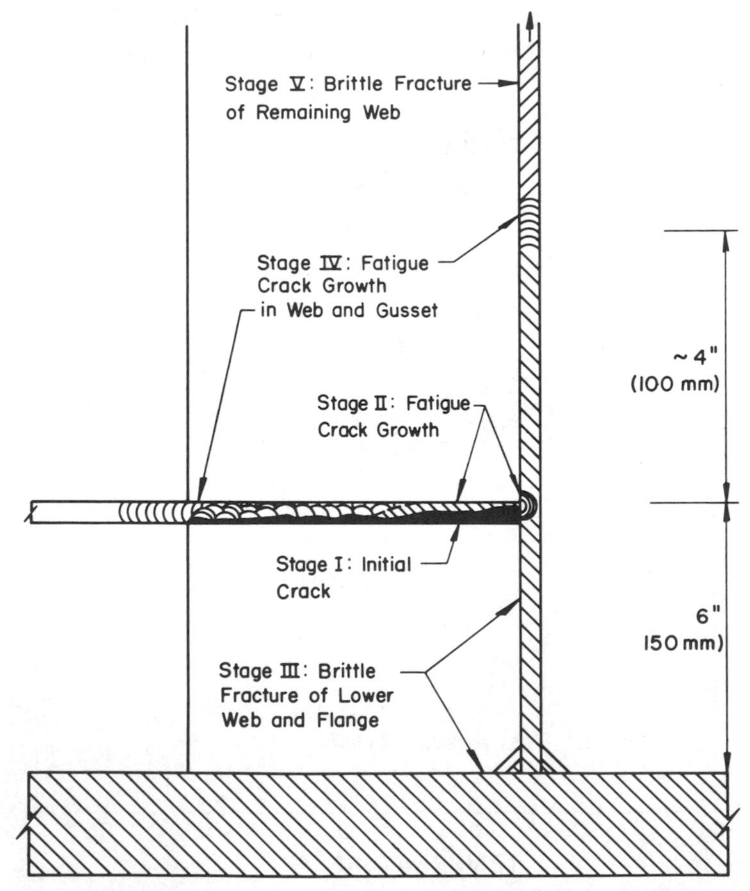

Figure 4.10 Schematic of stages of crack growth in gusset, web, and flange.

cleavage or brittle fracture occurred in the web after the crack had propagated into the web through the gusset plate–stiffener weld. This cleavage or brittle fracture extended down the web and into the flange. Figure 4.10 shows a schematic of the crack surface.

Fractographic examination of the crack surfaces at several locations with the transmission electron microscope corroborated the conclusions of the visual examinations. Figure 4.11 shows a fractograph of the crack surface in the web at the gusset. Fatigue crack growth striations are readily apparent at high magnification. The crack was also found to arrest about 4 in. (102 mm) up the web. Striations were detected in this region, and cleavage fracture on each side of it. This verified the visual observations made on the day the crack was discovered. The crack appears to have arrested because the lateral gusset plate prevented excessive crack opening and the crack tip was far removed from the residual tensile stress field at the level of the gusset plate welded connection.

The primary cause of failure was the large lack of fusion flaw in the weld between the gusset plate and the transverse stiffener near the backup bars. This resulted in fatigue crack growth in the web at the point of the gusset plate–stiffener and web intersection which later resulted in brittle fracture of the web plate and the bottom flange of the "east" girder.

Figure 4.11 Fatigue crack growth striations on the crack surface, 25,200×.

Probably the brittle fracture of the web occurred on the colder day, on February 9, 1975. Thus both fatigue and brittle fracture cracks were observed.

Failure Analysis

The initial flaw was from the lack of fusion in the single bevel groove weld between the gusset and transverse stiffener. The fact that this weld intersected with the stiffener and gusset plate welds to the web provided a direct path into the girder web. The resulting fatigue crack simultaneously grew as a semielliptical surface crack into the girder web and as an edge crack through the gusset plate–stiffener weld connection.

The stress intensity factor at the intersecting weld corner was defined as

$$K = F_e F_s F_w F_g \sigma \sqrt{\pi a} \tag{4.3}$$

where

$$F_e = \frac{1}{E(k)} \simeq 0.95 \tag{4.3a}$$

$$F_s = 1.211 - 0.186 \sqrt{\frac{a}{c}} \simeq 1.128 \tag{4.3b}$$

$$F_g = 2.64 \left[1 - 3.215 \frac{a}{t} + 7.897 \left(\frac{a}{t}\right)^2 - 9.288 \left(\frac{a}{t}\right)^3 + 4.086 \left(\frac{a}{t}\right)^4\right] \tag{4.3c}$$

$$F_w = \sqrt{\sec \frac{\pi a}{2t}} \tag{4.3d}$$

The initial crack size was taken as 0.39 in. (9 mm) and $t = 0.5$ in. (13 mm). The crack shape ratio c/a was 5 and kept constant, as it was observed at a second detail that the crack penetrated the web at the same time it propagated through the gusset plate–stiffener weld.

The number of cycles required to propagate the initial lack of fusion crack through the gusset plate weld and girder web plate thickness was estimated as

$$N = \int_{0.39}^{0.48} \frac{da}{3.6 \times 10^{-10} \Delta K^3} = 3.19 \times 10^6 \text{ cycles} \tag{4.4}$$

This is in good agreement with the service behavior and interval of life between 1968 and 1975.

Fracture appears to have developed as the fatigue crack penetrated the web and became a through crack. Since the crack was enlarging in a region of high tension residual stress, the critical stress intensity is given by

$$K = \sigma_Y \sqrt{\pi a} = K_c \tag{4.5}$$

where

$$\sigma_Y = 54 \text{ ksi } (373 \text{ MPa})$$

A crack size of 0.7 in. (18 mm) corresponds to a K_c value of 80 ksi $\sqrt{\text{in.}}$ (88 MPa $\sqrt{\text{m}}$). This is comparable to a crack detected in the floor beam north of the fracture 94 ft 6 in. (29 m) from pier 10. A core was removed from this area so that the crack surfaces could be examined. Figure 4.12 shows the exposed crack surfaces. The light areas represent metal that was

Figure 4.12 Exposed surface of crack at cored area adjacent to failure section.

fractured when the cracks were exposed. This shows that a web crack about $2\frac{1}{2}$ in. (63 mm) long existed. As discussed in [4.1], the upper 1 in. (25 mm) segment appeared to be created at the time of repair. The probable reason the larger $2\frac{1}{2}$ in. (63 mm) crack did not result in fracture of the girder was the higher toughness of the material during the summer of 1975. The lower segment of crack was slightly smaller than the crack that resulted in failure at the casualty section and hence was stable at the time of fracture which appears to have been in February 1975 when the temperature reached −22°F (−30°C).

Final fracture of the web occurred with further fatigue crack growth and yielding of the gusset plate which permitted the web crack to open, and this eventually led to the girder fracture sometime near May 7, 1975, the day the crack was discovered.

4.1.3 Conclusions

Conclusions Regarding Fracture and Failure

The fracture of the "east" main girder of the southbound lane of the Lafayette Street Bridge was due to the formation of a fatigue crack in the lateral bracing gusset plate to the transverse stiffener weld and web plate. This fatigue crack originated from a significant lack of fusion defect. At the failure cross section the partial penetration groove weld intersected the transverse stiffener–web weld and thus provided a direct path for the fatigue crack to penetrate into the web. Eventually, the fatigue crack precipitated a brittle fracture in part of the web and fractured the tension flange.

4.1.4 Actual Repair and Fracture Control

Other discontinuities due to lack of fusion were found to exist in other similar transverse stiffener–gusset plate welds in the two bridges. Therefore all gussets located in the regions of cyclic stress range and tensile stress were retrofitted to prevent other fatigue crack growth into the girder webs; also at this time inspections were made of other locations.

The scheme used to retrofit the structure is shown in Figure 4.13. Two vertical holes 1.25 in. (32 mm) were cut into the gusset plate and ground smooth at the corners of the gusset–transverse stiffener–web intersection. Two smaller vertical holes were drilled through the gusset plate and ground smooth at the tips of groove welds that join the gusset plate to the transverse stiffener. The set of holes adjacent to the web surface and the gusset were inspected with liquid penetrant. If no crack indication was observed, no further corrective action was taken. Where web cracking was detected, slotted holes were cut into the girder web with a 1.25 in. (32 mm) hole saw on two diagonal lines from the opposite side of the web

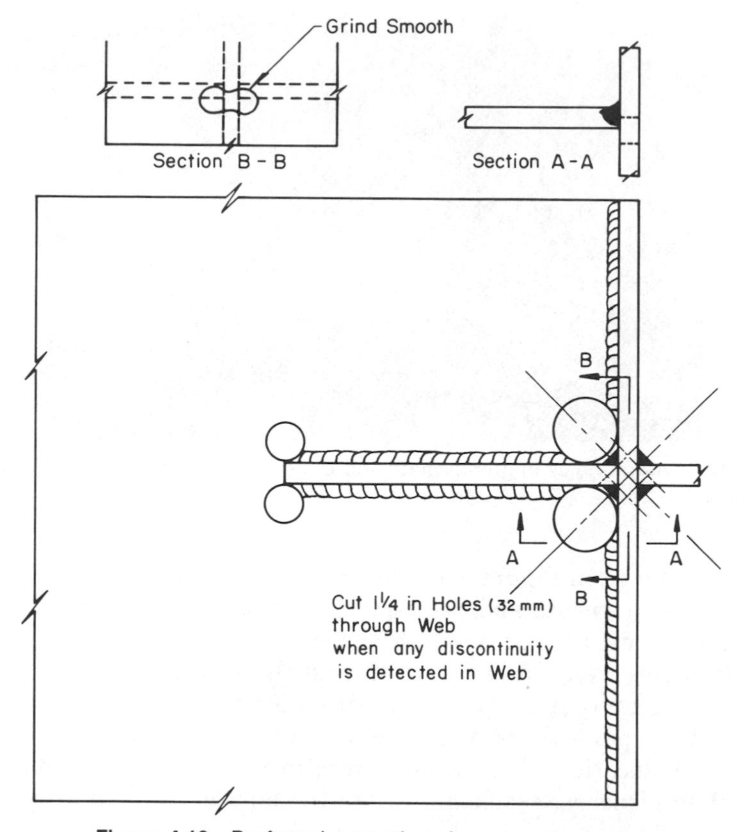

Figure 4.13 Preferred procedure for corrective action.

Figure 4.14 Retrofit holes in gusset plate and girder web. (*a*) Holes cut in gusset plate.

Figure 4.14 Retrofit holes in gusset plate and girder web. (*b*) Hole cut in girder web to remove crack.

surface, as shown in Figure 4.13. After each core was removed, the opening was ground smooth and the exposed web surface checked with liquid penetrant to ensure that a web crack did not extend beyond the hole. Then the holes were ground smooth and the area painted. Figure 4.14 shows photographs of a typical retrofitted area.

After the repairs there was a slight sag observed in the south facia girder at the location of the original fracture which was bolt spiced after the cracked girder was jacked up from the adjacent bridge.

REFERENCES

4.1 Fisher, J. W., Pense, A. W., and Roberts, R., Evaluation of Fracture of Lafayette Street Bridge, Proc. ASCE, *J. Struct. Div.* 103 (July 1977): 1339–1357.

4.2 Fisher, J. W., Pense, A. W., and Roberts, R., Investigation and Analysis of the Fractured Girder in Bridge No. 9800, T. H. No. 56 over Mississippi River in St. Paul, Minn., Minnesota Department of Highways (Materials, Research and Standards), October 1975.

4.3 Fisher, J. W., Bridge Fatigue Guide—Design and Details, AISC, Publication T112-11/77, 1977.

CHAPTER 5

Transverse Groove Welds

In the late 1940s and 1950s many groove welds placed into bridge structures were not well executed nor inspected following their installation. This lack of quality control has resulted in large number of lack of fusion defects, slag, and other discontinuities which have led to fatigue crack growth and fracture.

At least four different types of groove weld details have experienced crack propagation. One such condition was found to occur in the Aquasabon River Bridge in Ontario [5.1], where a rolled section was haunched by welding in an insert plate. A short length of vertical groove weld was thus attached perpendicular to the stress field. The embedded discontinuity as well as the poor quality of the detail produced crack propagation (see Figure 5.3).

Cracks have been found in the flange and web splices at groove welds in at least four bridges [1.1]. The flange groove weld cracks detected in the Quinebaug Bridge in Connecticut appear comparable to those observed in the Aquasabon River Bridge. Fatigue crack growth developed from large unfused regions of the weldment.

In two of the structures cracks were found in the groove welded splices of A514 steel tension members. In both cases part of the cracking was traced to weld repairs. Other cracks were found to be related to cold cracking which was apparently not detected at the time of fabrication. These structures were fabricated during 1969 to 1973. Only the cracks found in the I-24 Bridge were found to have experienced fatigue crack growth [5.2]. No evidence of fatigue crack extension was detected in the Silver Memorial Bridge [5.3] nor the Fremont Bridge [5.4].

Comparable cracks have been observed in the welds of groove welded cover plates. These welds were used to splice two different thickness plates that are then attached by fillet welds to either rolled sections or to riveted built-up girders. Four different bridges are known to have developed cracks at such details [1.1]. A review and summary will be given of the cracking that developed in the U.S. 51 Bridge at Peru, Illinois.

A third kind of groove welded detail that has experienced cracking occurs at splices in continuous longitudinal stiffeners. These attachments were considered secondary components, so no weld quality control has been imposed on the groove weld splice. The first such structure to experience cracking at this detail (see Figure 5.10) was the Quinnipiac River Bridge near New Haven, Connecticut [5.5]. A summary and review of this structure is provided later. At least three other bridges have developed cracks at such splices. One of these structures developed a crack during a cold winter prior to being placed into service. In every instance of cracking no weld quality control was imposed on the longitudinal stiffener splices. Some of these were in tension zones. As a result of cyclic stresses crack propagation occurred where there were large defects, and these eventually led to fracture.

The fourth groove weld detail that has resulted in cracking is the electroslag weld. The fracture of a flange on an I-79 Bridge at Neville Island at Pittsburgh in 1977 [5.6] has led to detection of significant flaws in at least five other bridge structures. Except for the I-79 structure, all discontinuities in the electroslag welded bridges were detected and retrofitted before significant crack propagation developed. The electroslag welds were found to have a lower level of fracture toughness than implied from qualification tests. Furthermore detection of the fabrication discontinuities was found to be difficult and not reliable. As a result of the I-79 failure and these other experiences, electroslag welds are not permitted in tension components of bridge structures [5.7].

5.1 FATIGUE-FRACTURE ANALYSIS OF AQUASABON RIVER BRIDGE

5.1.1 Description and History of the Bridge

Description of Structure

The Aquasabon River Bridge is located on the north shore of Lake Superior on Highway 17 (Trans-Canada Highway) 130 miles (210 km) east of Thunder Bay, Ontario. The structure was completed in 1948. The three-span continuous beam structure has a composite steel beam-reinforced concrete slab. It was designed by the Ontario Department of Highways to accommodate an H20 truck load [5.1].

The 200 ft (61 m) structure has three spans of 60 ft (18.3 m), 80 ft (24.4 m), and 60 ft (18.3 m), as shown on Figure 5.1. The superstructure con-

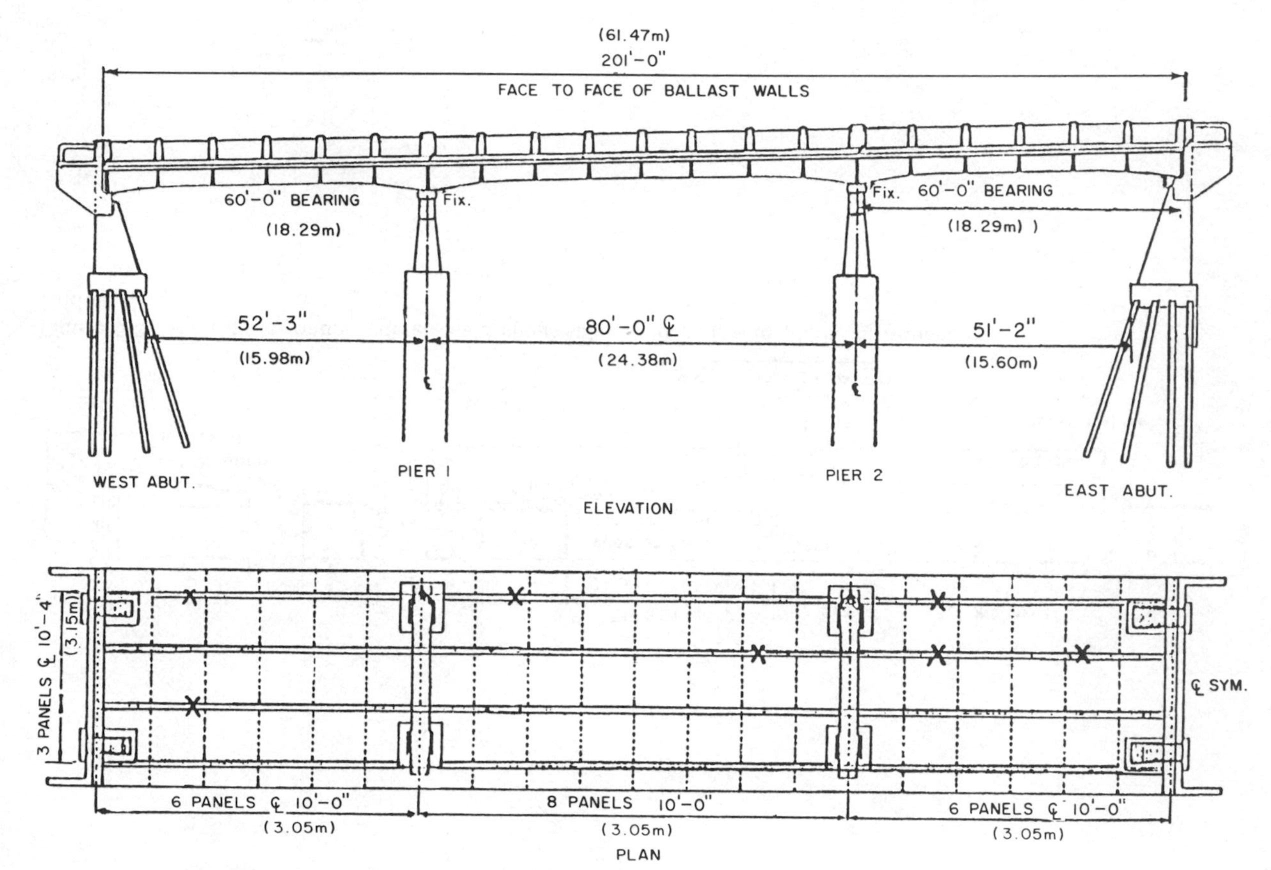

Figure 5.1 Plan and elevation of Aquasabon Bridge (crack locations are indicated with the mark X).

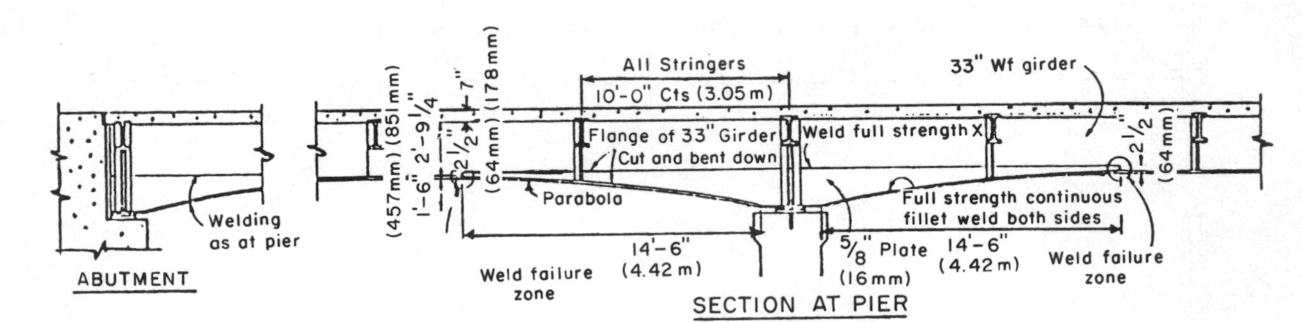

Figure 5.2 Main girder haunch detail, field shear splice and connector detail of Aquasabon Bridge.

sists of four longitudinal WF33 × 141 (84 cm) girders connected to 15 in. (384 mm) transverse floor-beam stringers. The steel structure supports a 7 in. (178 mm) reinforced concrete deck. The main girders are haunched both at piers and abutments. These haunches were fabricated by cutting the bottom flange from the web fillet and welding a $\frac{5}{8}$ in. (16 mm) parabolic insert plate into the web which resulted in a 51.25 in. (1.30 m) deep section at the piers and abutments (see Figure 5.2).

The main girders were field spliced at two points in the center span 22 ft (6.7 m) from each pier, as shown in Figure 5.2. The splice points were placed at the points of dead load contraflexure. The riveted splice consisted of $\frac{5}{8}$ in. (16 mm) flange plates on interior and $\frac{1}{2}$ in. (12.7 mm) plates on the exterior girders. All had two $\frac{3}{8}$ in. (10 mm) web plates for shear splices.

The reinforced concrete floor slab was connected to the girders by channel-type shear connectors of 11 × 3 in. (275 × 75 mm) welded at 1 ft 6 in. (0.457 m) intervals.

History of Structure and Cracking

The Aquasabon River Bridge was completed in 1948. In 1963 cracks were discovered at the vertical butt weld detail in three of the six haunch inserts of the north interior main girder, as shown in Figure 5.3*a* [5.1]. One of these cracks extended 44 in (1.12 m) into the girder web along a diagonal line starting from the vertical butt weld detail.

In 1973 the structure was subjected to rigorous testing to determine its safe load-carrying capacity. Prior to the 1973 tests, dye-penetrant and magnetic-particle inspection techniques were used on the repaired and other 21 remaining original welds. As a result four other weld cracks were discovered. The location of individual cracks are marked on Figure 5.1. One of these cracks is shown in Figure 5.3*b*.

The fatigue cracks that developed were in the main WF girders and are identified in Figures 5.1 and 5.3. These cracks propagated from large

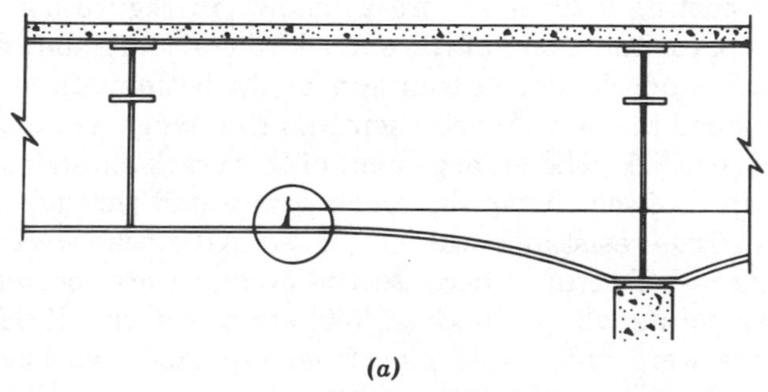

(a)

Figure 5.3*a* Location of weld crack in main girder (see Figure 5.3*b* for a photograph of circled area).

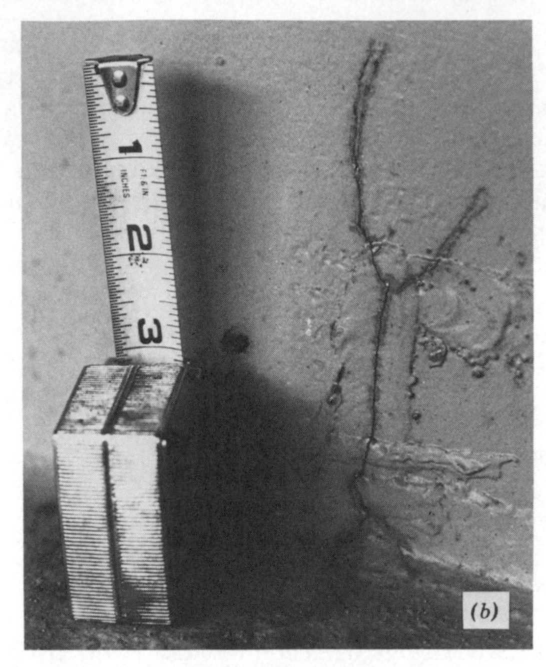

Figure 5.3*b* Typical crack at vertical groove weld detail (terminal point at haunch insert detail) (courtesy of Ontario Ministry of Transport and Communication).

initial weld imperfections or inclusions in the short transverse groove welds at the ends of the parabolic haunch inserts in the main girders (see Figure 5.3). Of 24 welded details, 7 had cracked by 1973. Additional cracks were detected since. By 1963 three of the six haunch inserts in the north interior girder had developed cracks. These were repaired by welding cover plates over the cracks. By 1973 four more welded details developed cracks. One of the cracks penetrated 7 in. (178 mm) up the web beyond the 2.5 in. (64 mm) transverse weld and had cracked about 65% of the bottom flange. Several of the web cracks were removed from the structure by cutting a circular core as shown in Figure 5.4. The crack surface and approximate size of the embedded crack are shown in Figure 5.5. With large imperfections residing near the bottom of the web, crack growth developed in the web and then into the flange as a part circular crack (see Figure 5.5). The enlargement of this crack extended it into the flange and up the web. After the crack penetrated through the flange, most of the fatigue resistance was exhausted. All cracks were discovered before the flanges fractured because the details were located near the contraflexure points where the dead load stress was small. Hence large fatigue cracks were able to develop from repeated live loads without brittle fracture of the remaining section.

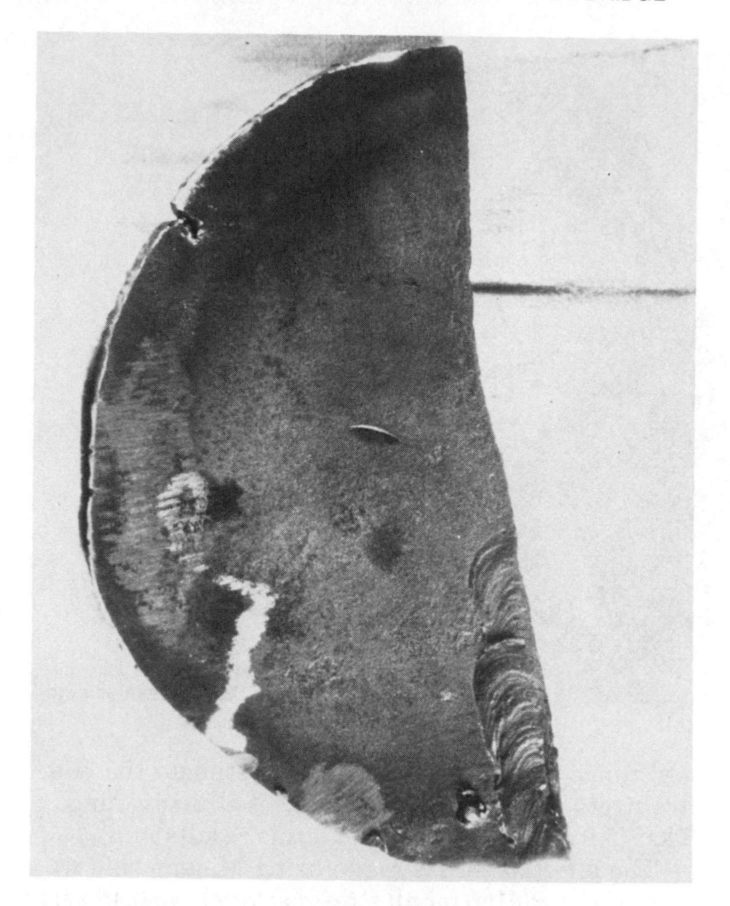

Figure 5.4 Circular cut-out section at haunch tip weld detail.

5.1.2 Failure Modes and Analysis

A rigorous field testing of the Aquasabon River Bridge was conducted in 1973. Two fully loaded vehicles, each with the gross weight GVW = 197 k (876 kN) distributed on five axles carrying the minimum payload of 153 k (680 kN), were used. The payload can be placed on the semitrailer to produce any tandem axle weight up to 110 k (490 kN). These loads were applied in four increments, while stresses in critical areas were monitored. Pier settlements were also measured.

Dynamic strains were recorded with the test vehicle loaded to a gross weight of 91 k (405 kN). The vehicle crossed the bridge at varying speeds between 30 to 40 mph (48 to 80 km/h). Based on the dynamic strain measurements and traffic conditions, a representative stress range histogram (S_r) was determined (see Figure 5.6).

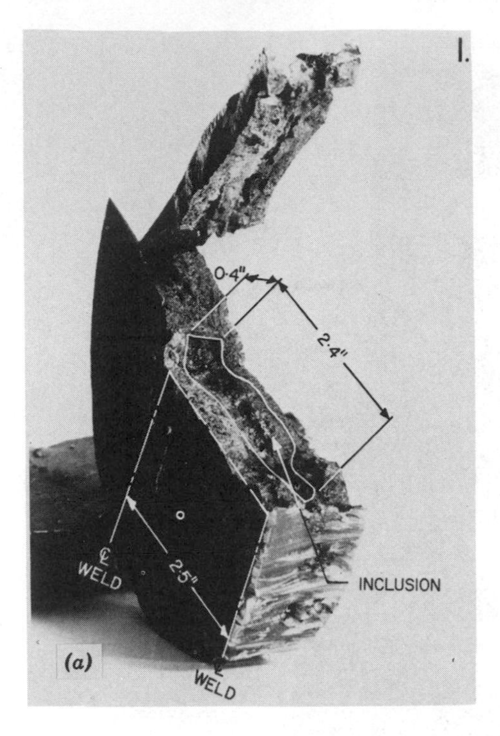

Figure 5.5*a* Details at weld inclusion.

The stress range histogram was used to estimate the effective root-mean-square stress range $S_{r\text{RMS}}$ and $S_{r\text{Miner}}$. If all stress range conditions above 0.85 ksi (5.9 MPa) are considered, this results in $S_{r\text{RMS}} = 1.92$ ksi (13.2 MPa). The effective stress range using Miner's rule was equal to $S_{r\text{Miner}} = 1.92$ ksi (13.2 MPa) for all stress cycles above 0.45 ksi (3.1 MPa).

A sample of vehicles crossing the bridge indicated that each vehicle produced from 6 to 15 stress cycles above 0.45 ksi (3.1 MPa). The smoothed histogram shown in Figure 5.6 indicates that 25 million stress

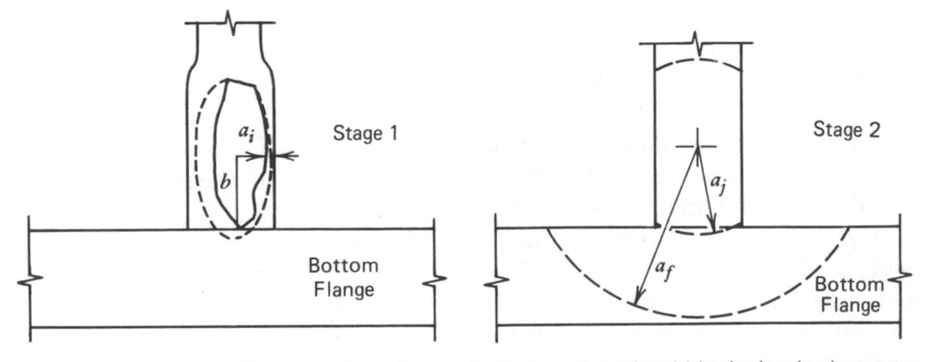

Figure 5.5*b* Assumed stages of crack growth (outer edge of weld inclusion is shown as dark line).

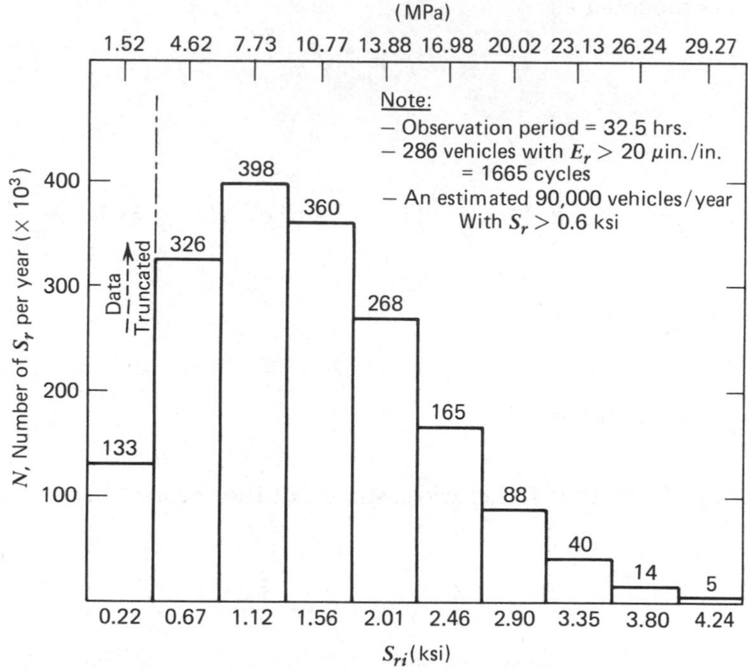

Figure 5.6 Stress range histogram (smoothed) of Aquasabon Bridge.

cycles exceeded 0.45 ksi (3.1 MPa) between 1948 and 1963. An additional 16 million cycles were accumulated between 1963 and 1973.

Temperature and Environmental Effects

It appears that temperature and environmental effects were not significant factors in the development of cracks in the short transverse groove welds of the Aquasabon River Bridge. The examination of the crack surface and the weld detail indicated that the cracks were all stable fatigue cracks.

This is reasonable considering the low dead load at the inflection points. As a result the maximum stress was small, and this permitted the large fatigue cracks that propagated into the flange and up the web to develop short of brittle fracture. During the second stage the crack grew as a penny-shaped circular crack through the flange and eventually developed into a through crack in the flange.

Failure Analysis

The fatigue crack propagation stages shown in Figure 5.5 were modeled by circumscribing the elliptical-shaped weld defect. The stress intensity factor was modeled as

$$K = F_e F_w \sigma \sqrt{\pi a} \tag{5.1}$$

where

$$F_e = \frac{1}{E(k)}, \qquad E(k) = \int_0^{\pi/2} [1 - k^2 \sin^2 \theta]^{1/2} d\theta \qquad k^2 = \frac{c^2 - a^2}{c^2}$$

$$F_w = \sqrt{\sec \frac{\pi a}{0.625}}$$

$$\frac{a}{c} = \text{constant} = 0.11$$

The penny shape that the crack assumed in the second stage of growth resulted in

$$K = \frac{2}{\pi} \sigma \sqrt{\pi a} \left(\sqrt{\sec \frac{\pi a}{4.8}} \right) \tag{5.2}$$

where a = crack radius and $a_i = c_f$ for stage 1. The crack growth equation was used to estimate the fatigue life for each stage of crack propagation:

$$N = \int_{a_i}^{2.35 \text{ in.}} \frac{da}{3.6 \times 10^{-10} \Delta K^3} \tag{5.3}$$

Table 5.1 Estimated Fatigue Life: $S_{r\text{Miner}}$ = 1.92 ksi (13.2 MPa)

Stage 1: Crack Growth through Weld			Stage 2: Crack Growth through Flange		
Initial Crack/Size a_i(in.)	Cycles of Stress	Years to Achieve	Initial Crack/Radius (in.)	Cycles of Stress	Years to Achieve
0.20 (5 mm)	14,050,000	9.4	1.40 (35.6 mm)	25,500,000	17
0.25 (6.4 mm)	2,664,000	1.8	1.42 (36 mm)	24,150,000	16
0.30 (7.6 mm)	40,500	0.3	1.45 (36.8 mm)	21,960,000	14.6
			1.50 (38 mm)	18,600,000	12.4

Several initial crack size conditions were examined in order to provide information on factors relative to crack propagation. The results of this examination are summarized in Table 5.1.

For the crack shown in Figure 5.5, it was estimated that the effective stress range was $S_{r\text{Miner}} = 1.92$ ksi (13.2 MPa) which occurred about 1,500,000 cycles each year. Approximately two to three years would be required before the crack penetrated into the bottom flange during the first stage of growth. An additional 15 to 20 years would be required before the crack propagated through the thickness of the flange. The results of the analysis are in good agreement with the actual field behavior of these details. The first cracked details were detected in 1963 after 15 years of service and fatigue cracking continued to develop through the 1970s.

5.1.3 Conclusions

The cracking that developed was caused by large imperfections that were fabricated in short transverse groove welds. The lengths of these welds were insufficient to produce sound connections.

The welded connection that caused the problem was difficult to fabricate. In 1948 equipment for nondestructive inspection was not available. Thus the weld quality was poor, and this permitted fatigue crack propagation to develop under the normal stress spectrum that the bridge was subjected to. The dynamic response of the structure also resulted in many stress cycles from passage of a single truck.

This case study demonstrates the necessity to provide sound groove welds perpendicular to the cyclic stress. It also suggests that short transverse groove welds are difficult to produce with sound welds and should be avoided.

5.1.4 Actual Repair and Fracture Control

The fatigue cracks discovered in 1963 in the transverse (or vertical) weld detail in the three of the six haunch inserts of the north interior girder were repaired by welding cover plates or insert plates into the cutout hole. Prior to the load tests in 1973 a thorough inspection of the repairs and remaining transverse weld details revealed four other weld cracks. The transverse weld area was cut out in a circular shape, and an insert was welded in its place. Where the crack penetrated the bottom flange, it was gouged out and filled with weld material at a slow rate of deposit, by using low-hydrogen-coated electrodes. All repaired surfaces were subsequently ground smooth and flush to eliminate stress concentrations. The probable adverse effects of the original repairs might be the possibility of new imperfections or inclusions fabricated into the repair welds. This was prevented by the installation of bolted flange splices to the entire haunch,

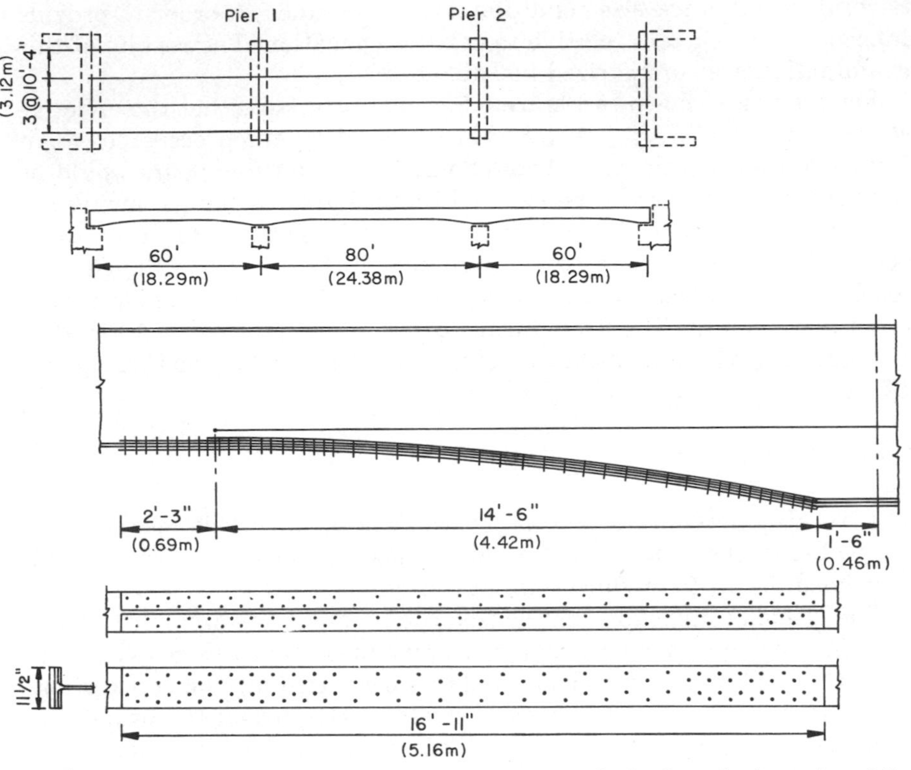

Figure 5.7 Retrofit detail (top and bottom reinforcing plates along the haunch and the hole drilled at the terminal point).

as cracks were later observed in the groove welds that connected the flange to the web plate insert. Figure 5.7 shows the flange splice detail.

This subsequent retrofit should offset and prevent any significant distress from developing in the structure.

5.2 FATIGUE-FRACTURE ANALYSIS OF QUINNIPIAC RIVER BRIDGE

5.2.1 Description and History of the Bridge

Description of Structure

The Quinnipiac River Bridge on I-91 near New Haven, Connecticut, is a four span structure over the Quinnipiac River and can be seen in Figure 5.8. Span 1 is of composite construction with WF beams and welded girders. Spans 2, 3, and 4 are noncomposite welded girders of a cantilever type with a suspended center span. The suspended part of the structure in

Figure 5.8 Profile of Quinnipiac River Bridge (courtesy of Connecticut Department of Transportation).

span 3 is 165 ft (50.3 m) long between the hinges that connect it to the anchor spans (spans 2 and 4, see Figure 5.9). The span lengths are 62 ft 11 in. (19.2 m) 112 ft (34 m) 220 ft (67 m), and 111 ft 1 in. (33.8 m) for the spans 1, 2, 3, and 4, respectively. The entire structure is on a skew. The cross sections of spans 2, 3, and 4 are composed of nine parallel welded girders supporting each roadway in between metal beam-type guard rails separating the traffic. The main girders have transverse X-type bracing and longitudinal stiffeners. The roadway is a $7\frac{3}{4}$ in. (197 mm) thick reinforced concrete deck.

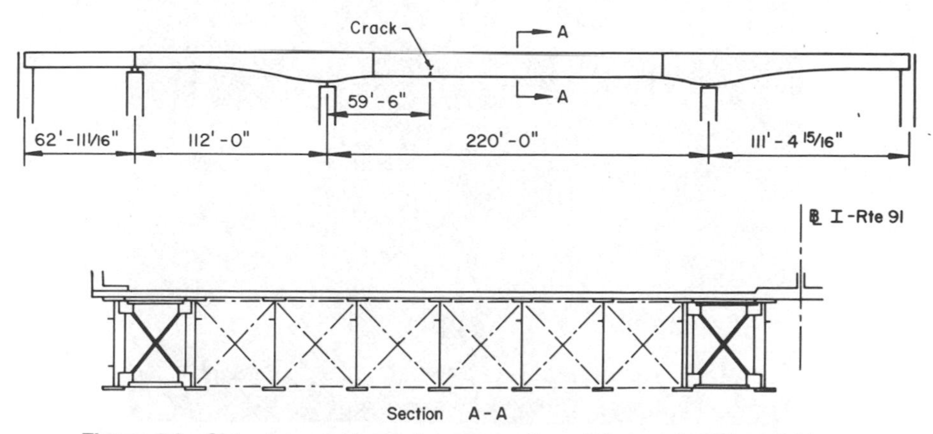

Figure 5.9 Side view and typical cross section of Quinnipiac River Bridge.

History of Structure and Cracking

The structure was opened to traffic in 1964. In November 1973 a large crack was discovered in the south fascia girder of the suspended span in the center portion of the bridge [5.5]. The bridge had experienced approximately nine years of service life at the time the crack was discovered. The location of the crack in the south facia girder of the suspended span is shown in Figure 5.9.

Figure 5.10 shows the crack that developed in the south fascia girder web. The crack propagated approximately to middepth of the girder and had penetrated into the bottom flange of the girder at the time it was discovered. The location of the crack is shown on the schematic profile of the bridge in Figure 5.9. It was approximately 59 ft 6 in. (18.1 m) from the left support of the noncomposite suspended span. A second crack was detected toward the midspan about 29 ft (8.8 m) from the cracked section.

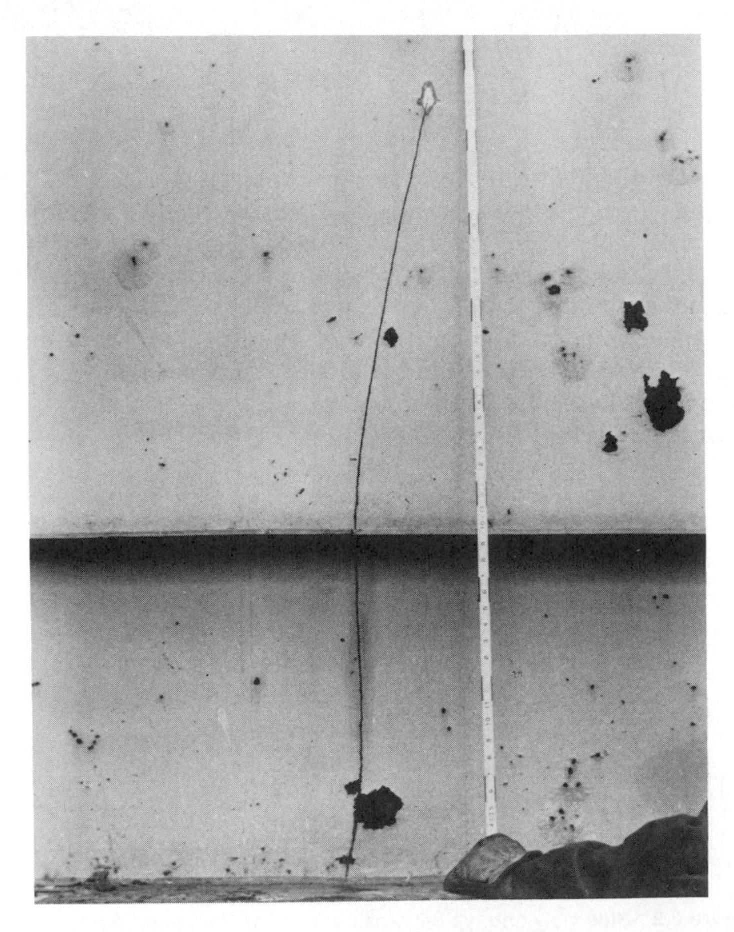

Figure 5.10 Crack in web of south fascia girder.

Figure 5.11 Crack in the longitudinal stiffener weld and partway through web.

This second crack had severed the stiffener but had not propagated through the web. Figure 5.11 shows this crack at the time it was detected.

A detailed study of the fracture surface was made on pieces (see Figure 5.12) that were removed from the section. This indicated that the fracture had initiated at the unfused butt weld in the longitudinal stiffener. The portions of the fracture surfaces in the vicinity of the stiffener groove weld were severely corroded from exposure to the environmental effects. The crack surface indicated that some crack extension probably developed from the unfused section of the butt weld across the thickness of the stiffener during transport and erection.

5.2.2 Failure Modes and Analysis

Cyclic Loads and Stresses

When originally analyzed, the stress range due to live load was approximated from the nationwide gross vehicle distribution [4.3]. The design stress range, S_r^D, in the bottom flange was equal to 4.35 ksi (30 MPa), and the effective stress range $S_{r\text{Miner}}$ was estimated as

$$S_{r\text{Miner}} = [0.35\alpha^3]^{1/3} S_r^D \tag{5.4}$$

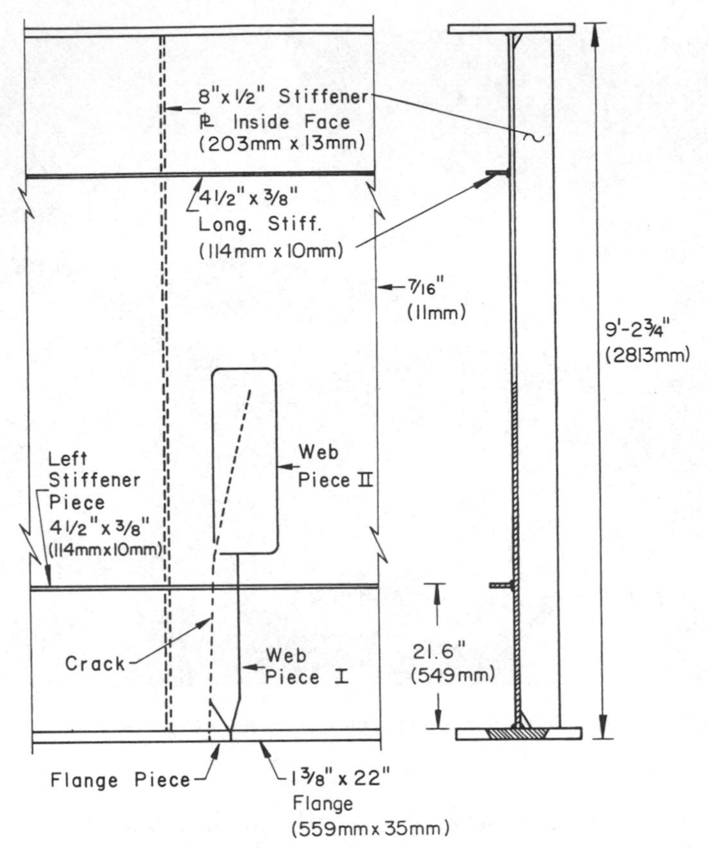

Figure 5.12 Schematic of girder showing sections removed for examination at crack.

with α assumed between 0.6 to 0.7; a value of 1.95 ksi (13.4 MPa) was used to assess the fatigue resistance. The corresponding stress at the longitudinal stiffener was equal to 1.17 ksi (8.1 MPa) [5.5].

During the summer of 1981 stress range measurements under traffic and controlled loads were obtained near the original failure section and near midspan where the second crack was found in 1973 (see Figure 5.11). Measurements were acquired on the tension flange and at the level of the longitudinal stiffener. Figure 5.13 shows the stress range histograms that were developed from the measurements. These indicated that the effective stress range from traffic was about 0.5 ksi (3.5 MPa) at the failure section and at the crack near midspan for the measurement period [5.8].

The average daily truck traffic (ADTT) crossing the bridge was estimated to be 4300. The strain measurements and the truck crossings at the time of the measurements indicated that about two stress range excursions that exceeded 0.4 ksi (2.8 MPa) were occurring for each truck crossing. Hence this produces about 3,200,000 random stress cycles per

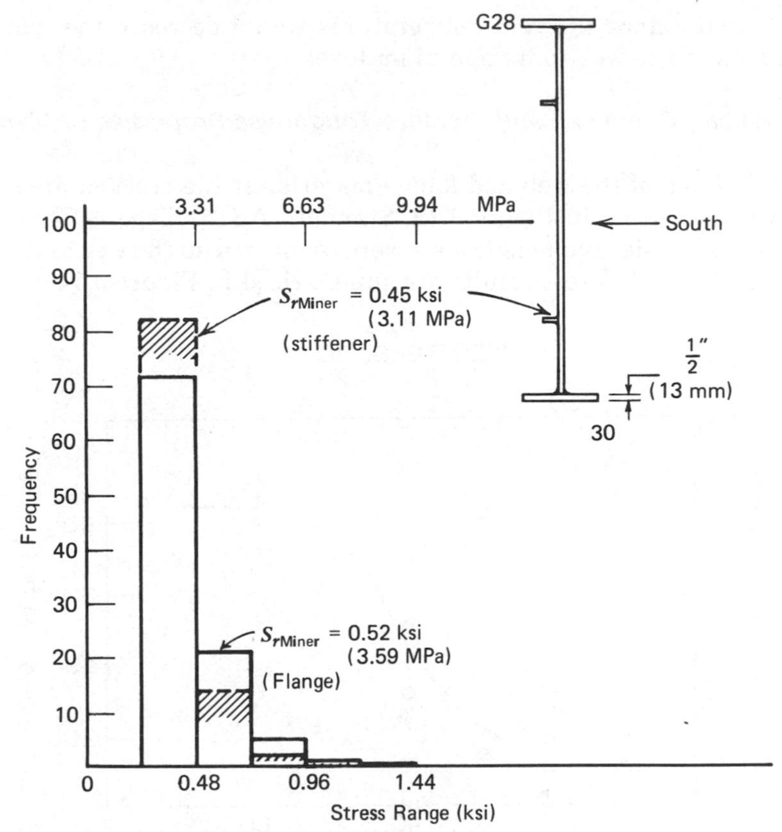

Figure 5.13 Stress range histograms for cracked section and near midspan.

year. Since the measured strains were extrapolated over the entire service period, a value of $S_{r\,\text{Miner}} = 0.6$ ksi (4.14 MPa) was used to assess the fatigue resistance.

The dead load stress in the bottom flange is 8.25 ksi (56.9 MPa). This results in a dead load stress at the stiffener of 4.95 ksi (34.1 MPa). The maximum live load stress in the bottom flange was estimated to be 3.3 ksi (22.8 MPa) and 2 ksi (13.8 MPa) at the longitudinal stiffener.

Temperature and Environmental Effects

Brittle fracture of the web probably occurred during freezing temperatures. Comparison of the fracture surfaces from three-point bend test specimens with the web crack indicated that the web had fractured when the ambient temperature was between 10°F (−18°C) and −10°F (−25°C). Brittle fracture initiated in a zone of high residual tensile stresses. Once the crack became unstable, it propagated through the web and was eventually arrested in the girder flange. It is highly probable that the brittle fracture occurred during the particularly cold winter of December 1972 to

March 1973, since the cold temperatures would decrease the material toughness of the web to its minimum level.

Mechanical, Chemical, and Fracture Toughness Properties of Material

Several pieces of the web and flange material at the cracked area were removed, as shown in Figure 5.12. Standard ASTM Type A Charpy V-notch (CVN) tests, and tensile tests were conducted on the web and flange material. The CVN test results are summarized in Figure 5.14.

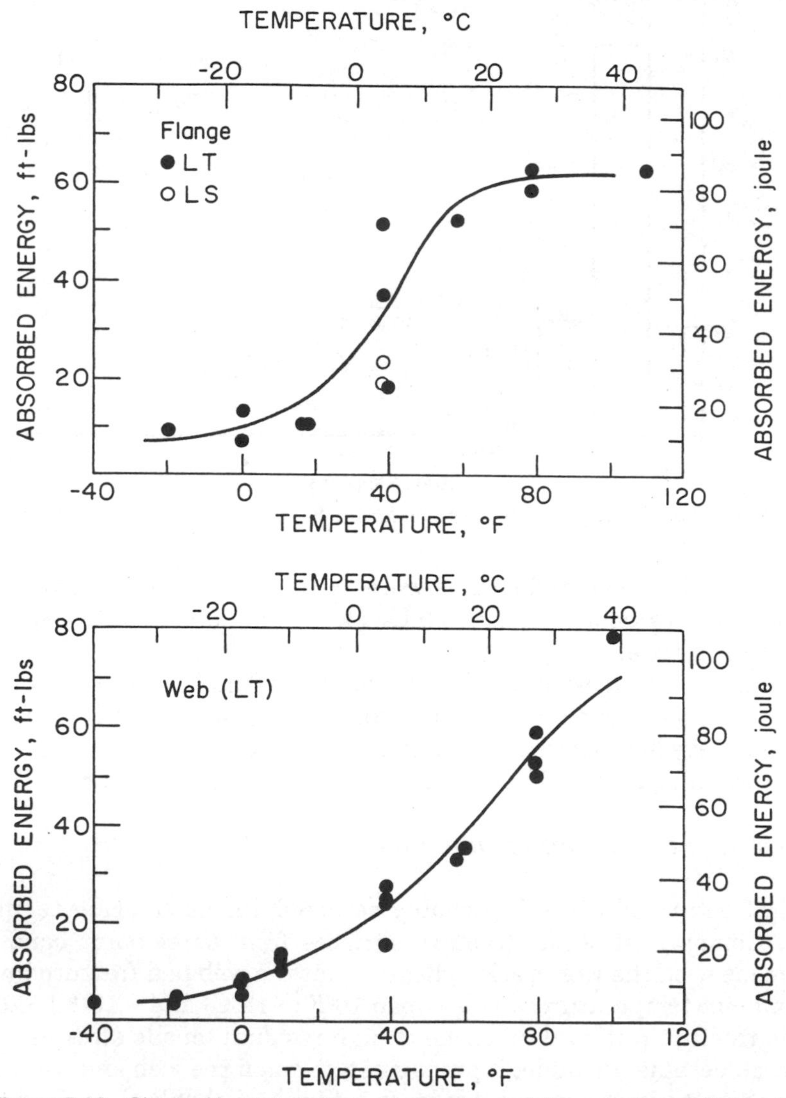

Figure 5.14 Charpy V-notch test results for web and flange adjacent to crack.

The tensile tests of the girder web provided a yield strength of 36.8 ksi (254 MPa) and an ultimate strength of 60.9 ksi (420 MPa). Both flange and web material satisfied the toughness requirements for group 2 of the 1974 Interim AASHTO specifications. The average CVN impact value for the web was 20 ft-lb (27 J) at 40°F (4°C). The flange provided 37 ft-lb (47 J) at 40°F (4°C).

Several three-point bend specimens were also fabricated from the web material. A notch with a 45° chevron front was machined at the center of the specimen to initiate fatigue crack growth.

The specimens were tested at a 0.1-sec loading rate and at temperatures of −40°F (−40°C), −20°F (−29°C), and 0°F (−18°C). From a comparison of the fracture surface of the web crack with the fracture surfaces of the K_c test specimens, an estimate of temperature at which brittle fracture occurred was made. On this basis it is believed that the fracture occurred at a service temperature between −10°F (−25°C) and 10°F (−14°C). It is apparent from Figure 5.14 that CVN energy absorption and consequently the dynamic toughness of the girder material decreased significantly at 0°F (−18°C). The estimated fracture toughness of the web steel from a J-integral analysis and K_J estimate is plotted in Figure 5.15. Also plotted in Figure 5.15 is the estimated dynamic fracture toughness

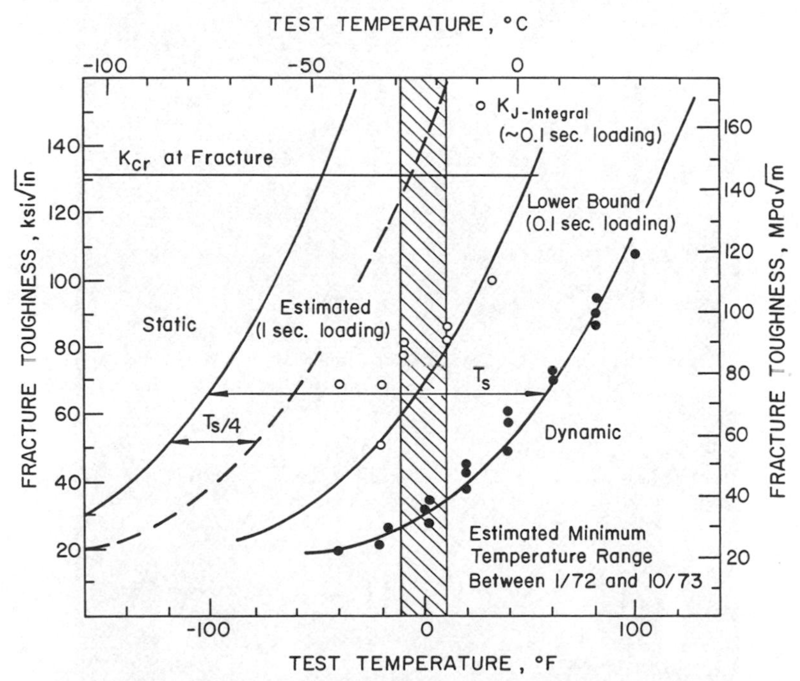

Figure 5.15 Fracture toughness of web estimated from Charpy V-notch tests and from three-point bend specimens in comparison to estimated K_{CR} for the structural crack when rapid extension occurred.

predicted from Barsom's correlation equation:

$$K_{Id} = [5E \text{ CVN}]^{1/2} \quad (\text{ksi} \sqrt{\text{in.}}) \tag{5.5}$$

The test data and the lower bound dynamic toughness indicate that the web material has a fracture toughness between 100 ksi $\sqrt{\text{in.}}$ (110 MPa $\sqrt{\text{m}}$) and 150 ksi $\sqrt{\text{in.}}$ (165 MPa $\sqrt{\text{m}}$) at the bracketed time of failure.

Visual and Fractographic Examination of Failure Surfaces

The fracture surfaces of flange-web connection, the longitudinal stiffener-web intersection (at which the crack originated), and the ends of the longitudinal stiffener are shown in Figures 5.16, 5.17, and 5.18, respectively.

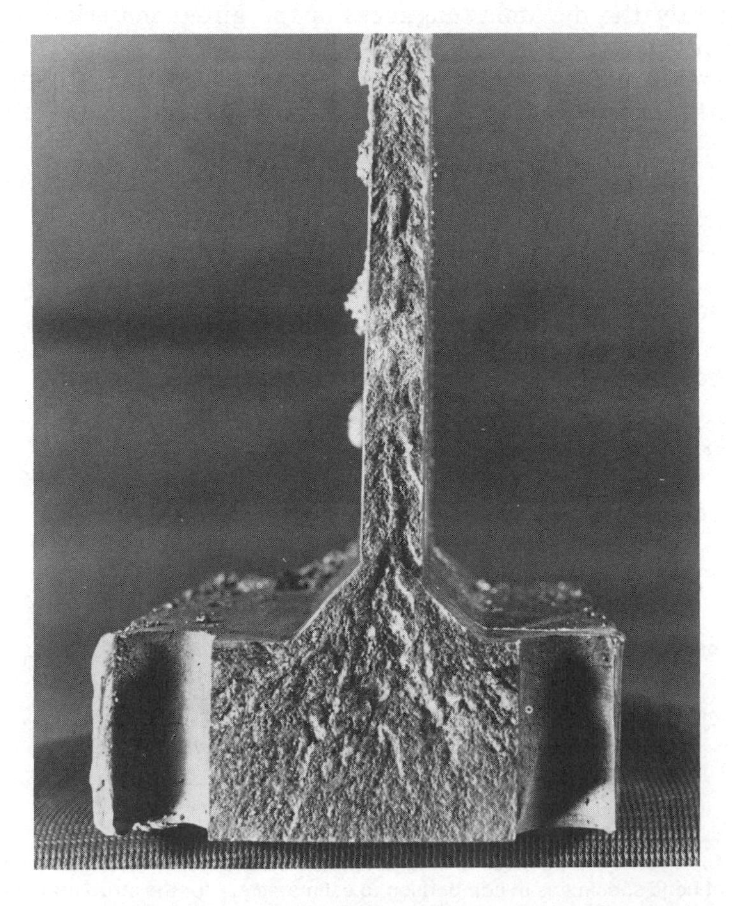

Figure 5.16 Fracture surface at flange-web junction.

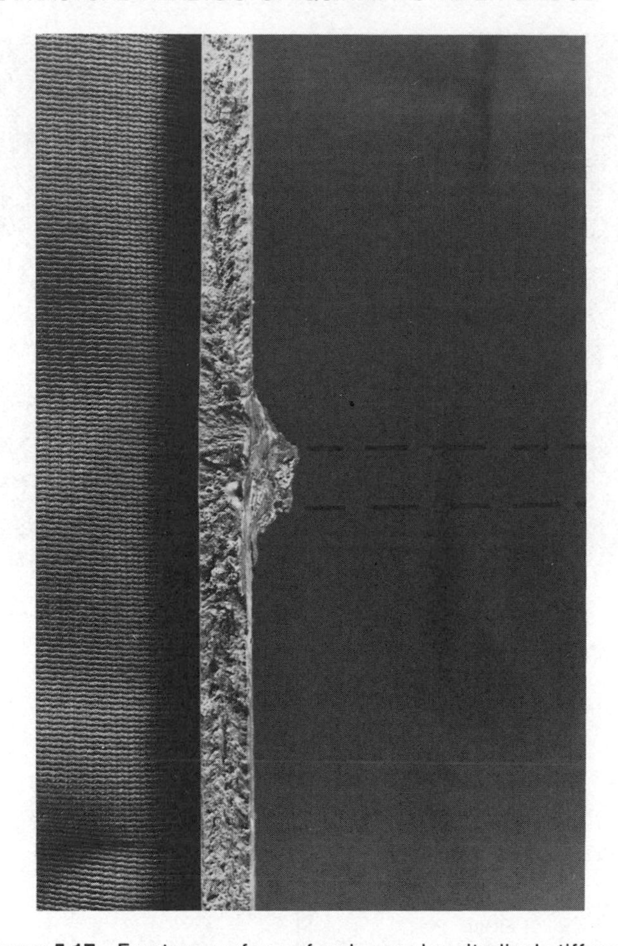

Figure 5.17 Fracture surface of web near longitudinal stiffener.

Visual examination of the fracture surfaces indicated that the failure had initiated at the web-stiffener intersection due to an unfused butt weld in the longitudinal stiffener. The web fracture surface indicated that brittle (cleavage) fracture occurred following the penetration of the web thickness by a fatigue crack. The cleavage fracture extended up as well as down the fractured web and penetrated some distance into the flange before it arrested. Several stages of crack growth and failure modes, starting from the unfused butt weld in the longitudinal stiffener–girder web interface, are indicated in a schematic manner in Figure 5.19.

Acetate surface replicas of the web–longitudinal stiffener intersection and near the flange-web intersection were made and examined with the transmission electron microscope. The examination of web-stiffener intersection confirmed that fatigue crack growth had occurred in that re-

Figure 5.18 Ends of longitudinal stiffener at crack.

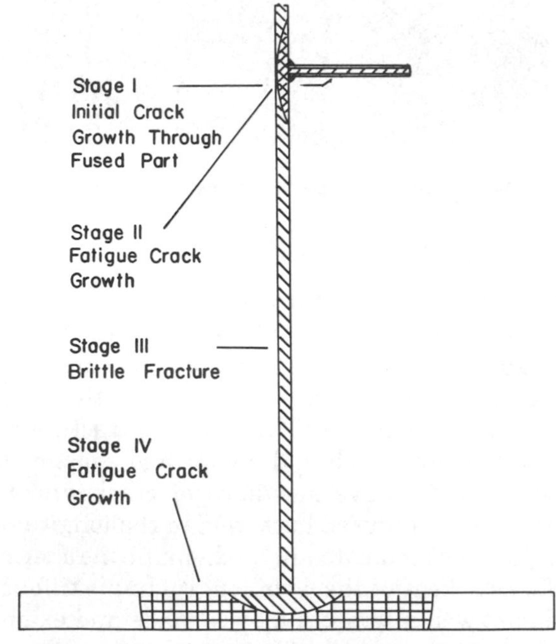

Figure 5.19 Schematic of crack growth stages.

Figure 5.20 Crack growth striations, 49,125×.

gion. Figure 5.20 shows the variable spaced crack growth striations in the web. Brittle (cleavage) fracture extended into the flange for about 1 in. (25 mm). Figure 5.21 shows the cleavage river pattern detected in the flange. Although the crack had obviously arrested and experienced fatigue crack propagation, no evidence of striations was found in the flange.

Failure Analysis

The primary cause of failure was the inadequate butt weld made across the width of the longitudinal stiffener. It is probable that some crack extension from the unfused section occurred during transport and erection of the girder. Several stages of crack growth based on the visual and fractographic examination of the fracture surface of the web are summarized in Figure 5.19.

The brittle fracture of the web developed from the initial fatigue crack. The crack became unstable and propagated rapidly down the web and arrested in the flange and near the middepth of the web. This appears to have occurred during the low temperatures in the winter of December 1972 to March 1973.

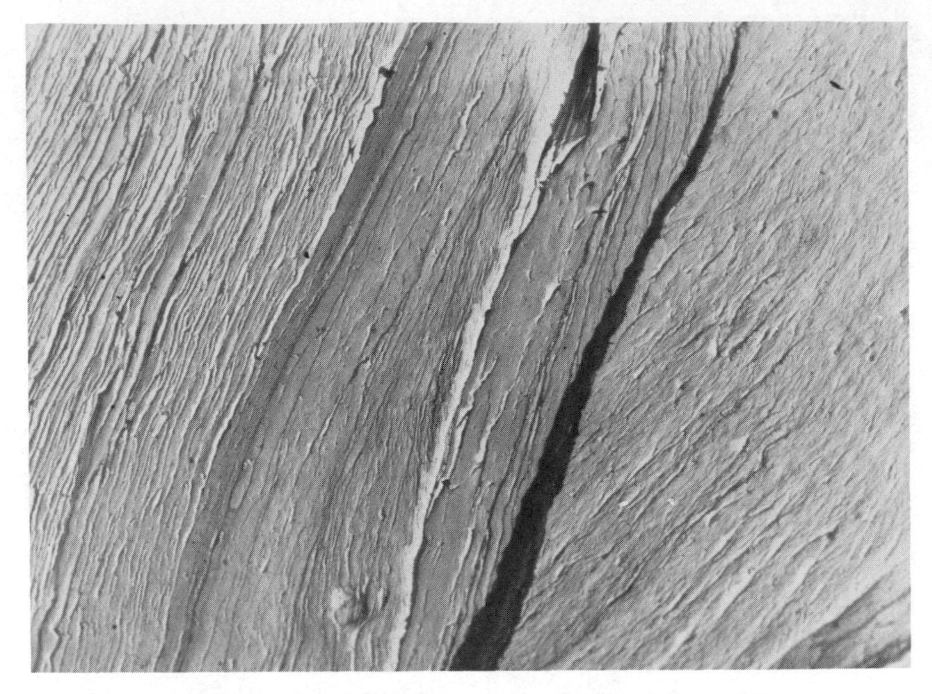

Figure 5.21 Cleavage in flange near bottom surface, 4300×.

The initial stage 1 behavior can be modeled as a through crack that may be symmetrical or eccentric in the stiffener cross section as shown in Figure 5.19. The symmetrical crack gives [5.8]

$$K = F_w \sigma \sqrt{\pi a} \tag{5.6}$$

where

$$F_w = \left[1 - 0.025 \left(\frac{2a}{t_s}\right)^2 + 0.06 \left(\frac{2a}{t_s}\right)^4\right] \sqrt{\sec \frac{\pi a}{t_s}} \tag{5.7}$$

where a = half the unfused thickness and t_s = thickness of the longitudinal stiffener.

When the crack is eccentric by an amount, e, with the mid-thickness of the plate, the stress intensity increases depending on the ratio $2e/t_s$. For a value $2e/t_s$ of 0.7:

$$K = \sigma \sqrt{\pi a} \left[1 + 0.86 \frac{a}{t_s} - 1.478 \left(\frac{2a}{t_s}\right)^2 + 2.307 \left(\frac{2a}{t_s}\right)^3\right] \tag{5.8}$$

The interval of life required to propagate through the plate thickness was estimated from

$$N = \int_{a_i}^{a_f} \frac{da}{3.6 \times 10^{-10} \Delta K^3} \tag{5.9}$$

This resulted in 8 to 80 million cycles for $2a = 0.25$ in. (6.4 mm) unfused thickness for Eqs. 5.6 and 5.8. This appeared to provide a reasonable bound for the behavior of the cracked section and at the second crack found near midspan which is shown in Figure 5.11. This analysis suggests that the failure section had a longitudinal stiffener welded joint with the worst-case condition at the time the bridge was placed in service. The crack near midspan is more nearly represented by the predicted behavior for stage 1. At the time of discovery both cracked sections had been subjected to about 30 million random variable stress cycles. The crack shown in Figure 5.11 had penetrated a short distance into the longitudinal welds and web plate.

The second stage of crack growth through the web was modeled as a flat circular crack with its center of radius at the tip of the stiffener. The stress intensity factor was taken as

$$K = \frac{2}{\pi} F_w \sigma \sqrt{\pi a} \tag{5.10}$$

where

$$F_w = \left(\frac{2b}{\pi a} \tan \frac{\pi a}{2b}\right)^{1/2} \tag{5.11}$$

For an initial crack size of $a_i = 4.5$ in. (114 mm), $b = 4.5 + 0.44 = 4.94$ in. (125.5 mm), and the stress range $= 0.6$ ksi (4.1 MPa), Eqs. 5.9 and 5.10 yielded 14,350,000 cycles for the crack to penetrate 95% of the web thickness. Hence it appears reasonable to assume that the crack instability in the web occurred after 23 million random variable stress cycles. The greatest uncertainty is with the first phase of crack growth which depends on the initial fabrication condition.

After the crack penetrates the web, a through-thickness crack results and the stress intensity factor is given by

$$K = \sigma \sqrt{\pi a} \tag{5.12}$$

Since two continuous fillet welds connect the longitudinal stiffener to the web, the web crack is in a zone of high residual tensile stress. Furthermore the longitudinal stiffener force is carried by the web alone. Therefore the web crack was assumed to be subjected to stresses at the yield point. At the low temperatures between December 1972 and March 1973, which were believed to be between −10°F (−23°C) and 10°F (−12°C), a yield stress of 45 ksi (310 MPa) was estimated [5.5]. Since the stress at the crack tip exceeds the elastic limit, the crack length $a = a' + r_y$ where a' is the physical crack length and r_y the plastic zone size. With the plastic zone size taken as

$$r_y = \frac{1}{2\pi}\left(\frac{K}{\sigma_Y}\right)^2 \tag{5.13}$$

Equation 5.12 becomes

$$K = \sigma\left[\frac{\pi a'}{1 - (\sigma/\sigma_Y)^2/2}\right]^{1/2} \tag{5.14}$$

With the average crack length $2a' = 2.8$ in. (71 mm), as shown in Figure 5.19 and a yield stress of 45 ksi (310 MPa), a stress intensity value of 133 ksi $\sqrt{\text{in.}}$ (147 MPa $\sqrt{\text{m}}$) results from Eq. 5.14. Figure 5.15 shows that the predicted critical stress intensity factor is in good agreement with the expected fracture resistance of the material.

Brittle fracture is arrested when the fracture toughness of the material is higher than the stress intensity factor. Once the crack becomes unstable, the dynamic fracture toughness controls the crack arrest condition. Figure 5.14 indicates that the absorbed energy at failure is about 10 ft-lb (14 J). Hence the estimated dynamic fracture toughness K_{Id} from Eq. 5.5 is 39 ksi $\sqrt{\text{in.}}$ (43 MPa $\sqrt{\text{m}}$). A second approximation is the empirical relationship

$$K_{Id} = (\sigma_Y + 24)\sqrt{0.5 \text{ in.}} \tag{5.15}$$

where σ_Y is the room temperature static yield point in ksi. Therefore $K_{Id} = 42.6$ ksi $\sqrt{\text{in.}}$ (46.8 MPa $\sqrt{\text{m}}$).

Brittle fracture was only arrested in the flange when the crack tip moved through the high tension residual stress field at the web-flange connection. At this stage the crack in the flange was semicircular in shape.

It was assumed that the tensile residual stresses due to welding were distributed over a semicircular area at the connection between the web and the flange with a radius of half the thickness of the web plus the weld leg size ($r = \frac{7}{32} + \frac{3}{8}$ in. $= 0.6$ in. or 15 mm). Over this area the stresses reach the yield stress. These residual tensile stresses are in equilibrium with compression residual stresses in the flange. The distribution of these compressive stresses was also assumed semicircular and their magnitude equal to one-fourth the yield point [5.9]. The residual stress distribution

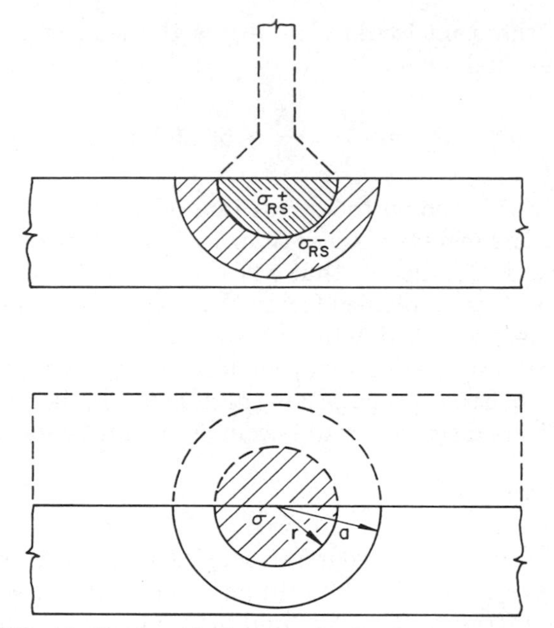

Figure 5.22 Idealized residual stress distribution in the bottom flange.

is shown schematically in Figure 5.22. The stress intensity for the loading conditions due to residual stress and applied loads can be estimated by superposition.

For a semicircular surface crack in a finite plate under uniform loading the stress intensity factor is

$$K = F_s F_w \frac{2}{\pi} \sigma \sqrt{\pi a} \tag{5.16}$$

and for a semicircular surface crack under uniform stress over a smaller circular area is given by [5.8]:

$$K = F_s F_w \frac{2\sigma}{\sqrt{\pi a}} [a - (a^2 - r^2)^{1/2}] \tag{5.17}$$

where r = radius of the residual tension zone.

The value of F_s in both Eqs. 5.16 and 5.17 was taken as 1.211. The back free surface correction F_w was defined by Eq. 5.11 and used for both loading conditions. The stress for uniform loading was taken as the sum of the dead load, maximum live load, and the residual compression stress:

$$\sigma_1 = \sigma_{\mathrm{LL}} + \sigma_{\mathrm{DL}} - \sigma_{rs}^- = 8.25 + 3.3 - \frac{45}{4} = 0.3 \text{ ksi} \quad (2.1 \text{ MPa})$$

The stress for the semicircular loading is the sum of the tension and compression residual stress:

$$\sigma_2 = \sigma_{rs}^+ + \sigma_{rs}^- = 45 + \frac{45}{4} = 56.25 \text{ ksi} \quad (388 \text{ MPa})$$

The influence of the web on the stress intensity factor is small, as illustrated in [5.9]. The increase is less than 10%. The superposition solution of Eqs. 5.16 and 5.17 indicates that a crack radius of just over 1 in. (25.4 mm) provides a stress intensity factor that is smaller than the dynamic fracture toughness of 40 ksi $\sqrt{\text{in.}}$ (44 MPa $\sqrt{\text{m}}$).

Further crack propagation by cyclic loading after crack arrest is governed by Eq. 5.16. Crack propagation through the remaining thickness of the flange until its discovery could occur in a year or less.

5.2.3 Conclusions

During its nine and a half years of service life, the cracked girder is estimated to have sustained 14.5 million random trucks which produced about 30 million stress cycles. The final brittle fracture and failure of the girder web resulted from an initial crack which started from an unfused butt weld in a longitudinal stiffener–girder web interface and enlarged from fatigue crack propagation. The crack condition at discovery in March 1973 indicated that the crack had developed in several stages as shown schematically in Figure 5.19. It seems probable that the web crack

Figure 5.23 Bolted splices at fractured section.

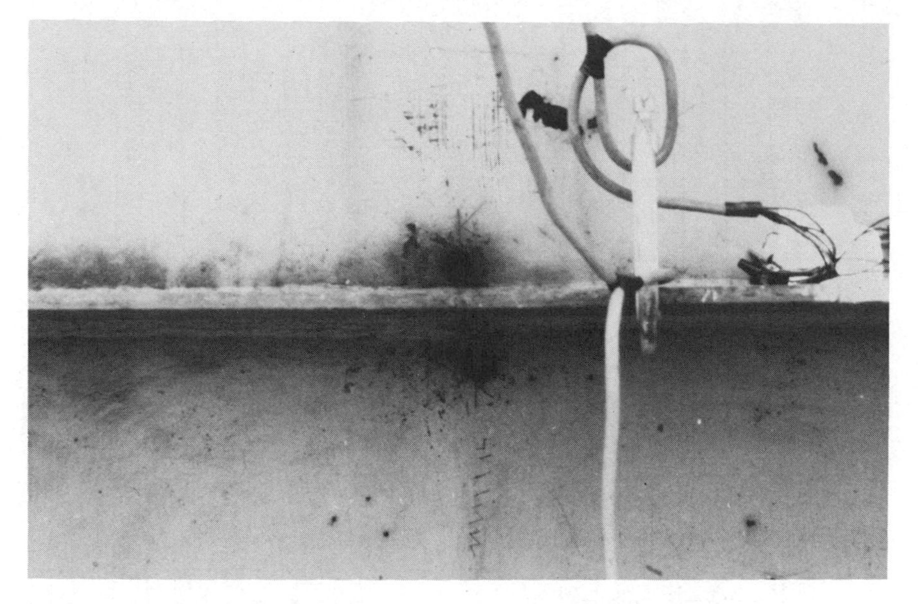

Figure 5.24 Drilled holes at stiffener crack.

instability occurred during the period of December 1972 to March 1973, when the material toughness of the web would be decreased considerably by low service temperatures.

5.2.4 Repair and Fracture Control

The cracked south facia girder was repaired by Conn-DOT using bolt splice plates following the removal of the crack segments. Figure 5.23 shows the bolted splice detail used to repair the cracked girder. In addition holes were drilled in the web in order to isolate the crack shown in Figure 5.11 which was propagating into the girder web. Figure 5.24 shows the retrofit holes and the cracked stiffener.

5.3 FATIGUE-FRACTURE ANALYSIS OF U.S. 51 OVER THE ILLINOIS RIVER AT PERU

5.3.1 Description and History of Bridge

Description of Structure

The U.S. 51 Bridge over the Illinois River at Peru is a 2292 ft (699 m), thirteen-span structure with the three main truss spans consisting of a cantilever truss 1007 ft (307 m) long. The suspended center span of the

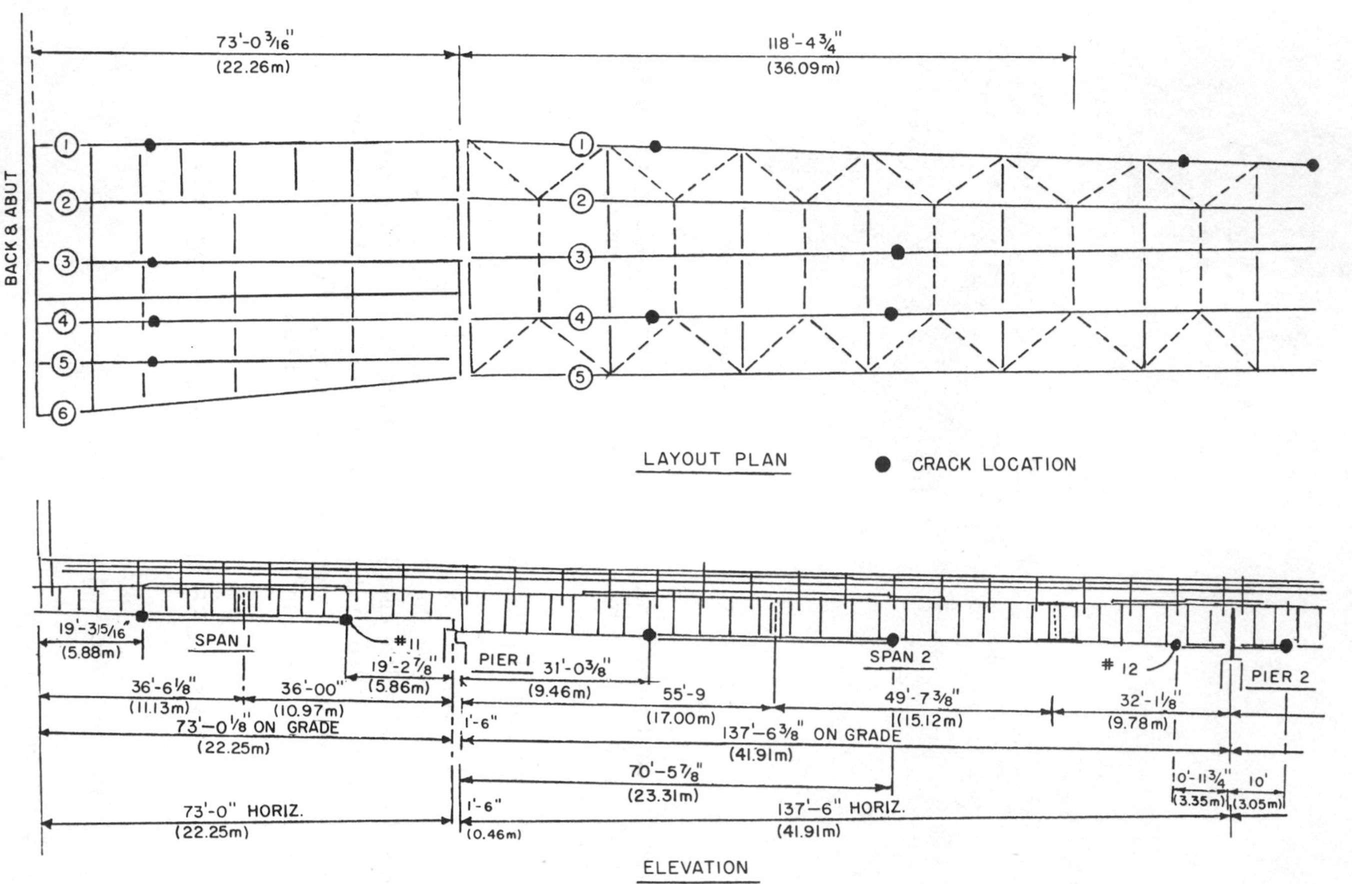

Figure 5.25 Plan and elevation of spans 1 and 2.

truss is 477 ft (145 m), and the two side spans are 265 ft (81 m). The approach spans consist of two 75 ft (23 m) single-span units, one two-span and two three-span units, all with spans of ±137.5 ft (±42 m). Except for spans 1, 2, and 3, each span is constructed with four riveted plate girders spaced 10 ft 3 in. (312 cm) apart. Span 1 has six girders, and spans 2 and 3 have five girders. Figure 5.25 shows the plan and elevation of a typical 137.5 ft (42 m) span with five girders. Each girder was 6 ft 10.5 in. (209.5 cm) deep back to back of the 8 × 6 × $\frac{7}{8}$ in. (203 × 152 × 22 mm) flange angles. The web was supplied from $\frac{7}{16}$ in. (11 mm) plate. Welded cover plates were attached to the riveted plate girders at the piers and at midspans, as illustrated in Figure 5.26. Near midspan the transitions were provided between a $\frac{15}{16}$ in. (24 mm) and a $\frac{1}{2}$ in. (12.1 mm) cover plate. Over the support at pier 2 the groove weld was at a transition between a $2\frac{3}{8}$ in. (60 mm) plate and $1\frac{1}{4}$ in. (32 mm) plates.

The girder webs were fabricated from A7 steel. The flange angles and cover plates were fabricated from A373 steel.

The bridge is one of the few early girder bridges that combined riveted girders with welded cover plates.

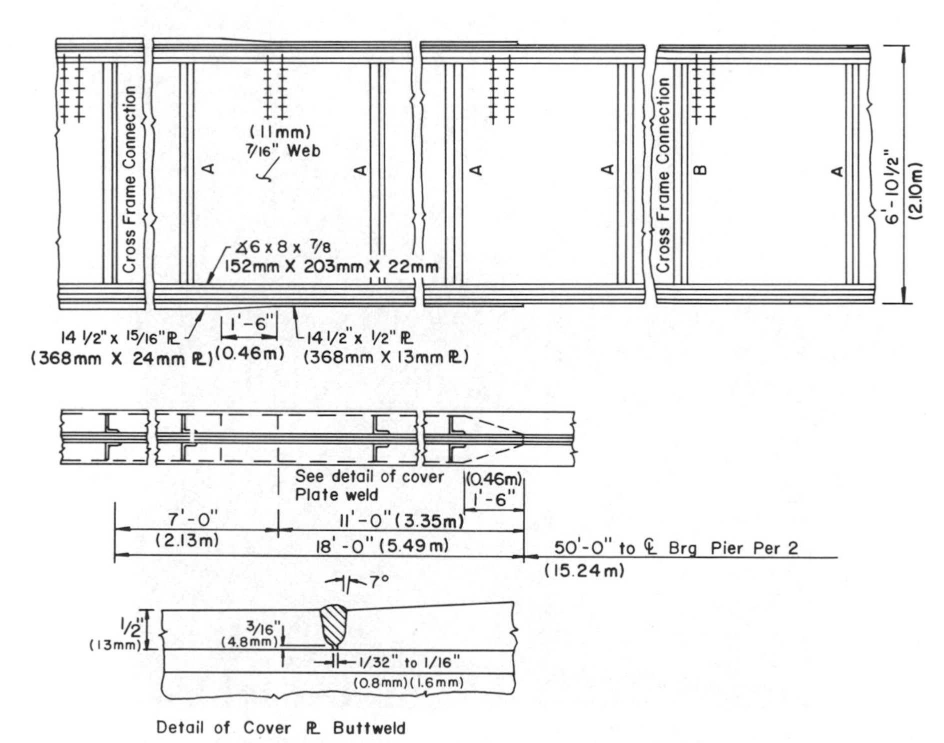

Figure 5.26 Typical welded cover plate detail.

History of Structure and Cracking

The structure was built in 1958. On July 22, 1980, painters reported a cracked cover plate at a transition weld which is shown in Figure 5.27. As of January 30, 1981, district maintenance personnel had located 13 transverse cracks in the bottom cover plate groove welds, including two locations in the negative moment region adjacent to pier 2. Eleven of the locations can be seen in Figure 5.25. Cover plates were cracked in two at two locations, one including the outside leg of a flange angle, as shown in Figure 5.28. Eight girder cover plates were cracked between 2 and 4 in. (5 and 10 cm) from each edge of the cover plate. At three locations grinding revealed full-width subsurface cracking, as illustrated in Figure 5.29. Four out of six girders were cracked near the quarter point of the north approach span (see Figure 5.25).

Figure 5.27 Cracked groove weld in cover plate (courtesy of Illinois Department of Transportation).

Figure 5.28 Crack extending into flange angle (courtesy of Illinois Department of Transportation).

Figure 5.29 Ground groove weld showing crack (courtesy of Illinois Department of Transportation).

5.3.2 Failure Modes and Analysis

Location of Cracks and Stages of Crack Growth

Of the 13 transverse cracks detected in the cover plate groove welds, two were in the compression flange of the east girder in the negative moment regions adjacent to pier 2. The cover plates were cracked completely at two locations. The cracks all appeared to be fatigue cracks, which had propagated from large internal flaws in the cover plate groove welds.

Ultrasonic testing of suspected weld defects in spans 11, 12, and 13 (corresponding to the riveted plate girders numbered 1, 2, and 3, respectively) was also carried out. These tests revealed a $\frac{3}{4} \times \frac{7}{16}$ in. (19 × 11 mm) crack in the outstanding leg of an 8 × 6 × $\frac{7}{8}$ in. (203 × 152 × 22 mm) bottom flange angle near pier 12. All cover plate butt welds were ground smooth and tested with dye-penetrant to locate subsurface cracks, as illustrated in Figure 5.29.

Cyclic Loads and Stresses

The 1979 estimated annual daily traffic crossing the bridge was 10,700 vehicles per day. The 1980 estimated annual daily truck traffic was 1950 multiunit vehicles per day, including occasional 120 kip (534 kN) permit loads.

The design stress range in span 1 was found to equal 10 ksi (69 MPa) at the groove weld near the 0.26L point from the north approach for HS20 loading. The stress ranges in span 2 were found to equal 8.5 ksi (58.6 MPa) and 11.4 ksi (78.6 MPa) at the groove welds in the positive moment region near the 0.23L and 0.55L points where cracks had been detected. The design stress range adjacent to pier 2 in the compression flange was 4.4 ksi (30 MPa).

The net section properties were used to estimate stresses in the tension flanges of the positive moment areas, and the gross section was used for the compression flange in the negative moment regions.

The actual service stresses were not available. However, an estimate of the effective stress range can be made by assuming that the daily truck traffic corresponds to the nationwide distribution [4.3]. The effective stress range becomes

$$\begin{aligned} S_{r\,\text{Miner}} &= \alpha S_r^D[\Sigma \gamma_i \phi_i^3]^{1/3} \\ &= \alpha(0.35)^{1/3} S_r^D \\ &= 0.5(0.705) S_r^D = 0.35 S_r^D \end{aligned} \tag{5.18}$$

Hence an effective stress range of 3 to 4 ksi (21 to 28 MPa) is probable if the factor α is taken as 0.5.

Assuming one stress cycle for each vehicle passage and that the total truck traffic has not significantly changed between 1958 and 1980 would result in 15.7 million random variable stress cycles corresponding to the Miner effective stress range.

Mechanical, Chemical, and Fracture Toughness of Material

No information is available on the material other than that it was supplied to the requirements of A7 and A373 steel. This suggests that a yield strength of about 35 ksi (242 MPa) is probable. The temperature shift between dynamic and static fracture toughness would be ~162°F (90°C). Available data suggest that a fracture toughness of 100 ksi $\sqrt{\text{in.}}$ (110 MPa $\sqrt{\text{m}}$) would not be unreasonable for such low yield point material at the probable minimum service temperature of −20°F (−29°C). Since no fractures are apparent at the cracked sections, this seems to be a reasonable assumption.

Failure Analysis

The observed behavior of the cover plate groove weld splices indicates that the cracks have developed from flaws embedded at the root of the single-U groove welds. Figure 5.30 shows a schematic of the probable defect configuration. As can be seen in Figure 5.26, a $\frac{3}{16}$ in. (5 mm) root gap was used with no evidence of back gouging. It is probable that these welds were made using the beam as a backup bar. Hence the probable size of the initial crack a_i is likely to be from $\frac{1}{4}$ to $\frac{1}{2}$ in. (6 to 12 mm) depending on the thickness of the plates at the groove weld. Since the

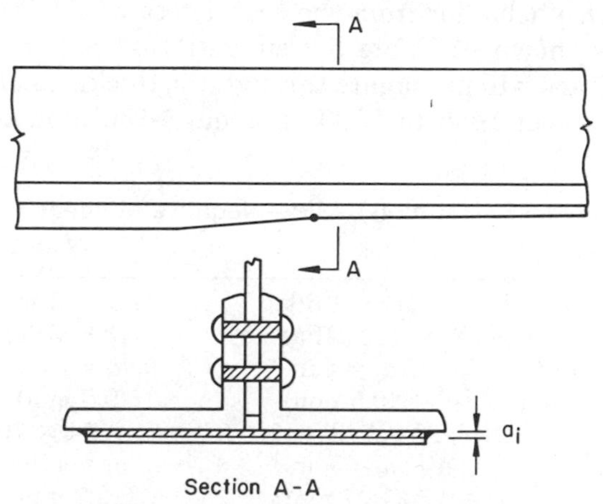

Figure 5.30 Idealized root crack in groove weld.

Table 5.2 Estimated Life—Positive Moment Regions

t_f	$\Delta\sigma = 2$ ksi (13.8 MPa)	$\Delta\sigma = 3$ ksi (20.7 MPa)	$\Delta\sigma = 4$ ksi (27.6 MPa)
$\frac{1}{2}$ in. (12.7 mm)	$a_i = \frac{1}{4}$ in. (6 mm) $N = 16 \times 10^6$	$a_i = \frac{1}{4}$ in. (6 mm) $N = 4.75 \times 10^6$	$a_i = \frac{1}{4}$ in. (6 mm) $N = 2 \times 10^6$
$\frac{3}{4}$ in. (19 mm)	$a_i = \frac{1}{4}$ in. (6 mm) $N = 51.6 \times 10^6$	$a_i = \frac{1}{4}$ in. (6 mm) $N = 15.28 \times 10^6$	$a_i = \frac{1}{4}$ in. (6 mm) $N = 6.45 \times 10^6$

cover plate is not connected to the flange angles, except along the edges, the applicable stress intensity factor for an edge crack is

$$K = 1.12\sigma\sqrt{\pi a}\left(\sqrt{\sec\frac{\pi a}{2t_f}}\right) \tag{5.19}$$

where t_f is the thickness of the thinner plate at the groove weld splice. The predicted cycles of stress are given by

$$N = \int_{a_i}^{t_f} \frac{da}{3.6 \times 10^{-10}\Delta K^3} \tag{5.20}$$

Appearing in Table 5.2 is a list of cycles required to propagate the initial crack through the plate thickness in the positive moment regions where the estimated effective stress range was 2 to 4 ksi (13.8 to 27.6 MPa).

The stress concentration that results from the reinforcement on the groove weld and the taper of the thicker plate do not significantly affect the crack growth behavior from the root defect.

The results shown in Table 5.2 suggest that a $\frac{1}{4}$ in. (6 mm) initial unfused root crack will propagate through the thickness of the cover plate during the interval 1958 to 1980 as about 16 million trucks cross the

Table 5.3 Estimated Life—Negative Moment Regions

t_f	$\Delta\sigma = 1.5$ ksi (10 MPa)	$\Delta\sigma = 2$ ksi (13.8 MPa)
$1\frac{1}{4}$ in. (32 mm)	$a_i = \frac{3}{8}$ in. (9.5 mm) $N = 74.6 \times 10^6$	$a_i = \frac{3}{8}$ in. (9.5 mm) $N = 34 \times 10^6$
	$a_i = \frac{1}{2}$ in. (12.7 mm) $N = 41.3 \times 10^6$	$a_i = \frac{1}{2}$ in. (12.7 mm) $N = 18.9 \times 10^6$

structure. The effective stress range was likely between 2 and 3 ksi (13.8 and 20 MPa) at the positive moment region welds.

In the negative moment region the design stress range was 4.4 ksi (30 MPa). This corresponds to an effective stress range between 1.5 ksi (10 MPa) and 2 ksi (13.8 MPa) if Eq. 5.18 is used to estimate the stress cycles. Equations 5.19 and 5.20 yield the life estimates for the $1\frac{1}{4}$-in. (32-mm) groove weld splices given in Table 5.3.

This evaluation suggests that larger initial crack conditions existed in the thicker groove welded splices.

5.3.3 Conclusions

1. The primary cause of the cracking was the poor quality of the groove welds at the thickness transitions in the welded coverplates. The poor quality welds resulted from the use of a single-U groove weld which appears to have been made with the beam as a backup. Hence lack of fusion and slag would exist at the weld root and create a large initial crack condition.
2. All defects have enlarged as a result of fatigue crack propagation. At a number of joints the crack tip had not penetrated through the outside surface. Several subsurface cracks were exposed by grinding the surface and removing $\frac{1}{16}$ in. (1.6 mm).
3. A reasonable estimate of the cycles required to crack the cover plate was provided by modeling the defects as edge cracks. Assuming the nationwide gross vehicle weight distribution was applicable to the bridge structure, lack of fusion between $\frac{1}{4}$ in. (6 mm) and $\frac{1}{2}$ in. (12.7 mm) would provide fatigue life estimates in agreement with the truck traffic volume.

5.3.4 Repair and Fracture Control

1. All groove welds that were found or suspected to be defective were retrofitted with bolted splice plates. A typical splice is shown in Figure 5.31. The repair work was carried out by Illinois-DOT day labor crews. Those groove welds that were found to be acceptable by nondestructive inspection were ground smooth and left as constructed.
2. The contact surfaces between the splice plates and the existing cover plates were ground smooth to ensure proper fitting. The fillet welds that connected the cover plate to the angles were removed by grinding for a length of several inches. At the groove weld the cover plate was ground back up to $\frac{1}{4}$ in. (6 mm) to ensure that no connection existed between the cover plate and the flange angles at the cracks. This measure would prevent subsequent crack propagation from penetrating into the flange angles.

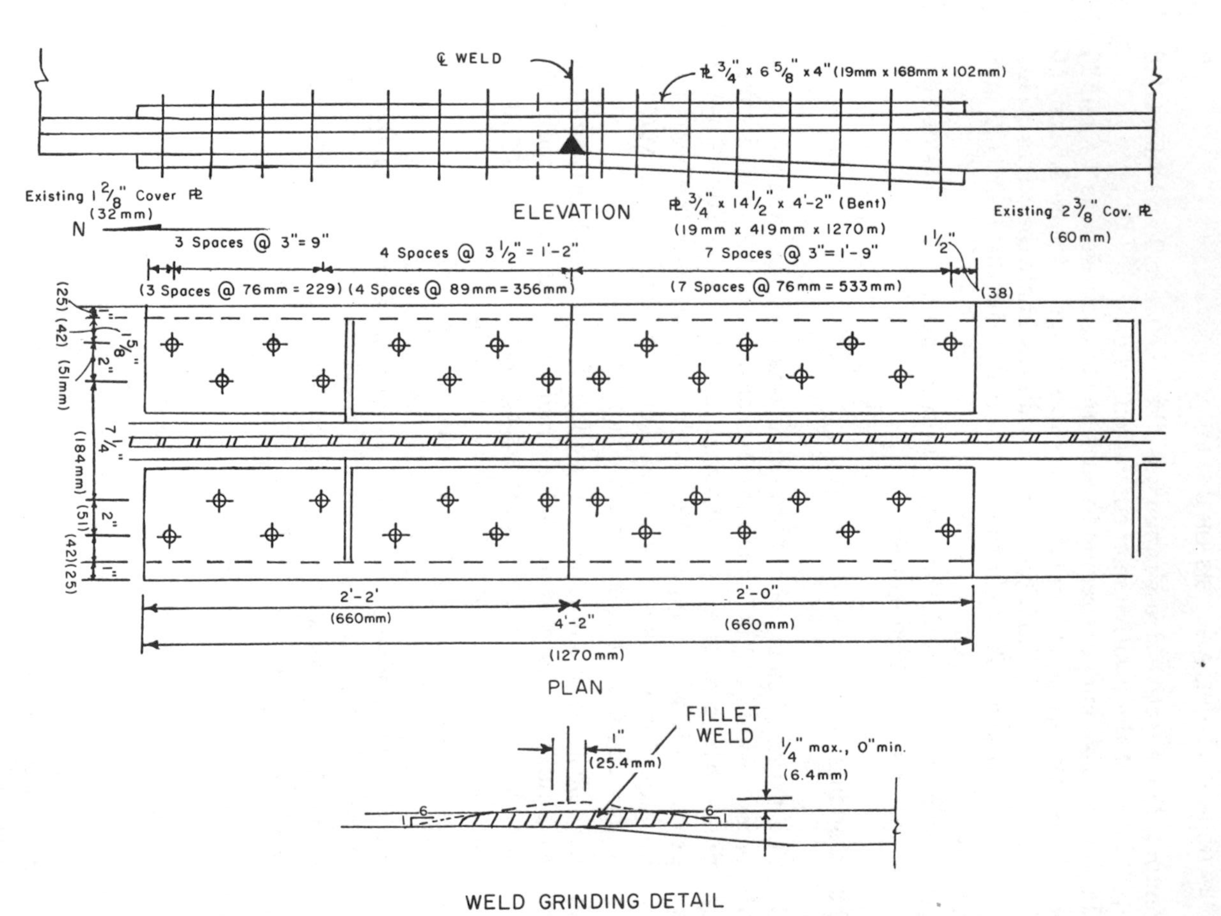

Figure 5.31 Typical repair splice and removal of fillet weld.

3. The outside splice plates were bent to conform to the beveled transition at the groove welded joints, as illustrated in Figure 5.31. Additional bolts were added to compensate for the forces induced by the bent plate.

REFERENCES

5.1 King, J. P. C., Csagoly, P. F., and Fisher J. W., Field Testing of Aquasabon River Bridge in Ontario, Transportation Research Record No. 579, 1976, pp. 48–60.

5.2 Fisher, J. W., and Pense, A. W., Evaluation of the Cracking of the I-24 Bridge over the Ohio River near Paducah, Kentucky and Recommendations for Its Retrofit, Fisher, Fang & Assoc. Report for Illinois-DOT, November 1980 (limited distribution.)

5.3 Frank, K. H., and Colwell, A. B., Evaluation of Silver Memorial Bridge—Phase III. Weld Metal Core Evaluation, PMFSEL, Univ. of Texas at Austin, July 1981, PMFSEL-WV-2 (limited distribution).

5.4 Hanson, J. M., Koob, M. J., and Bresler, G., Post-Construction Evaluation Study Fremont Bridge (Willamette River) I405-Bridge 2529, Wiss, Janney, Elstner and Assoc. Inc., December 31, 1981 (limited distribution).

5.5 Fisher, J. W., Pense, A. W., Hausammann, H., and Irwin, G. R., Analysis of Cracking in Quinnipiac River Bridge, *J. Struct. Div.*, ASCE, 106 (April 4, 1980).

5.6 Engineering News Record, Cracked Girder Closes I-79 Bridge, *Engineering News Record,* February 10, 1979.

5.7 Federal Highway Administration Notice N 5040.23, Electroslag Welding, March 1977.

5.8 Roberts, R., Fisher, J. W., Irwin, G. R., Yen, B. T., Bellenoit, J. R., Rohr, D., and Durkee, M., Improved Quality Control Procedures for Bridge Steels, Report FHWA-RD-82, Federal Highway Administration, Washington, D.C., 1982.

5.9 Roberts, R., Fisher, J. W., Irwin, G. R., Boyer, K. D., Hausammann, H., Krishna, G., Morf, U., and Slockbower, R. E., Determination of Tolerable Flaw Sizes in Full Size Welded Bridge Details, Report FHWA-RD-77-710, Federal Highway Administration, Washington, D.C., 1977.

CHAPTER 6

Web Penetrations

After cracking was discovered in the rigid frame bents of the Dan Ryan elevated transit structure in Chicago in 1978, an examination was made of other structures with comparable details. In the Dan Ryan structure, the bottom compression flanges of the longitudinal girders which frame into the steel box bents pass through slots in the box girder web. The flange plates were groove welded at the web slots, but not very well. Large unfused areas were found to exist in the web at the tips of the flange plates.

At least two other bridge structures have experienced fatigue cracking at similar lack-of-fusion discontinuities. In both cases the details were retrofitted before significant damage developed; the faulty details were isolated by installing vertical slots ending in holes. In the Girard Point pier caps the flange plates were connected to the web slots by exterior fillet welds [3.5]. Fatigue tests carried out on a detail, which was inserted into a slot cut into a rolled beam web and then fillet welded to one side, verified the low fatigue resistance of this type of lack-of-fusion condition. In the Metro pier cap vertical reaction brackets were passed through the box girder webs. These brackets were groove welded to the webs using backup bars, producing a cracklike condition and lack of fusion adjacent to the flange which caused significant crack growth in a short period of time. Crack growth was prevented from entering the tension flange by isolating the vertical joint with 3 in. (75 mm) holes.

6.1 FATIGUE-FRACTURE ANALYSIS OF THE DAN RYAN RAPID TRANSIT STEEL BOX BENTS

6.1.1 Description and History of the Structure

Description of the Structure

The Chicago Transit Authority's (CTA) Lake–Dan Ryan Rapid Transit Line provides direct service between the city center and the south and west sectors of the city. The tracks ascend from the median of an expressway onto a viaduct which carries them through a curve of 400 ft (120 m) radius to a smaller junction curve that joins them with the old North–South elevated structure at 17th and State Streets. Two views of the structure are shown in Figure 6.1.

General features of the viaduct at the location of the cracked bents (piers) 24, 25, and 26 are shown in Figures 6.2 and 6.3. The superstructure consists of four continuous plate girders that carry a cast-in-place concrete deck and two tracks. The rails of the two tracks are supported on wooden ties resting on a ballast. The bents supporting the four continuous I-beam plate girders have a cross section shaped like a box; that is, it consists of two box shaped column legs and a horizontal box member, as shown in Figure 6.3. The stringer bottom flanges pass through flame-cut slots near the bottom of the side plates or webs of the box girders. The stringer flange was inserted through these slots and then connected to

Figure 6.1*a* General view of Dan Ryan Rapid Transit Structure (courtesy of Chicago Department of Public Works).

Figure 6.1*b* View of cracked bents looking west from Clark Street (arrows are pointing to locations of fracture) (courtesy of Chicago Department of Public Works).

the supporting box girder web by groove welds all around the flange surface.

The stringers at the crack locations have a track curvature of 400 ft (120 m) in the horizontal plane, and they intersect the boxes at different angles (approximately 45°). With the exception of bent 35, the bents were fabricated with portions of the superstructure girders welded within the boxes. The girder stubs projected 3 to 4 ft (0.9 to 1.22 m) from the sides of the boxes and are connected to the continuous girders by either a pinned link or a bolted connection (see Figure 6.1*a*).

History of the Structure and Cracking

The structure was designed in 1967 in accordance with the criteria and procedures established by the American Railway Engineering Association (AREA), the American Welding Society Specifications, and the CTA design policy for elevated railways. The construction of the Dan Ryan Line started in April 1968, and it was completed and opened to traffic in September 1969. On an average weekday, 467 trains (in both directions) pass over this viaduct. The trains vary in length from eight cars during the rush hours to two cars during the night [6.2, 6.3].

On January 4, 1978, a crack was discovered in one of the steel box girder bents near 18th and Clark Streets. Subsequent inspection of the structure verified the existence of this crack in bent 24 and also revealed that bents 25 and 26 were cracked as well [6.1]. A typical crack is shown in Figure 6.4. The most recent inspection prior to the discovery of the

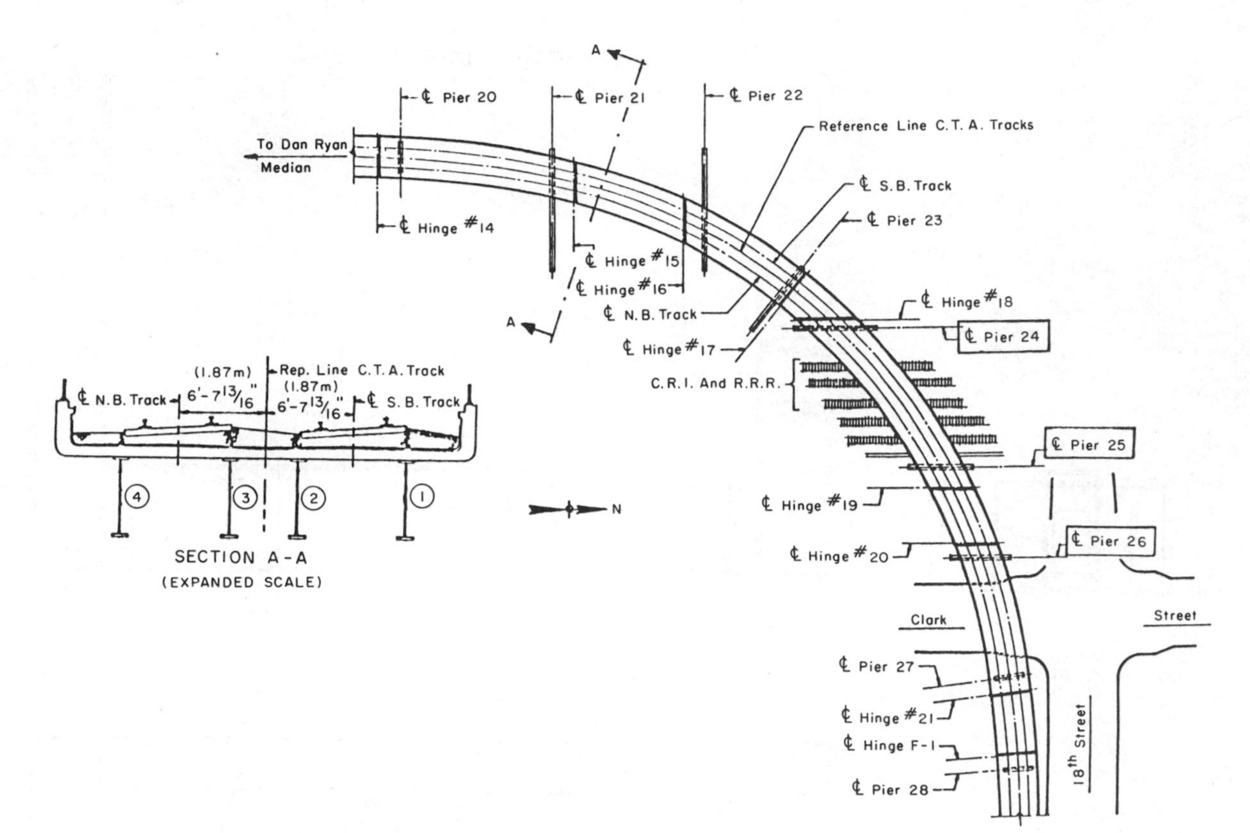

Figure 6.2 Schematic plan and locations of cracked bents (bents 24, 25, and 26).

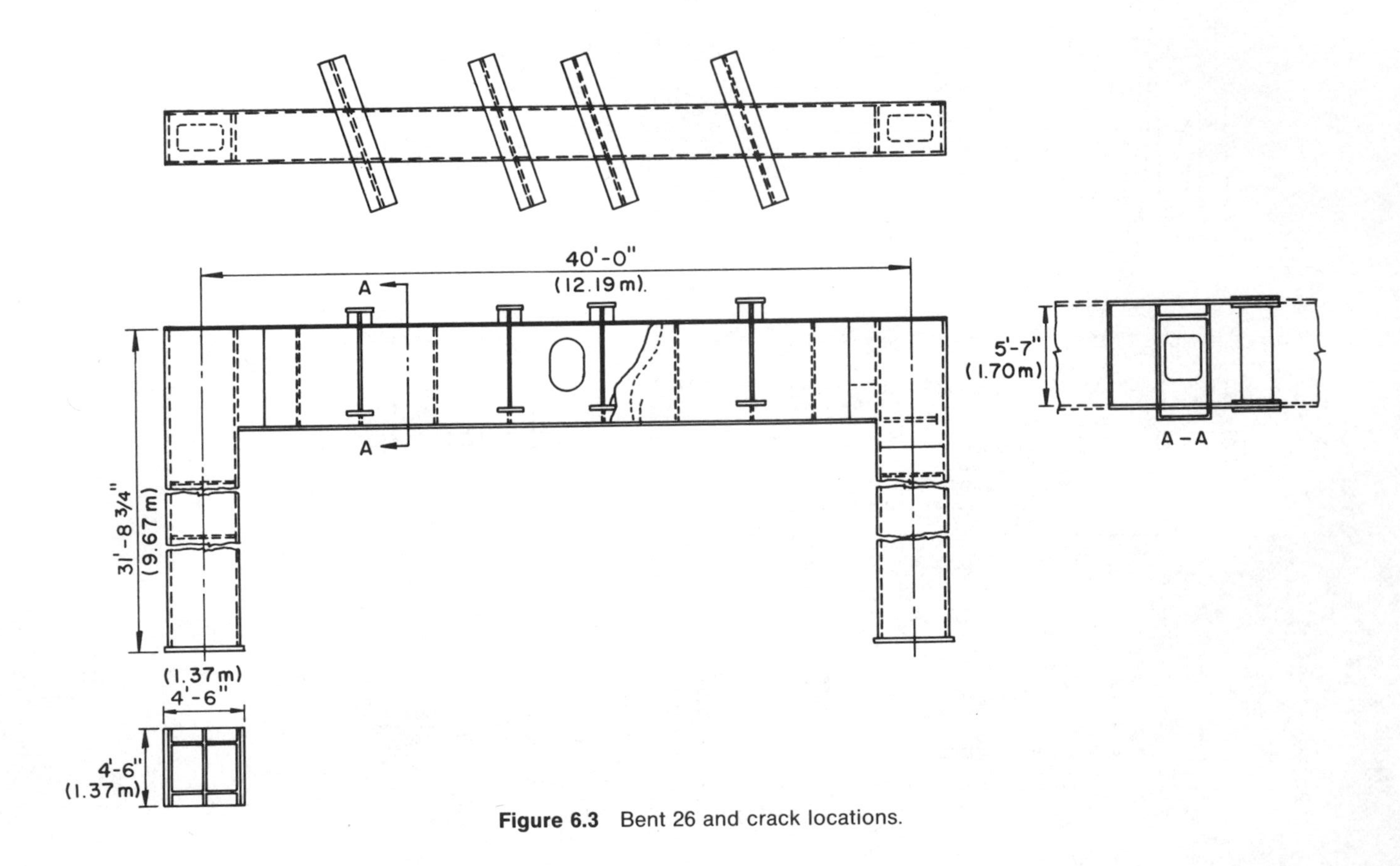

Figure 6.3 Bent 26 and crack locations.

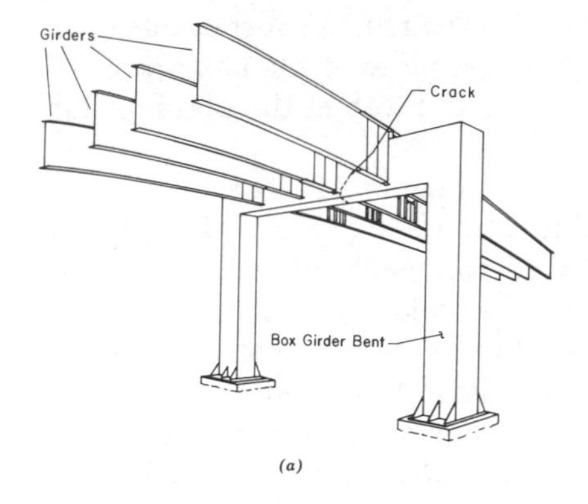

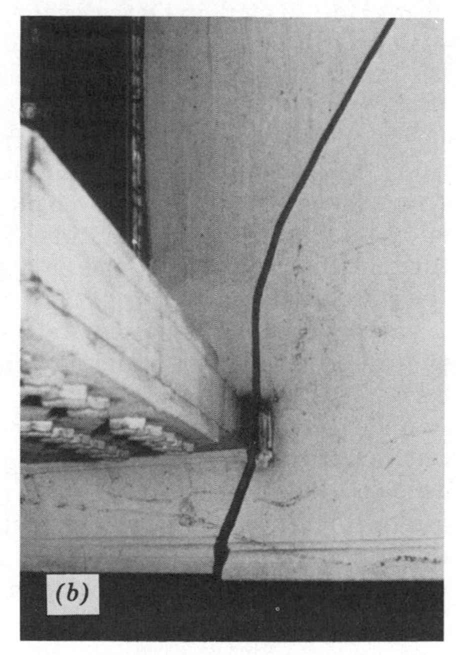

Figure 6.4 Fracture of the box girder (bent 26). (*a*) Schematic showing location of fracture (bent 26); (*b*) close-up view of cracked box girder web.

cracks was conducted in July 1976, and no significant cracks were found and reported at that time.

Initial examination of the fractures in bents 24, 25, and 26 indicated that they originated from fatigue cracks at or near the welded junctions where the bottom flanges of stringers intersect the side plates of a steel box bent (see Figure 6.4).

Prior to the discovery of the fractures, temperatures in the Chicago area were extremely low on several occasions. There were five days when the temperatures dropped below 0°F (−18°C), and the coldest temperature recorded at Midway Airport was −7°C (−22°C) in December 1977.

6.1.2 Failure Modes and Analysis

Location of Cracks

The field examination and measurements revealed that the locations, shapes, and patterns of the cracks were similar. A common feature of the cracks in bents 24, 25, and 26 was their close proximity to the vertical edge of the bottom flange of the stringers [6.1, 6.2]. In all three bents the cracks completely separated the box girder bottom flange plate and both side plates or webs. The crack openings at the bottom flange measured about $\frac{3}{4}$ in. (20 mm). In two cases the cracks were arrested at the edge of or with only little penetration into the top flange plate of the box girder of the bents. In bent 24 the crack extended into the web of the plate girder framing into the bent.

The initial field examination of fractures in bents 24, 25, and 26 indicated that they initiated at the welded junction of a plate girder bottom flange tip and the box girder web plate. Chevron markings were observed on the fracture surfaces pointing toward the welded junction. The fracture surfaces on bents 25 and 26 appeared to be lightly rusted, whereas the fracture surfaces on bent 24 were more heavily rusted.

The trajectory of the crack in the box girder of each bent was influenced by tension, shear, and flexural stresses, as can be seen in Figure 6.4.

The examination of the fracture surfaces in each bent showed that a brittle fracture occurred in the box girder web plates after fatigue cracks developed from the unfused welds at the edges of the bottom flange of girders that intersected the horizontal box girder. In the fractured bents the intersections were connected by partial penetration groove welds that resulted in a large unfused region in the box girder web. Typical stages of

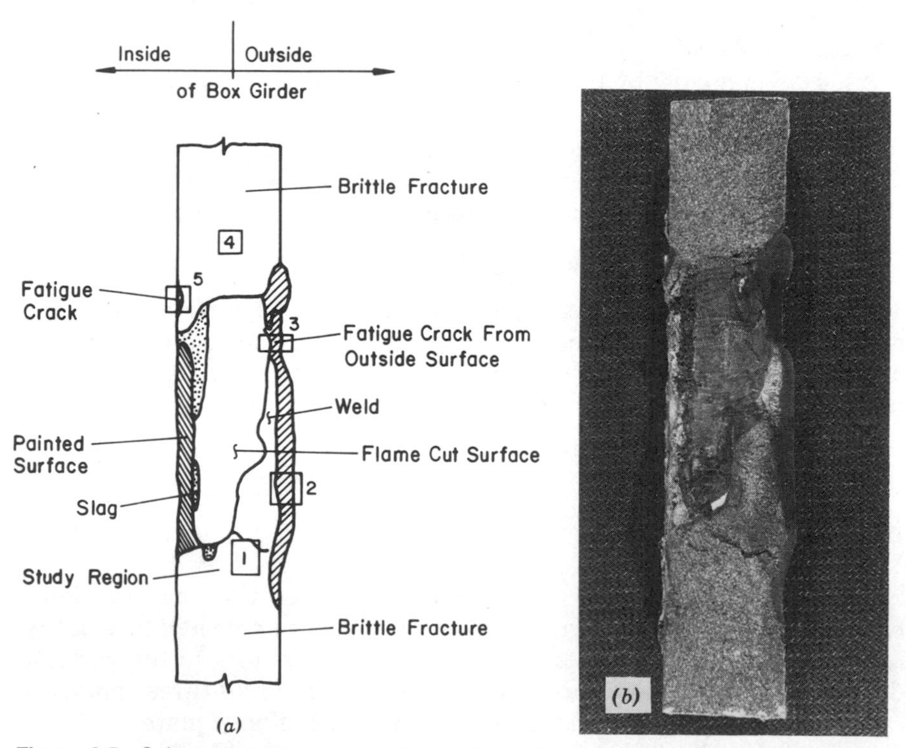

Figure 6.5 Schematic of fracture surface at bent 26. (*a*) Stages of fatigue crack growth; (*b*) fracture surface of box girder web.

crack growth in the web of one of the intersections in bent 26 are shown in Figure 6.5.

Cyclic Loads or Stresses

Following the discovery of cracking in the bents of the Dan Ryan Rapid Transit, the bents were analyzed using the finite element method for computing stresses. The analysis indicated that the stresses were generally less than the values allowed by the governing specifications of the American Railway Engineering Association. The maximum design shear stress in the web plate on the east side of bent 24 was 17.27 ksi (119 MPa) which exceeded the value of 12.5 ksi (86 MPa) currently allowed by these specifications [6.2].

The design cyclic stress range during passage of two trains varied from 3.6 ksi (25 MPa) to 8.7 ksi (60 MPa) in the three bents at the edge of the stringer flange [6.2]. With variable amounts of impact and multiple presence of two trains on occasion, it was estimated that an effective stress range of 2.2 ksi (15 MPa) to 3 ksi (21 MPa) was reasonable for trains that crossed the structure.

Figure 6.6 shows the stress-time variation that occurs in the box girder web as four- or six-car trains cross the structure. The trains vary in length between two and eight cars. Most stress cycles during service are produced by one train crossing the structure at a time. Altogether, an estimated 1.4 million trains crossed the structure prior to discovery of the cracked bents. As illustrated in Figure 6.6, passage of a train results in one primary stress range cycle for each train. Smaller stress cycles also exist but appear to be negligible.

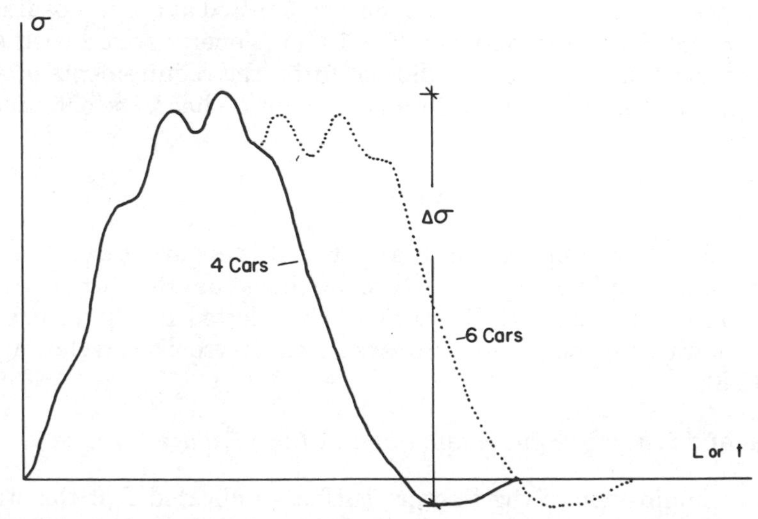

Figure 6.6 Stress range in box web plate at crack location, Dan Ryan Viaduct.

Temperature and Environmental Effects

As mentioned earlier, the temperatures in the Chicago area were known to be extremely low prior to the discovery of the fractures. During the 1977 to 1978 winter season there were five occasions when the temperatures dropped below 0°F (−18°C), and the coldest temperature recorded at the Midway Airport in December 1977 was −7°F (−22°C). Low temperatures were also experienced in Chicago on January 3, 1978, at the time the cracks in the three box girders of the bents were discovered [6.1, 6.2].

Mechanical, Chemical, and Fracture Properties of the Material

Several sample pieces were removed from the fractured box girders [6.2]. Tensile tests, chemical analysis, Charpy V-notch tests (CVN), compact tension fracture tests (CT), and metallographic and fractographic examinations were conducted on these samples [6.2, 6.3].

The web plate tensile tests provided a yield strength (0.2% offset) of 33.7 ksi (235 MPa) and an ultimate strength of 69.6 ksi (480 MPa). Machining and testing was done according to ASTM Designation E8. Both tensile and chemical tests indicated that the material used in the structure conformed to the ASTM requirements for A36 steel.

Charpy V-notch tests were performed on the specimens in accordance with ASTM Designation E23. The results are summarized in Figure 6.7. The results indicated that all plates tested satisfied the impact requirements of the AASHTO specifications (15 ft-lb or 20 J, at 40°F, or 4°C) for a minimum service temperature of −29°F (−34°C) for zone 2. The transition temperature behavior for the material ranges between +40°F (4°C) and +70°F (+21°C).

Compact tension fracture specimens were tested at a 1-sec loading rate between −30°F (−34°C) and +30°F (−1°C) in general accord with ASTM Designation E399. The results did not fulfill the requirements of ASTM E399, and therefore the stress intensity factor K_J had to be obtained by a J-integral approximation:

$$K_J = (EJ)^{1/2} \tag{6.1}$$

The results are summarized in Figure 6.8. It is apparent from the figure that a significant increase in fracture toughness develops when temperatures are greater than 10°F (−12°C). Also plotted in Figure 6.8 is the dynamic fracture toughness estimated from Barsom's correlation equation (2.8).

Visual and Fractographic Examination of the Failure Surface

Visual examination of the fracture surfaces indicated that the stringer bottom flange tips were partially welded to the box girder side plate.

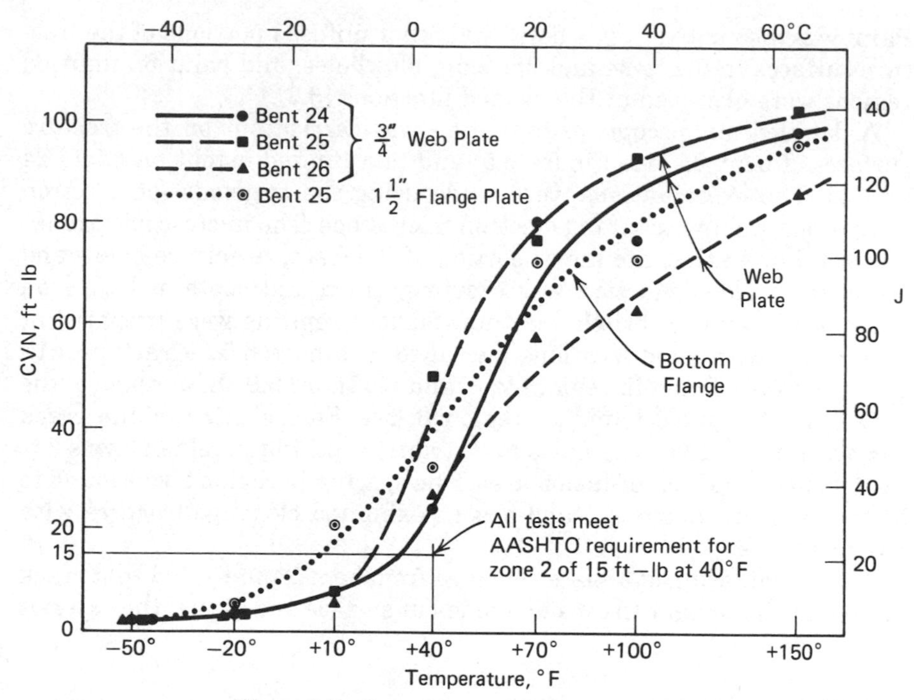

Figure 6.7 Comparison of CVN test results.

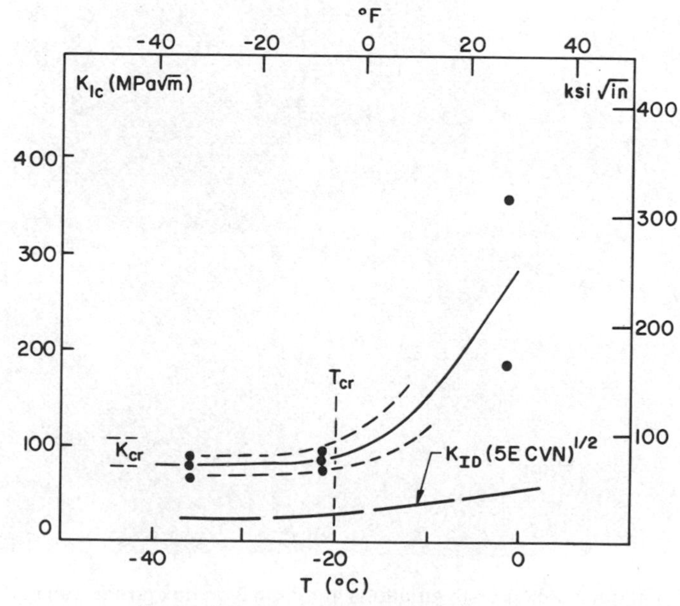

Figure 6.8 K_{Ic} versus test temperature.

Paint was also noted in gas holes and other unfused portions of the fracture surfaces in bents 24 and 26. Slag, blowholes, and paint on unfused regions were observed at the welded junctions [6.2].

A detailed microscopic examination was carried out on the fracture surfaces of bent 26 (see Figure 6.5) and to a limited extent on bents 24 and 25. The examinations were made using the transmission electron microscope and the scanning electron microscope. The microscopic examinations showed that the major portion of the fracture surface below and above the welded junction was coarse grained and contained chevron markings, typical of brittle fracture. Surface replicas were prepared at each of the numbered locations identified in Figure 6.5*a*. Crack growth striations were found in regions 2, 3, and 5. Figure 6.9 shows some of the striations that were found in region 3 (see Figure 6.5*a*) of the crack surface from bent 26. The measured striation spacing in region 3 was 2×10^{-6} in. near the lack-of-fusion area. The spacing in region 2 was found to be 8×10^{-7} in. Regions 1 and 4 both exhibited cleavage fracture with river patterns.

The examination of a piece removed from bent 24 indicated that crack growth initiated near the inside flame-cut surface and also at the exterior

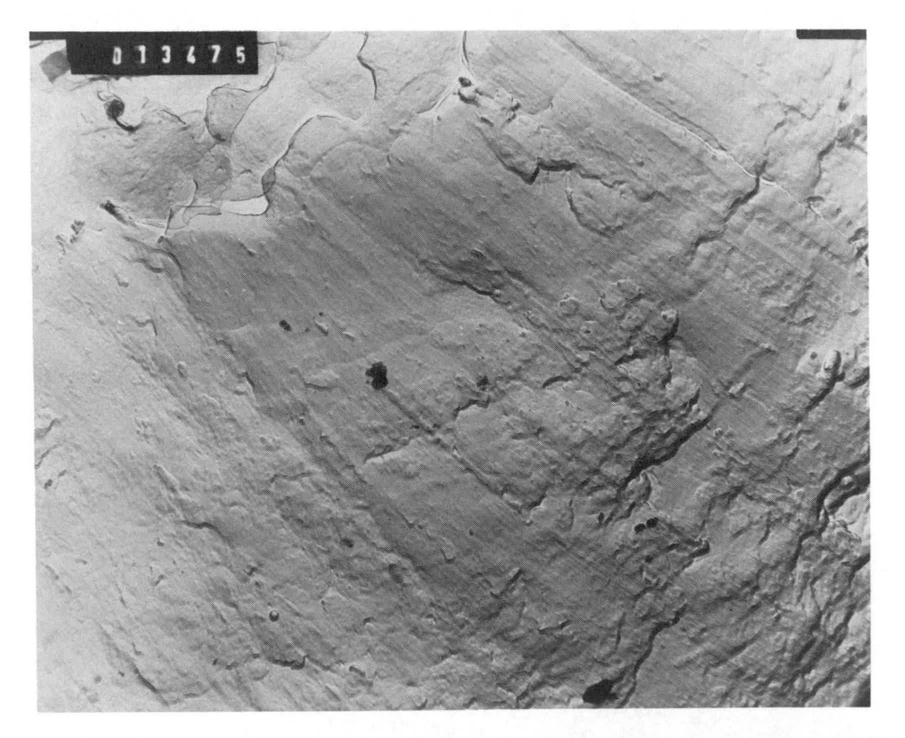

Figure 6.9 Fatigue crack growth striations in region 3 on box girder web surface at bent 26; 17,000×.

weld surface and extended toward the outside. Similar observations were made for bent 25.

The microscopic examination established that fatigue crack growth striations existed at the weld toe area as well as in the regions adjacent to the initial lack-of-fusion defects in the web at the plate intersections. Hence a significant amount of fatigue crack growth occurred on each surface of the web plate. It was estimated that the fatigue crack extension from the outside surface varied from $\frac{1}{8}$ in. (3.2 mm) to $\frac{1}{4}$ in. (6.4 mm). Similar amounts of crack extension occurred from the inside lack-of-fusion areas on all three bents. The estimated vertical length of the fatigue-extended cracks was 3 in. (76 mm) for bents 24 and 25 and 2 in. (51 mm) in bent 26. This crack condition included the vertical height of the flame-cut web and the additional extension due to fatigue cracking.

The metallographic study did not indicate any unusual conditions in the steel web plates. The grain structures were typical for A36 steel plate.

Failure Analysis

Two modes of fatigue crack extension were observed at the flange tips of the plates inserted through the web. Crack growth occurred from the outside web surface at the weld toe and from the unfused interior areas adjacent to the inside web surface. A schematic of the general crack condition is shown in Figure 6.10.

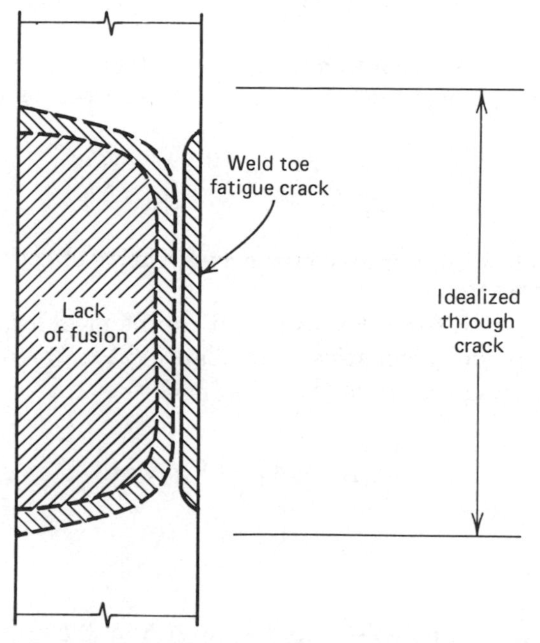

Figure 6.10 Idealized cracks in box girder web.

The stress intensity factor for crack propagation at the outside web surface at the weld toe was estimated from the relationship [6.4]:

$$K = F_e F_s F_w F_g \sigma \sqrt{\pi a} \tag{6.2}$$

where

$$F_e = \frac{1}{E(k)} \simeq 1.0 \quad \text{as} \quad c \gg a$$

$$F_s = 1.211 - 0.186 \sqrt{\frac{a}{c}} \simeq 1.0$$

$$F_w = \left(\frac{2t'_w}{\pi a} \tan \frac{\pi a}{2t'_w} \right)^{1/2}$$

$$t'_w = t_w + \frac{D}{\sqrt{2}} - \text{lack of fusion} \simeq 0.4 \text{ in. (10 mm)}$$

$$F_g = \frac{K_{tm}}{1 + 0.88a^{0.576}} \quad \text{(in.)}$$

The stress gradient correction factor F_g was derived from a finite element analysis of the stress field at the weld toe [6.4] which indicated that $K_{tm} = 6$.

The number of cycles required to propagate an initial defect of 0.03 in. (0.75 mm) to the observed depth of $\frac{1}{8}$ in. (3.2 mm) was established to be

$$N = \int_{0.03}^{0.125} \frac{da}{C \Delta K^3} = \frac{10^{10}}{3.6} \int_{0.03}^{0.125} \frac{da}{\Delta K^3} \tag{6.3}$$

This yielded 1.65×10^6 variable stress cycles at an effective stress range $S_{re} = 2.2$ ksi (15 MPa).

Crack propagation from the inside out can be modeled as a cruciform joint which is shown schematically in Figure 6.11. The stress intensity factor for this configuration is given by [6.5]:

$$K = \left(A_1 + A_2 \frac{a}{w} \right) F_w \sigma \sqrt{\pi a} \tag{6.4}$$

where

$$w = t_p \left(\frac{1}{2} + \frac{H}{t_p} \right) \quad \text{and} \quad t_p = 1.5 \text{ in. (38 mm)}$$

$$A_1 = 0.528 + 3.287\frac{H}{t_p} - 4.361\left(\frac{H}{t_p}\right)^2 + 3.696\left(\frac{H}{t_p}\right)^3$$

$$-1.875\left(\frac{H}{t_p}\right)^4 + 0.415\left(\frac{H}{t_p}\right)^5$$

$$A_2 = 0.218 + 2.717\frac{H}{t_p} - 10.171\left(\frac{H}{t_p}\right)^2 + 13.122\left(\frac{H}{t_p}\right)^3$$

$$-7.755\left(\frac{H}{t_p}\right)^4 + 1.783\left(\frac{H}{t_p}\right)^5$$

$$F_w = \left[\sec\frac{\pi a}{2w}\right]^{1/2}$$

The cyclic life is provided by

$$N = \int_{a_i}^{a_f}\frac{da}{C\Delta K^3}$$

$$= \frac{1}{C\Delta\sigma^3}\int_{0.7}^{0.85}\left[\frac{1 + 2H/t_p}{[(A_1 + A_2(a/w)][\pi a \sec(\pi a/2w)]^{1/2}}\right]^3 da \quad (6.5)$$

Equation 6.5 predicted 1.3 million stress cycles. An approximation to the integral equation, termed I, in the expression for N is given by

$$I = \left[0.71 - 0.65\left(\frac{2a_i}{t_p}\right) + 0.79\left(\frac{H}{t_p}\right)\right]^3\frac{1}{\sqrt{t_p}} \quad (6.6)$$

Therefore the cyclic life N is

$$N = \frac{I}{C\Delta\sigma^3} \quad (6.7)$$

Since the weld is from one side, the web plate thickness must be taken as $2 \times \frac{3}{4} = 1\frac{1}{2}$ in. (37 mm). For ratios

$$\frac{2a_i}{t_p} = \frac{2 \times 0.70}{1.5} = 0.933 \quad \text{and} \quad \frac{H}{t_p} = 0.10$$

Equation 6.6 becomes $0.004967\sqrt{\text{in.}}$ The fatigue life predicted by Eq. 6.7 is

$$N = \frac{0.00497}{3.6 \times 10^{-10}\Delta\sigma^3} = \frac{13.8 \times 10^6}{\Delta\sigma^3} \simeq 1.3 \times 10^6 \text{ cycles} \quad (6.8)$$

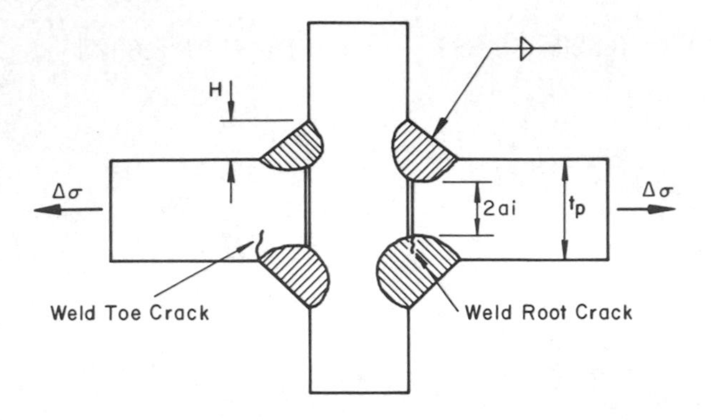

(a) Welded Both Sides

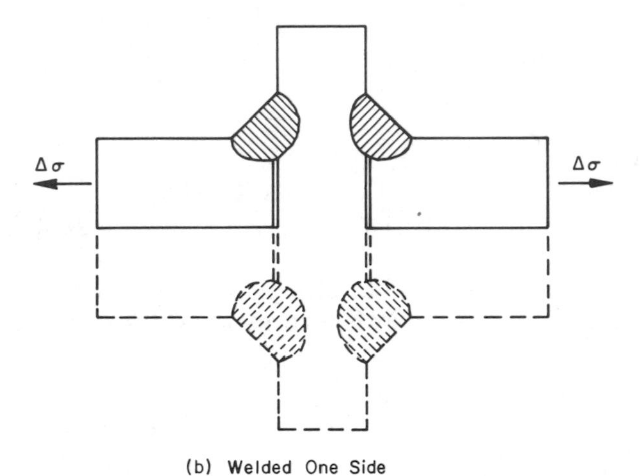

(b) Welded One Side

Figure 6.11 Cruciform model. (*a*) Welded both sides; (*b*) welded one side.

Hence significant growth would be expected from both the inside and outside surfaces at the details.

All three bents experienced cleavage and rapid crack extension after the fatigue cracks had propagated through the weld metal and resulted in through cracks at the flange tip in the box girder webs. Small ligaments between the inside lack-of-fusion and the external weld toe cracks were ignored. The stress intensity factor for the through thickness cracks becomes

$$K = \sigma\sqrt{\pi a} \tag{6.9}$$

For a crack length $2a = 3.0$ in. (76 mm) and yield point stress as a result of weld shrinkage, load, and detail geometry, the stress intensity factor becomes 74 ksi $\sqrt{\text{in.}}$ (81 MPa $\sqrt{\text{m}}$). A comparison with Figure 6.8 indicates crack instability is very probable as this is equal to the lower bound fracture toughness of the web plate at −10°F (−23°C).

6.1.3 Conclusions

Conclusions Regarding Fracture and Failure

A detailed examination of the fracture surfaces from the cracked steel pier bents of Dan Ryan Elevated structure showed that these fractures were all caused by fatigue cracking. The cracks originated from poor quality welds at the edges of stringer bottom flanges where they intersect box bents.

During fabrication slots for the stringers flanges were flame cut into box girders. Subsequently, flange plates were inserted through these slots and welded to the box bent girder web. This connection created a high stress concentration at the weld adjacent to the stringer flange and large unfused cracklike embedded defects. Thus at very low levels of the stress range the joint was very sensitive to fatigue crack propagation. These fatigue cracks appear to have reached a critical condition at the low temperatures that occurred in Chicago in late December 1977 and early January 1978. Thus all these factors—fatigue cracks, poor quality welds, high stress concentrations, and low temperatures—contributed to the brittle fractures of the box girders which were discovered in early January 1978.

6.1.4 Repair and Fracture Control

Immediately after the discovery of cracks in box girder web plates on January 3, 1978, the Department of Public Works of the City of Chicago and the Chicago Transit Authority started construction of temporary support structures under the box girders at bents 21 through 26 [6.2].

These temporary structures consisted of jacking beams placed on both sides of each box girder to support its longitudinal plate girder. The jacking beams in turn were supported by a steel falsework tower, with bracing resting on timber mudsill and crushed stone. The temporary support structures carry the weight of the tracks and trains until a permanent repair and retrofit is carried out.

In order to minimize the severe stress concentration and high residual stresses at the intersection of the girder flange and box girder webs at other locations that had not experienced crack instability, the details were retrofitted by cutting holes and sawing between them to create a "dumbbell"-like geometry. Figure 6.12 shows a schematic and photograph of the retrofit holes and sawcut.

One in. holes were also drilled at the ends of all cracks in the affected bents to minimize further growth pending a permanent repair.

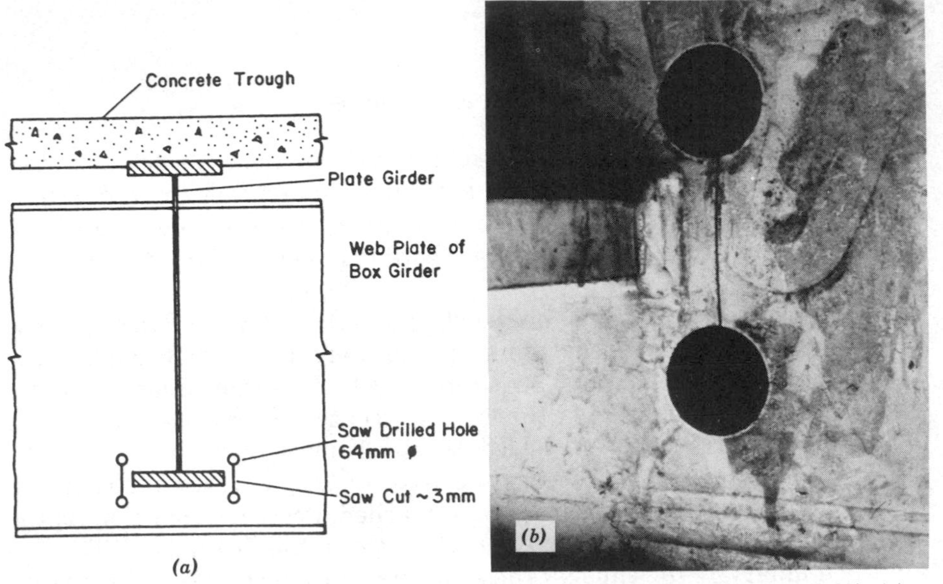

Figure 6.12 Typical retrofit detail at bent 26 (same as bents 21, 22, 24, and 25). (*a*) Schematic showing retrofit; (*b*) photograph of retrofitted box side plate (or web).

REFERENCES

6.1 Wiss, Janney, Elstner and Associates, Inc., Inspection of Junctions of Girder Flanges and Pier Side Plates on Piers Nos. 21 through 26 on the Dan Ryan, Wiss, Janney, Elstner and Associates, Inc., Consulting Engineer, Northbrook, Ill., 1978 (limited distribution).

6.2 The Technical Committee, Final Report on Causes of Fracture in Bents Nos. 24, 25, 26 of Dan Ryan Rapid Transit, Department of Public Works, Chicago, Ill., January 1979 (limited distribution).

6.3 Fisher, J. W., Hanson, J. M., Hausammann, H., and Osborn, A. E. N., Fracture and Retrofit of Dan Ryan Rapid Transit Structure, Final Report, 11th Cong., IABSE, August 31–September 5, 1980, Vienna.

6.4 Norris, S. N., and Fisher, J. W., Fatigue Behavior of Welded Web Attachments, *J. Constructional Steel Res.* 1 (January 1981).

6.5 Frank, K. H., and Fisher, J. W., Fatigue Strength of Fillet Welded Cruciform Joints, *J. Struct. Div.*, ASCE, 105 (September 1979):1727–1740.

CHAPTER 7

Welded Holes

Several bridges have developed cracks because of misplaced holes that were subsequently filled with weld. This has resulted in lack of fusion and other cracklike defects at each welded hole that eventually experienced significant crack growth and fracture.

At least three multibeam bridges developed major cracks in one or more girders as a result of such misplaced welded holes. The Illinois I-57 overpass at Fariña was one such structure, and it was analyzed and evaluated [7.1]. Misplaced welded holes for the diaphragms between the rolled sections resulted in the cracking of one of the beams. Similar conditions were found in two structures in Iowa.

In every instance cracks were observed to originate at several holes. These often linked and resulted in fracture of the section. Only the Illinois structure experienced a crack that turned and traveled down the length of the beam.

A similar condition was found to exist in a cover-plated plug welded chord of the Burned River Truss Bridge in Ontario [7.3]. The cover plate was attached to a rolled section with plug-welded holes. This created a lack-of-fusion and defect condition that was similar to that of the weld-filled holes. The crack eventually caused the chord to fracture.

7.1 FATIGUE-FRACTURE ANALYSIS OF COUNTY HIGHWAY 28 BRIDGE OVER I-57 AT FARINA, ILLINOIS

7.1.1 Description and History of the Bridge

Description of Structure

County Highway 28 Bridge over I-57 is located north of Farina, Illinois, in Fayette County. The structure is a skewed four-span composite–rein-

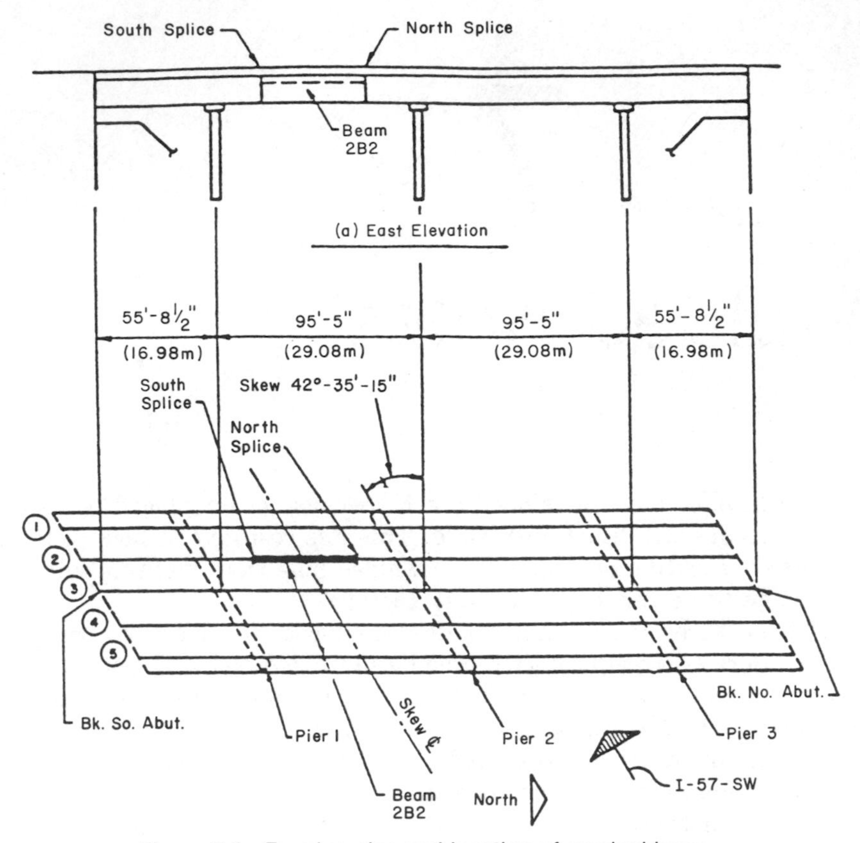

Figure 7.1 Framing plan and location of cracked beam.

forced concrete slab steel-beam bridge. It was completed and opened to traffic in 1968. The bridge primarily carries local traffic over Interstate I-57.

The general plan and elevation of the structure is given in Figure 7.1. A typical cross section is shown in Figure 7.2. The bridge has a 32 ft (9.75 mm) wide 7 in. (178 mm) thick reinforced concrete deck supported by five continuous W36 × 150 steel beams of ASTM A36-67 steel. It has two 95 ft 5 in. (29 m) main spans and two 55 ft 8½ in. (17 m) side spans. The concrete slab is connected to the main girders with shear studs with different spacings depending on their location along the span.

The entire structure was constructed in accordance with the "AASHO Standard Specifications for Road and Bridge Construction", January 2, 1958; the supplemental specifications dated January 3, 1966; and the 1968 specifications of Illinois-DOT.

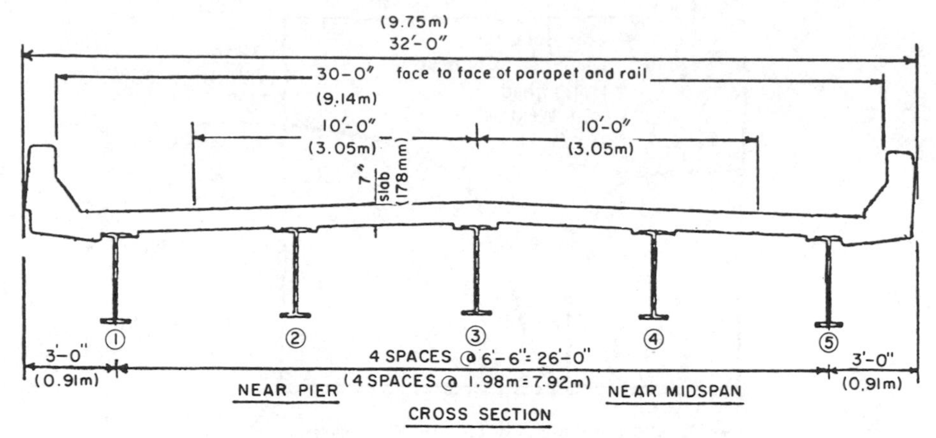

Figure 7.2 Cross section of structure.

History of Cracking

On March 18, 1977, a large crack in the south side of the first interior girder (girder 2) of the second span was discovered by a construction technician while driving under the bridge [7.1]. The fracture in the stringer originated at one of the four horizontal rivet holes that were mispunched and later improperly repaired and filled with weldments.

During the winter of 1976 to 1977 very low temperatures of −20°F (−29°C) were experienced in the area where the bridge was located [7.1].

7.1.2 Failure Modes and Analysis

Location of Fracture

The location of the fracture in the south interior beam is shown in Figure 7.3, and a detailed schematic drawing is given in Figure 7.4.

There were two separate fractures which started from two holes filled with weldment immediately under the clip angle supporting the diaphragm. Both cracks were initiated from weld-filled holes with large slag inclusions. Both cracks propagated through the flange of interior beam 2B2 at a common termination point. The cracks also extended up and turned longitudinally along the web at about middepth. One crack extended about 1 ft (0.305 m), while the other crack extended horizontally approximately 15 ft (4.57 m), reaching the next diaphragm, as illustrated in Figure 7.4. Photographs showing the crack surfaces and the weld-filled holes are given in Figures 7.5 and 7.6.

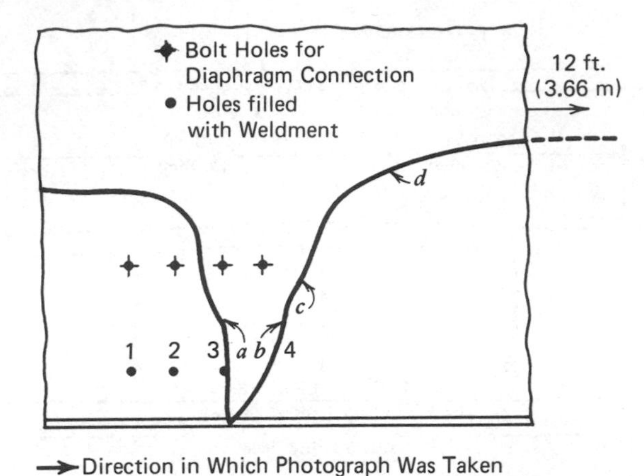

Figure 7.3a Schematic of fracture.

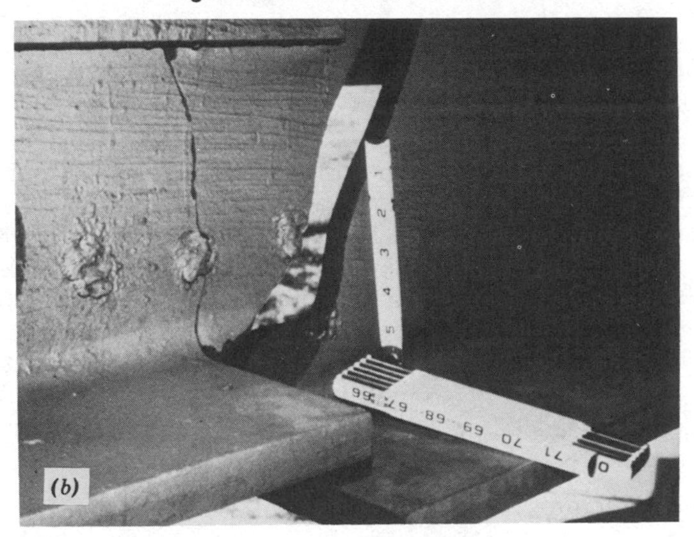

Figure 7.3b Photograph of cracks at the welded bolt holes 3 and 4 (courtesy of Illinois Department of Transportation).

The visual examination of the fracture surfaces shown in Figures 7.5 and 7.6 indicated that the failure was primarily a brittle fracture along the entire length of the cracks.

There was very little ductility in the fracture. The brittle fracture probably occurred during the extremely low temperatures recorded in the area during the winter of 1976 to 1977. Close examination of the plug welds showed the existence of herringbone patterns near the slag and voids in welds [7.1]. Thus the fracture likely initiated from fatigue-sharpened cracks originating from slag and voids in the plug welds. During low temperatures the brittle fracture resistance was decreased. The crack

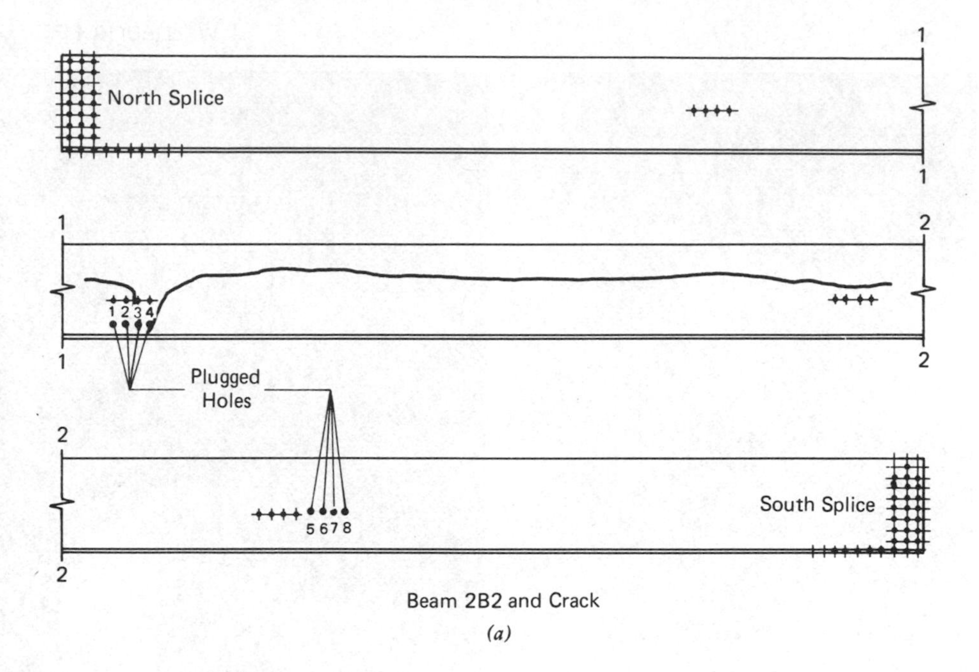

Beam 2B2 and Crack

(a)

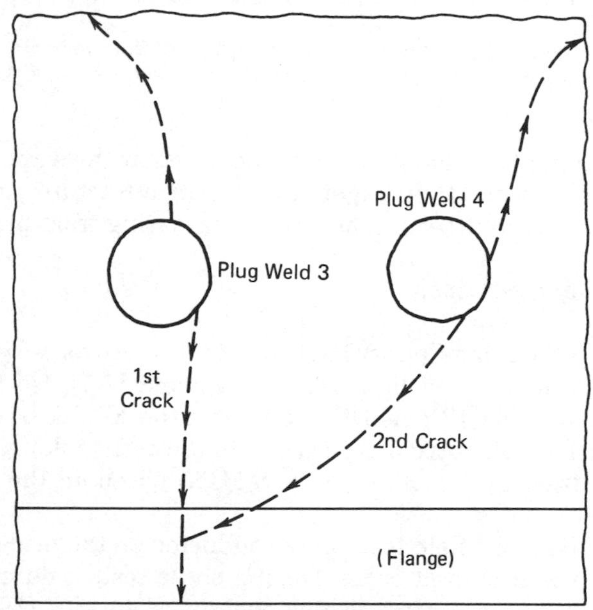

Dotted Line Indicates Crack Position
Arrows on Dotted Line Indicate Direction of Crack

Crack Initiation and Propagation from Plug Welds

(b)

Figure 7.4 Location of fracture and initiation and propagation of cracks. Beam 2B2 and crack showing crack initiation and propagation from plug welds.

Figure 7.5 Fracture surfaces at welded holes (courtesy of University of Illinois). (*a*) Photograph from direction a, hole 3; (*b*) photograph from direction b, hole 4.

originating from plug weld 4 (probably occurred at the same time as the cracks from plug weld 3) propagated in a brittle fracture mode into the bottom flange and intersected the crack originating from plug weld 3.

Cyclic Loads and Stresses

The normal traffic using County Highway 28 Bridge over I-57 is composed mainly of automobile and light farm equipment [7.1]. No truck traffic counts were available. Illinois-DOT estimated the ADT to be 400 vehicles in 1976 [7.2]. The computed design stress in the bottom flange adjacent to the origin of fracture is 15.2 ksi (104.8 MPa) based on the HS20 truck loading [7.2].

A series of live load field tests were conducted on the bridge, following the repair of the fractured beam. During these tests a dump truck was loaded with gravel to a gross vehicle weight of 44 kips (199 kN). This truck was driven across the bridge in order to determine the live load response of the structure.

A series of strain gauges were installed on replacement beam 2B2 at several locations, including the fracture initiation point (for strain gauge locations, see the schematic in Table 7.1). These gauges were used to estimate the service live load stresses [7.1]. Results of the field tests are

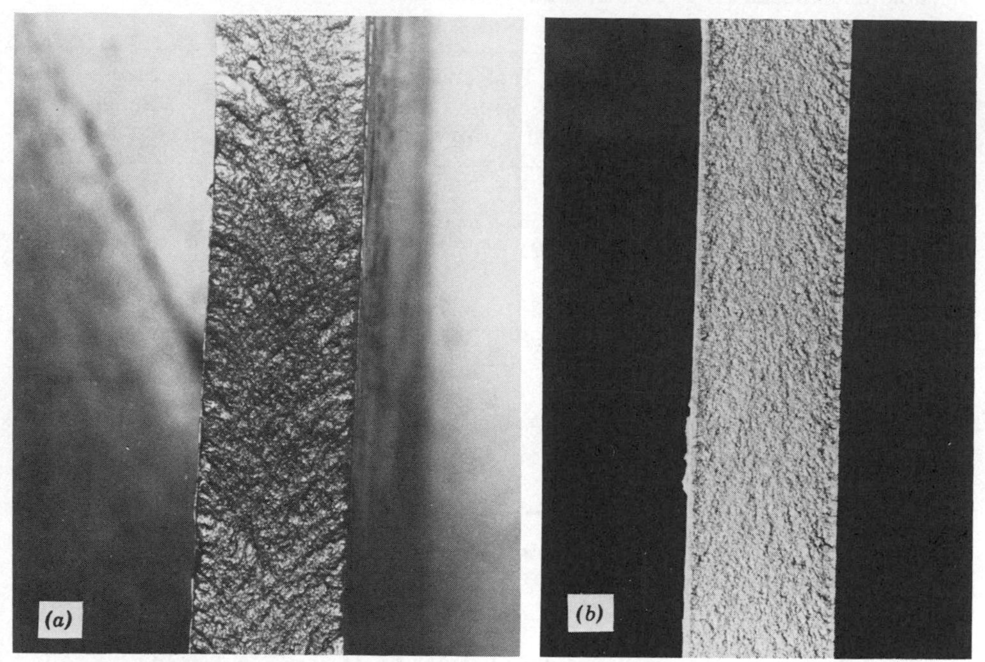

Figure 7.6 Fracture surfaces of the beam web (courtesy of University of Illinos). (*a*) Photograph from direction c; (*b*) photograph from direction d.

summarized in Table 7.1. The measured stress ranges in beam 2B2 were all quite low; the maximum stress range was 3.46 ksi (23.9) MPa). The ASTM A36-67 steel used in the beam has a yield point of 43.5 ksi (300 MPa). Therefore live load tensile stresses observed in the beam, even with the addition of the dead load stresses, are well below the design stress. Welding residual stresses from the weld-filled holes would induce stresses near the yield point at each hole.

Temperature and Environmental Effects

The fractures of the south side interior girder in the bridge over I-57 were discovered in the wake of the very cold 1976–77 winter when temperatures dropped as low as −20°F (−29°C). Aside from the low service temperature, no other significant environmental effects (such as corrosion) were apparent.

Mechanical Chemical, and Fracture Properties of the Material

The lower T-shaped portion of beam 2B2 was removed from the bridge by flame cutting between two bolted splices, as shown in Figure 7.7. Specimens fabricated from this portion of beam 2B2 were used to conduct

Table 7.1 Measured Strains and Computed Stresses in Beam 2B2 Under Test Truck

Truck Run	Gauge Location	Tension Stress ksi	Tension Stress (MPa)	Compression Stress ksi	Compression Stress (MPa)	Stress Range ksi	Stress Range (MPa)
1. *Crawl*							
Centered on beam	44	2.56	(17.65)	0.34	(2.34)	2.90	(19.99)
line 2, southbound	34	2.21	(15.24)	0.28	(1.93)	2.49	(17.77)
2. *Crawl*							
Right wheels over	44	2.28	(15.72)	0.34	(2.34)	2.62	(18.06)
beam line 2,	34	1.98	(13.65)	0.28	(1.93)	2.26	(15.58)
southbound							
3. *30 mph*							
Centered on beam	44	2.84	(19.58)	0.62	(4.28)	3.46	(23.86)
line 2, southbound	34	2.47	(17.03)	0.56	(3.86)	3.03	(20.89)

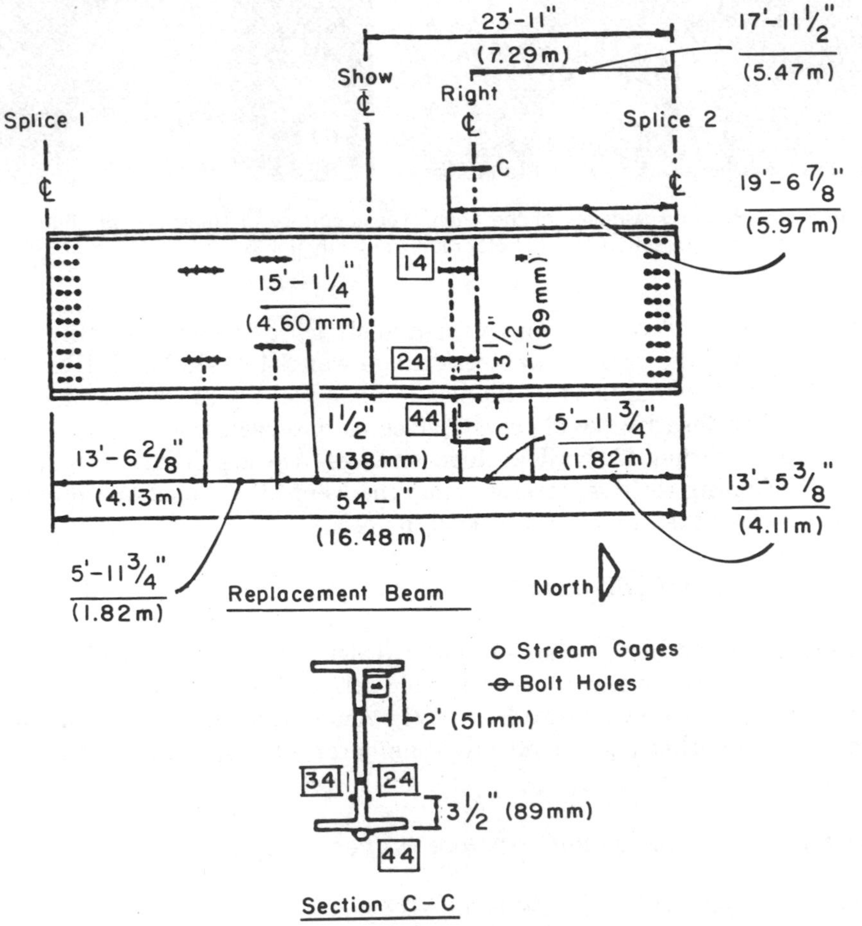

Strain Gage Location for Field Test

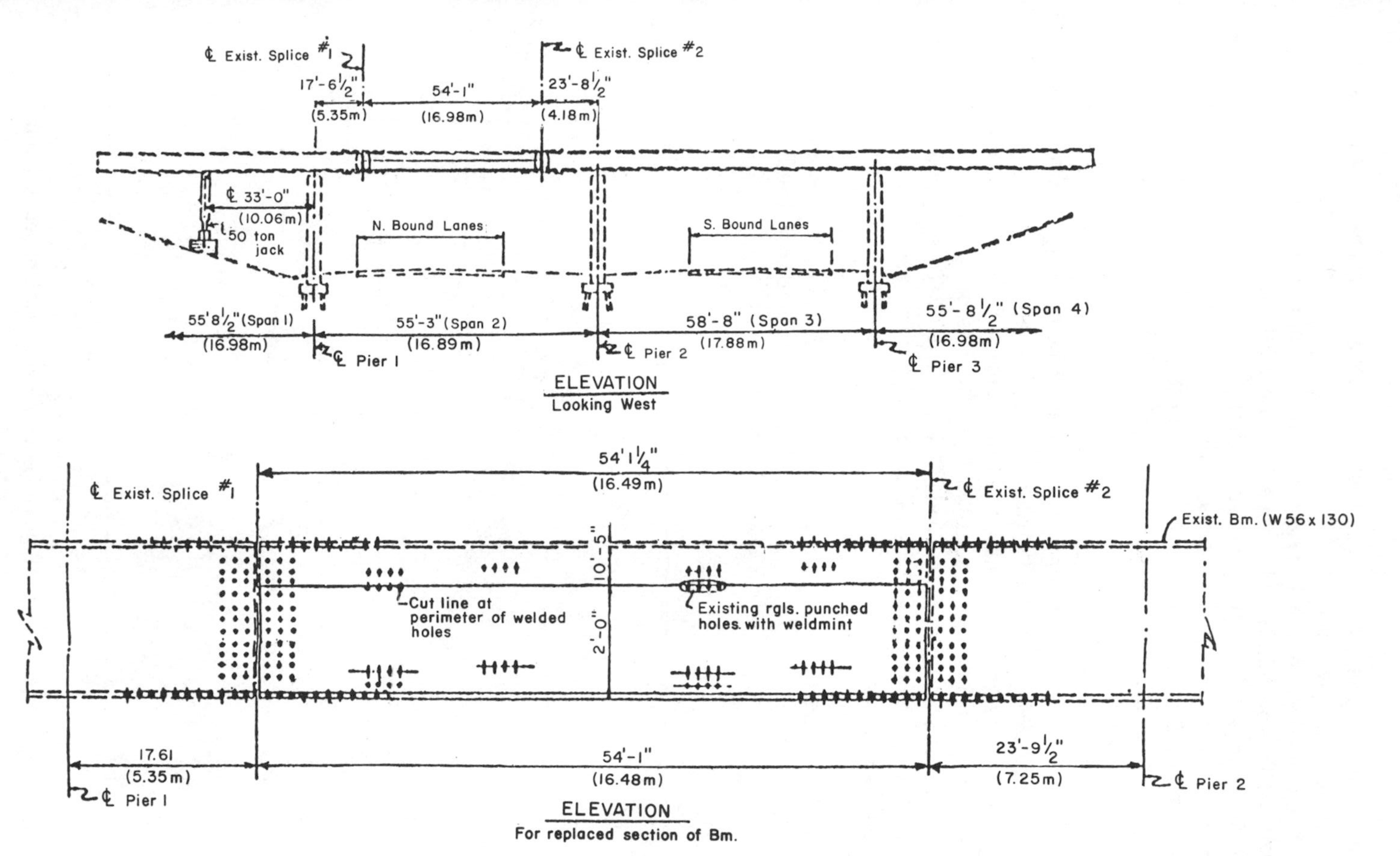

Figure 7.7 Retrofit detail (replacement to beam 2B2).

tensile tests, Charpy V-notch tests, fracture toughness tests, and for chemical analysis [7.1].

The chemical analysis indicated that the composition of the steel was well within the requirements specified for A36 steel.

The tensile test results indicated that the web material had an average yield point of 43.5 ksi (300.0 MPa) and an ultimate strength of 71.3 ksi (491.6 MPa). Similar results were obtained for the flange. The tensile properties of the steel satisfied the specification requirements for A36 steel.

Charpy V-notch tests were carried out in accordance with ASTM standard A370. Four sets of standard size specimens were cut from the web and flange of the fractured girder.

The test results are summarized in Figure 7.8. Both longitudinal and transverse specimens were tested from the beam web. All of the specimens tested at 40°F (4°C) exceeded the 15 ft-lb (20 J) requirement of the AASHTO specifications for A36 steel. The web specimens were designated as follows:

> Web specimens "WL" with their lengths parallel to the direction of rolling of the beam (the notch being perpendicular to the direction of rolling).

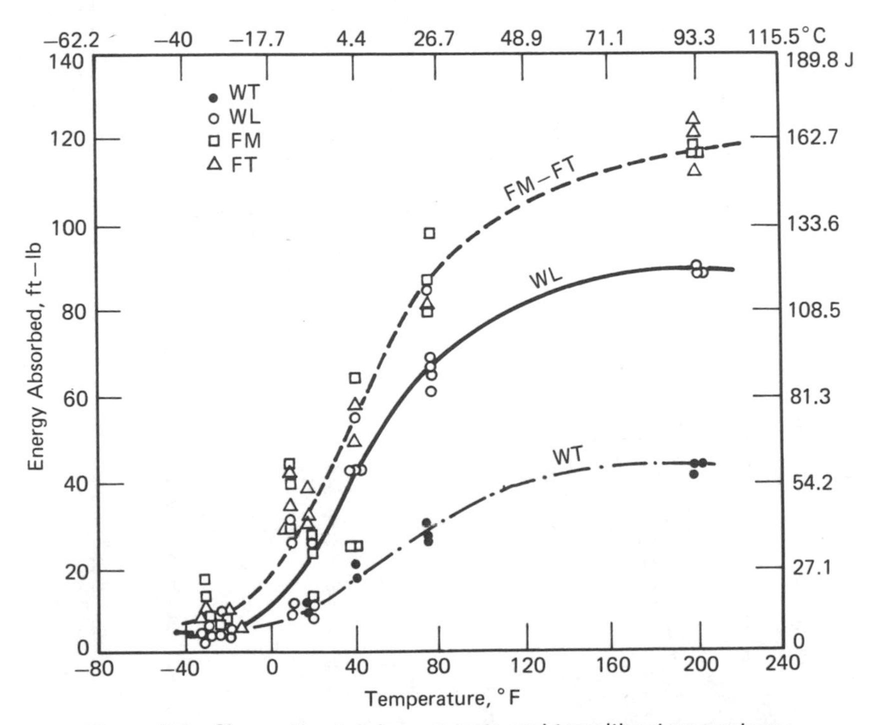

Figure 7.8 Charpy V-notch impact tests and transition temperature.

Web specimens "WT" with their lengths transverse to the direction of rolling (the notch being parallel to the direction of rolling).

Two sets of flange specimens were cut with their lengths parallel to the direction of rolling of the beam and their notches parallel to the surface of the flange. The flange specimens were located as follows:

Flange specimens "FT" at the top of the flange.
Flange specimens "FM" at the middepth of the flange.

In addition to the Charpy V-notch tests, fracture toughness tests were conducted on bend specimens in order to establish the toughness of the steel. Since A36 steel was used in the beam, valid K_{Ic} tests were not possible. Static fracture toughness values, K_Q, for the steel were estimated from the maximum load. No load-deflection data was acquired; hence a more accurate determination using the J-integral was not possible. These tests provided K_Q values at the maximum load of about 49 ksi $\sqrt{\text{in.}}$ (53.9 MPa $\sqrt{\text{m}}$) at −20°F (−29°C). A single test at −40°F (−40°C) yielded 50 ksi $\sqrt{\text{in.}}$ (55 MPa $\sqrt{\text{m}}$). These results were compatible with the estimated K_c values provided by the Barsom-Rolfe correlation equation and the intermediate rate temperature shift. Figure 7.9 shows the estimated dynamic fracture toughness and the K_Q values which are underestimates since the deformation was not measured nor the K_J calculated.

In addition to the Charpy V-notch and K_Q tests, a series of tensile tests were conducted on rectangular specimens containing plug welds [7.1]. The test specimens were approximately 3 in. (76.2 mm) wide. The specimens were machined such that the plug weld was centered on the width of the specimen. Three tests were carried out at −20°F (−29°C) and one at −40°F (−40°C). The results are summarized in Table 7.2.

All of these tests failed at nominal stresses on the gross section that were well in excess of the yield point. Obviously, none of the plug welds in

Table 7.2 Tensile Test of Specimens with Plug Welds

Specimen Number	Gross Area in.^{2}	(mm^2)	Specimen Test Temperature °F	(°C)	Tensile Strength kips	(kN)	Stress on Gross Area ksi	(MPa)
10a	1.842	(1188)	−20	(−29)	100.97	(449.1)	54.82	(378.0)
10b	1.822	(1175)	−20	(−29)	105.13	(467.6)	57.70	(397.8)
10c	1.815	(1171)	−20	(−29)	114.9	(511.1)	63.31	(436.5)
10d	1.807	(1166)	−40	(−40)	99.6	(443.0)	55.12	(380.0)

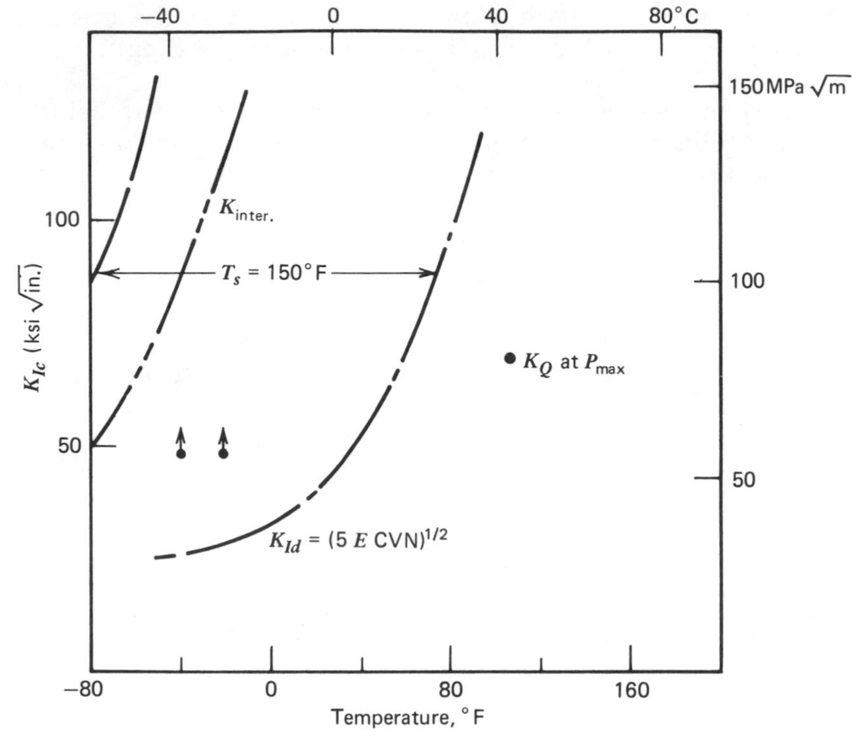

Figure 7.9 Fracture toughness of the beam web.

these specimens had initial flaws that were as large as those at the failure section. Failure occurred at stresses about midway between the yield point and tensile strength of the base plate.

Visual and Fractographic Examination of the Crack

Visual observations indicated that the plug welds initiated the fracture of beam 2B2.

Photographs of parts of the fracture surfaces are shown in Figures 7.5 and 7.6. These indicated that a small shear lip existed along the outer edges (plate surfaces) over most of their length. Figure 7.3*b* demonstrates a similar condition. No spalled paint exists at the plug weld and for a short distance each side of the plug welds. Therefore the spalled mill scale confirms the plastic behavior associated with the shear lips. The herringbone surface patterns and the fine texture of the surface near the plug welds shown in Figure 7.5 also exhibit the brittle fracture characteristic [7.1]. The slag and other voids which initiated the unstable crack extension are also visible in Figure 7.5. Notice that the herringbone patterns in Figure 7.6 also point toward the plug welds.

Failure Analysis

The cracklike defects observed in the web at the weld-filled holes were modeled as through thickness cracks of length $2a = 1$ in. (25 mm). The stress intensity factor for this case is given by

$$K = \sigma\sqrt{\pi a} \tag{7.1}$$

The amount of fatigue crack extension at such a defect will be very small, considering the number of trucks and pieces of farm equipment that are likely to have crossed the structure during its nine years of service. Assuming that 10 to 20% of the ADT count results in measurable stress cycles, that would yield 130,000 to 260,000 variable stress cycles during its service life. If the measured stress range on the web for the test truck is used to estimate the crack extension (i.e., $\Delta\sigma \simeq 3$ ksi, or 21 MPa) the amount of crack size increase can be estimated as

$$\begin{aligned} N = 260{,}000 &= \int_{0.5}^{a_f} \frac{da}{3.6 \times 10^{-10}\Delta K^3} \\ &= \frac{10^{10}(2)}{3.6\pi^{3/2}\Delta\sigma^3}\left(\frac{1}{\sqrt{0.5}} - \frac{1}{\sqrt{a_f}}\right) \end{aligned} \tag{7.2}$$

Equation 7.2 results in a final crack size of 0.505 in. (12.8 mm). The crack extension under cyclic loading is 0.005 in. (0.13 mm) and would not be enough to be detected. It is probable that even less extension would occur because the number of stress cycles and the magnitude of the effective stress range are likely to be less than the values used in Eq. 7.2. Hence it is reasonable to assume that no observable crack extension was seen adjacent to the initial defect condition. Cyclic loading served to "sharpen" the defect and create a more cracklike condition.

Crack instability will occur when the critical stress intensity is reached. As can be seen in Figure 7.9, the minimum fracture toughness at −20°F (−29°C) is between 75 ksi $\sqrt{\text{in.}}$ (83 MPa $\sqrt{\text{m}}$) and 100 ksi $\sqrt{\text{in.}}$ (110 MPa $\sqrt{\text{m}}$). The stress intensity factor can be determined from the superposition of the stress intensity factor for a uniform stress due to dead and live loads and a uniform "pressure" stress due to the residual tensile stress acting on the crack. The uniform stress contribution is provided by Eq. 7.1. A plastic zone size correction $a = a' + r_y$ was introduced, and this results in

$$K_1 = \sigma\left[\frac{\pi a'}{1 - (\sigma/\sigma_Y)^2/2}\right]^{1/2} \tag{7.3}$$

where

$$r_y = \frac{1}{2\pi}\left(\frac{K_c}{\sigma_Y}\right)^2$$

For an assumed dead and maximum live load stress of 10 ksi (69 MPa) and a yield point of 50 ksi (345 MPa) adjusted for temperature and strain rate, Eq. 7.3 yields a value $K_1 = 12.7$ ksi $\sqrt{\text{in.}}$ (87 MPa $\sqrt{\text{m}}$).

The residual tension yield stress field acting on the crack surface results in

$$K_2 = \sigma_{rs}^{+} \sqrt{\pi a} \tag{7.4}$$

which is identical to a uniformly stressed sheet. Without the plastic zone size correction, this yields a value of $K_2 = 62.7$ ksi $\sqrt{\text{in.}}$ (69 MPa $\sqrt{\text{m}}$). If the same criteria is introduced for the plastic zone size, the stress intensity K_2 can be estimated as 88.7 ksi $\sqrt{\text{in.}}$ (97.6 MPa $\sqrt{\text{m}}$). Hence the stress intensity $K = K_1 + K_2$ is between 75.4 ksi $\sqrt{\text{in.}}$ (82.9 MPa $\sqrt{\text{m}}$) and 101.4 ksi $\sqrt{\text{in.}}$ (111.5 MPa $\sqrt{\text{m}}$). An examination of Figure 7.9 suggests that the fracture resistance of the material is between 85 ksi $\sqrt{\text{in.}}$ (93.5 MPa $\sqrt{\text{m}}$) and 110 ksi $\sqrt{\text{in.}}$ (121 MPa $\sqrt{\text{m}}$). Hence the fracture toughness of the material will be exceeded, and crack instability occurs as the natural crack tip is sharpened by repeated loads.

7.1.3 Conclusions

The analysis of the fracture of a rolled beam in County Highway 28 overpass at I-57 near Farina, Illinois, enabled the following conclusions to be made about the condition of the structure:

1. The maximum live load stresses measured in the cracked beam were 3.46 ksi (23.86 MPa) which was far below the AASHTO allowable stresses for such a member.
2. The chemical and mechanical properties of the steel were well within ASTM-A36 specifications.
3. The steel met the Charpy V-notch requirements for service temperatures down to −30°F (−34°C) which was lower than the temperatures experienced by the fractured beam.
4. Fatigue sharpened the natural cracks, and brittle fracture resulted from the presence of plug-welded holes. These welded holes resulted in large cracklike discontinuities that were susceptible to crack propagation. Radiographs of other weld-filled holes showed plug welds with slag inclusions and voids.

7.1.4 Repair and Fracture Control

The repair of County Highway 28 Bridge over I-57 was carried out by removing the fractured section of beam 2B2 by flame cutting the beam web longitudinally about 10 in. (254 mm) below the top flange for the entire length between splices.

A new T-shaped section that matched the section removed from the beam was field-welded horizontally to the remaining section of the web and then bolted to the adjacent beam segments, using the existing splice plates. The repaired portion of the beam is shown in Figure 7.7. The T-shaped section removed from the structure was used to fabricate test specimens in order to establish the chemical, physical, and fracture toughness characteristics of the steel [7.1]. Fractographic studies were also carried out on the crack surfaces.

No adverse behavior is anticipated from the repaired member. Nondestructive tests were carried out on the longitudinal field weld. Furthermore it is located near the neutral axis, so no appreciable cyclic stress is likely to ever occur at that level.

REFERENCES

7.1 Handel, W., and Munse, W. H., Investigation of Interstate I-57 Bridge Beam Brittle Fracture, University of Illinois, Structural Research, Series No. 477, March 1980.

7.2 Illinois Department of Transportation, Personal communication from J. B. Nolan to J. W. Fisher, dated March 23, 1981.

7.3 Csagoly, P. F., Fatigue Considerations in the Ontario Highway Bridge Design Code, Proc. Symposium on Fracture Control in Engineering Structures, Laurentian University, Sudbury, August 1979.

CHAPTER 8

Continuous Longitudinal Welds

Several structures have had transverse cracks detected in the longitudinal weldments after the structures were in service. In the United States three tied arch bridges were found to have such cracks in the corner welds of the tie girders which were welded built-up box sections [5.2, 8.1, 8.2]. Two of these structures were located at the same bridge site, and it was necessary to close these structures until repairs were completed, as large cracks were detected in the transverse groove welds as well [8.1]. The source of cracking in all three structures was found to be hydrogen-related cold cracking that had occurred at the time of fabrication. All of the affected components were fabricated from A514 steel.

A similar type of cracking was also detected in the Entella River Bridge in Italy. This structure was a continuous box girder structure which had portions of the welded trapezoidal box fabricated from A514 steel as well [8.3].

In all of these structures some degree of fatigue crack propagation was detected. In the tie girder boxes crack growth was detected near the crack tip. In the box girder structure new cracks were found to occur after other cracks were removed by grinding [8.3].

8.1 FATIGUE-FRACTURE RESISTANCE OF THE GULF OUTLET BRIDGE

8.1.1 Description and History of the Bridge

Description of Structure

The Gulf Outlet Bridge is a three-span truss with a tied arch suspended span. The side spans are 273 ft (83.2 m) long, and the center span is 702 ft (214 m). The tied arch suspended span is 546 ft (166.4 m) long (see Figure 8.1). Each panel of the truss structure is 39 ft (11.9 m) long. The two main

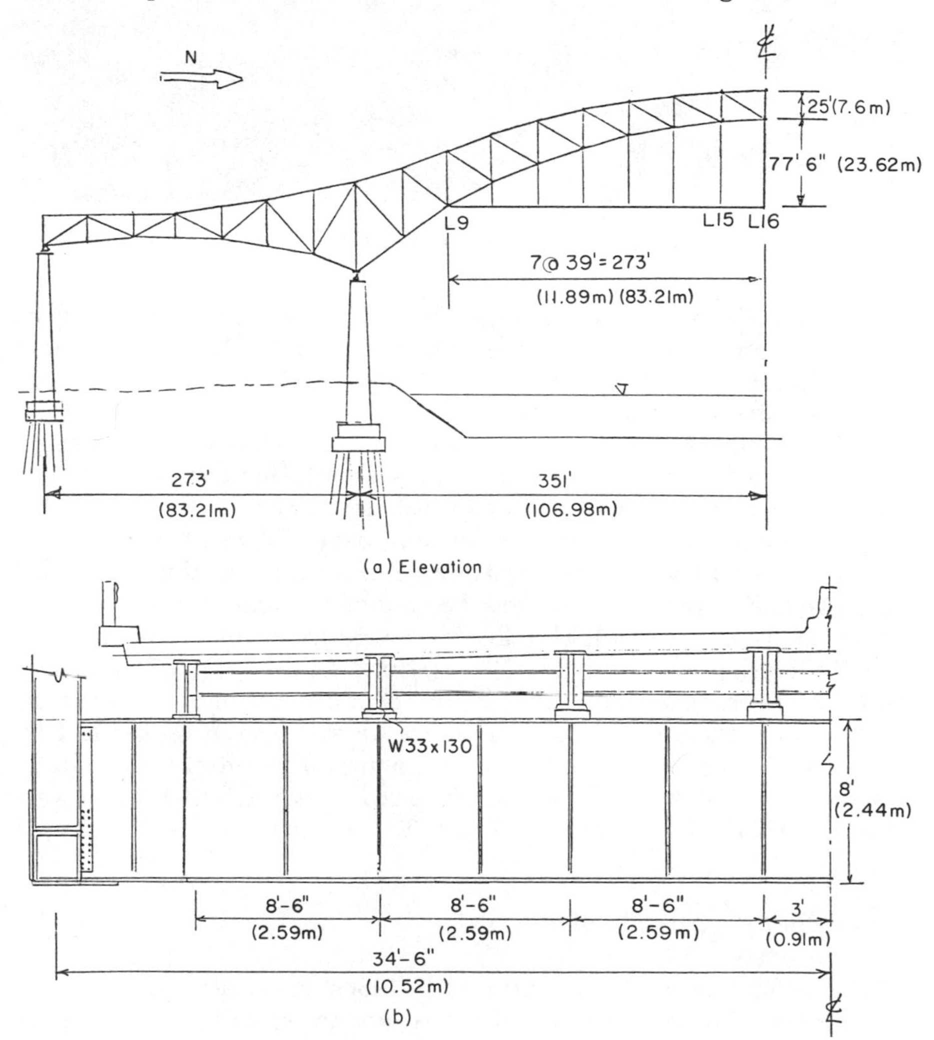

Figure 8.1 Elevation and cross section of the Gulf Outlet Bridge. (*a*) Elevation; (*b*) cross section showing floor beam and stringers.

Figure 8.2 Overview of the Gulf Outlet Bridge.

trusses are spaced at 75 ft (22.9 m). Figure 8.2 shows an overview of the structure and the suspended span. The box tie girder for the tied arch suspended span was 25 in. (635 mm) × 28 in. (711 mm) in cross section and fabricated from A514 steel.

The structure spans the Mississippi River Gulf Outlet shipping channel near New Orleans, Louisiana. Three-span welded plate girders frame into the structure on the north and south approaches.

The tie girders each contain a pin link near midspan between panel points L15 to L16 which were used as an erection aid. The link connection is shown in Figure 8.3 and the floor-beam–hanger connections to the floor beam can be seen as well. The tie girder webs were $\frac{3}{4}$ in. (19 mm) plates, and the flange plates were $\frac{5}{8}$ in. (16 mm) thick. At the pin link connections the web plates were increased to $3\frac{1}{2}$ in. (89 mm) thickness to permit the pinned connection. The web plate thickness was also increased to 1 in. (25.4 mm) thickness at each floor-beam connection to offset the bolt holes necessary for the floor-beam and hanger connections. Fillet welds were used to form the tie girder box section. Ten in. pins were used at the pin links.

History of the Structure and Cracking

The structure was designed in the early 1960s. Fabrication and erection was completed in the 1964 to 1965 period, and the structure was placed in service in October 1965.

Figure 8.3 Tie girder showing pin link connection.

As a result of concern with A514 steel weldments, a detailed inspection was carried out on the tie girder web plate groove welds in 1978. Ultrasonic and radiographic examinations were made. No significant discontinuities were detected in the groove welded connections. However, near the pin plate groove welds, transverse cracking was inadvertently detected in the box section corner fillet welds as the radiography film lapped over those connections.

Most of these cracks did not extend to the outside face of the fillet welds and were not apparent on the surface until a surface layer was removed by grinding. At the pin plate and at the bolted splices in the tie girders, fillet welds were placed inside the box section as well to assist in transferring the forces in the web plates to the flange plates. Figure 8.4 shows typical cracks in an interior box corner weld after surface grinding and application of liquid penetrant.

In 1979 several of these defects were removed from the box corner welds by coring in order to make a detailed evaluation of the causes of cracking and to determine whether or not fatigue crack growth had occurred.

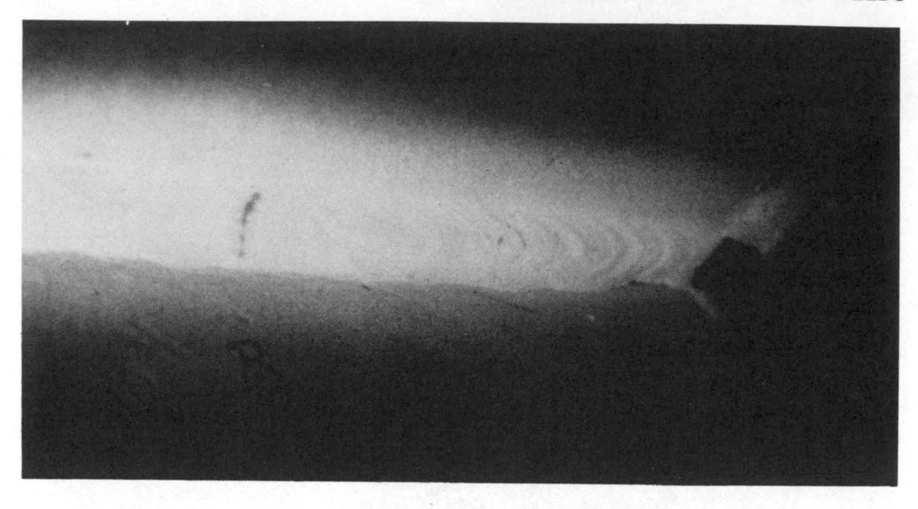

Figure 8.4 Transverse cracks in longitudinal fillet weld.

8.1.2 Crack Conditions and Analysis

Location of Cracks

Cracking was initially detected in the east truss (see Figure 8.1) at the pin link connection in member L15–L16. The cracks were not visible on the weld surface nor defined by liquid penetrant. A segment of the box corner was removed with a hole saw in order to permit an examination of the crack. Figure 8.5 shows the type of segment removed as well as the ground and polished surface of the weld with the crack.

In the region where the crack was detected, it was apparent that the original automatic submerged arc weld had been built up to a larger size by manual weld passes. The increased weld size was necessary to accommodate the transfer of force from the box webs at the pin plate connection into the balance of the section. All cracks were detected in regions where such manually made weld passes existed. This included the interior and exterior box corner welds in the region of the pinned connections and those sections at bolted splices in the tie girder where similar reinforcement weld passes were made.

Since most cracks did not extend through the surface layer of the weld, it was necessary to surface grind the weld in order to expose the crack and permit a reliable method of detection with the dye-penetrant. All of the manual weld passes were surface ground and tested in this manner. Several of the automatic submerged arc welds were also examined, but no cracks were detected in that type of weldment.

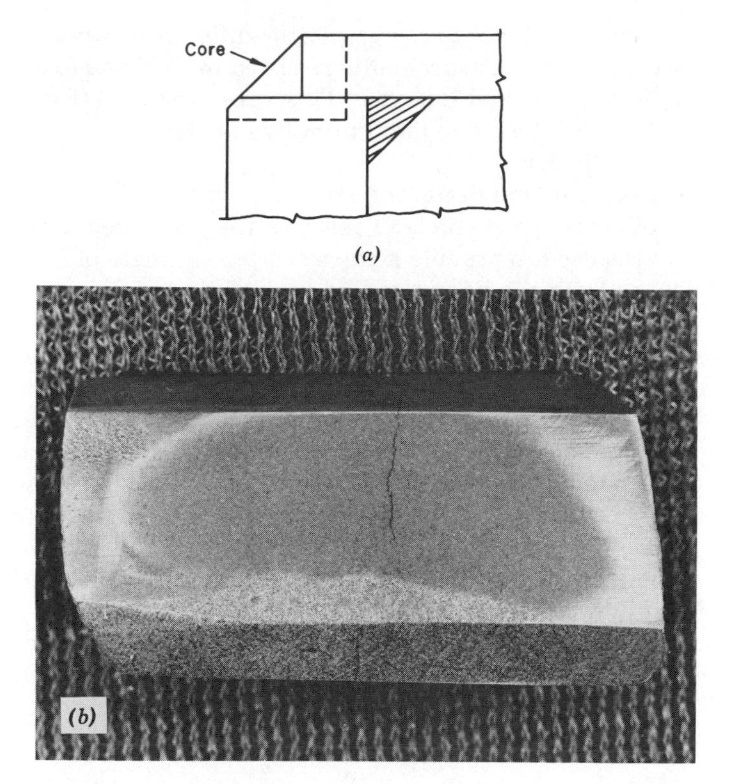

Figure 8.5 Detail showing core removed from box corner weld. (*a*) Schematic of typical core; (*b*) polished and etched surface of weld showing crack.

Cyclic Loads and Stresses

No live load stress measurements are available on this structure. The calculated dead load stress in the tie girder is 36.6 ksi (252 MPa). The dead load stress was assumed to be uniformly distributed over the cross section. Also in panel L15–L16 the pinned linkage minimizes the possibility of bending due to deflection.

The design live load stress range which included an impact factor of 7.5% was 6.9 ksi (48 MPa). Based on the design calculations and measurements on other long span bridges [5.3, 5.4], the actual stress range from traffic was known to be much less. Assuming the truck traffic using the structure corresponds to the nationwide distribution [4.3], the effective stress range was estimated as

$$S_{r\,\mathrm{Miner}} = 0.7\alpha S_r^D \tag{8.1}$$

where α is a reduction factor to correct for the difference in calculated and measured stresses. The measurements reported in [5.3] and [5.4] indicate that $\alpha \simeq 0.35$ is a reasonable value. This results in an effective stress range of 1.7 ksi (12 MPa). The maximum live load stress will not exceed $0.6S_r^D = 4.1$ ksi (28 MPa).

Because of its span and grade the structure will likely experience less than one stress cycle per vehicle. Therefore the estimated annual daily truck traffic was used to provide a conservative estimate of the variable stress cycles to which the tie girder had been subjected. It was estimated that between 1965 and 1979 five million trucks crossed the bridge.

The residual stresses due to welding the box corners were estimated using a finite element discretization of the section [8.2]. This procedure is described in [8.4]. It was assumed that the automatic submerged arc welds at the outside corners were made first and that the additional manual weld pass sequence could be simulated by initial passes inside the box and final passes on the outside corners [8.2]. This resulted in the residual stress distribution shown in Figure 8.6. The box corner welds were made with E70 electrodes, so a yield strength of 80 ksi (550 MPa) was used for the weldment.

The structure is located in zone 1 for a minimum service temperature of 0°F (−18°C). The actual minimum service temperature at the site is 20°F (−7°C). Aside from the effect of temperature on fracture toughness, no other environmental effects are probable.

Mechanical, Chemical, and Fracture Properties of the Material

Sample cores were removed from the $\frac{5}{8}$ in. (16 mm) flange plates, the $\frac{3}{4}$ in. (19 mm) web plates and the $3\frac{1}{2}$ in. (89 mm) web plates of the east and west trusses of member L15–L16. The base plate material was used for tensile tests (0.252 specimens), Charpy V-notch tests, and compact tension fracture tests. Chemical analysis were also carried out on each core sample. Generally the cores were sliced into $\frac{1}{2}$ in. (12 mm) and 1-in. (25 mm) thick segments to evaluate the properties through the thickness.

The chemical content of each plate was found to be consistent with the requirements of A514 grade H. Some heats were found to be slightly higher in alloy content. This would increase the carbon equivalent and decrease weldability.

The tensile tests indicated that the $\frac{3}{4}$ in. (19 mm) plate had a yield point of 122.9 ksi (848 MPa) and a tensile strength of 132.9 ksi (917 MPa). The $\frac{5}{8}$ in. (16 mm) plate had a yield point of 117.3 (809 MPa) and a tensile strength of 124.5 ksi (859 MPa). The $3\frac{1}{2}$ in. (89 mm) plate had a yield point of 108.1 ksi (746 MPa) and a tensile strength of 122.8 ksi (847 MPa).

The Charpy V-notch tests are summarized in Figure 8.7. This shows that only the $\frac{5}{8}$ in. (16 mm) flange plates met the zone 1 requirements of 25 ft-lbs (34 J) 30°F (−1°C). The $3\frac{1}{2}$ in. (89 mm) web plates were found to

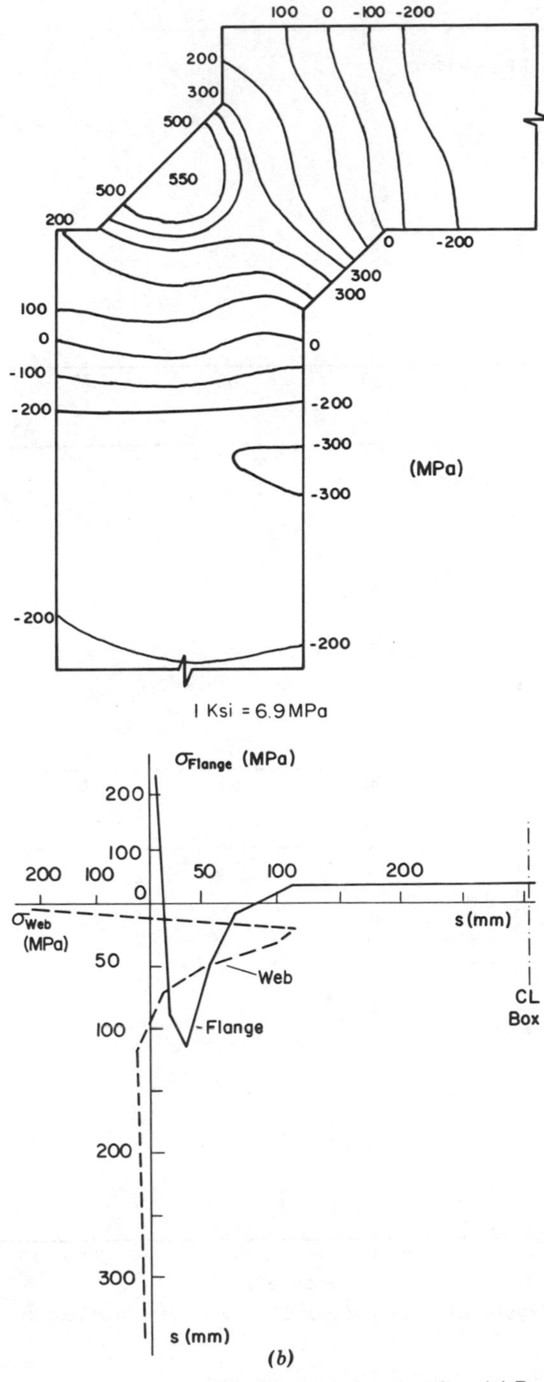

Figure 8.6 Predicted residual stress distribution in box section. (*a*) Residual stress distribution predicted at box corner; (*b*) average residual stress distribution on box section.

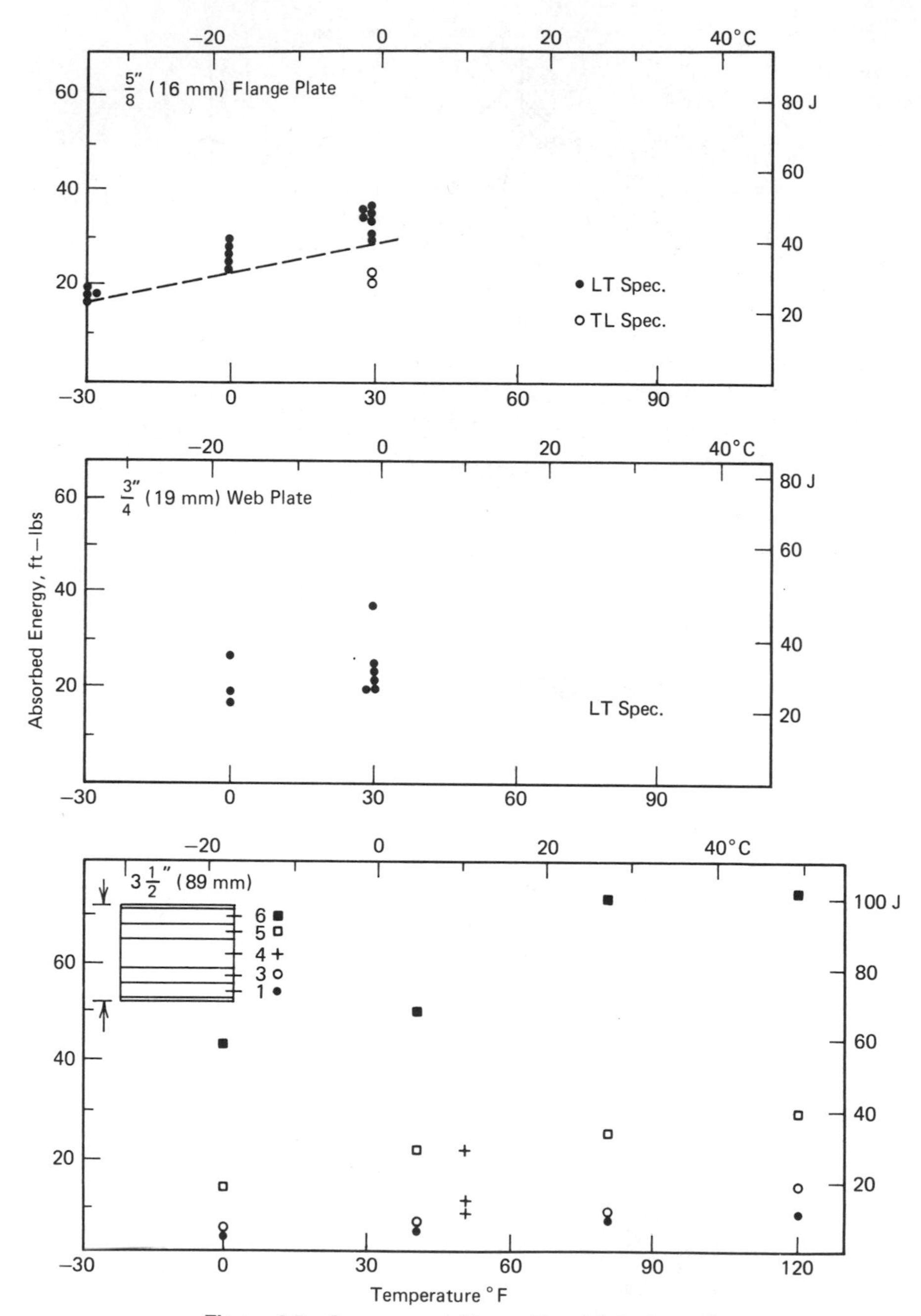

Figure 8.7 Summary of Charpy V-notch test results.

provide a significant difference in absorbed energy depending on their location through the thickness. The crack tip was located near slice 4 in the $3\frac{1}{2}$ in. (89 mm) plate.

Three compact tension fracture tests were carried out on the mid-thickness slice from the sample cores. In addition, 13 precracked Charpy V-notch tests were made to correlate with the compact tension. These tests results were all carried out at a 1 sec loading and are summarized in Figure 8.8. Also shown is the estimated dynamic fracture toughness K_{Id} provided by the Barsom correlation equation.

The test results indicate that the minimum fracture toughness of the $3\frac{1}{2}$ in. (89 mm) plate at 20°F (−7°C) is about 55 ksi $\sqrt{\text{in.}}$ (60 MPa $\sqrt{\text{m}}$) for 1 sec loading. The $\frac{3}{4}$ in. (19 mm) web plate and the $\frac{5}{8}$ in. (16 mm) flange plate provide a dynamic fracture toughness of the same magnitude. Their static fracture toughness is much higher (i.e., 75 to 80 ksi $\sqrt{\text{in.}}$ or 88 MPa $\sqrt{\text{m}}$). Since the loading rate is much lower than 1 sec, it is probable that the minimum fracture toughness of the $3\frac{1}{2}$ in. (89 mm) plate was 60 ksi $\sqrt{\text{in.}}$ (66 MPa $\sqrt{\text{m}}$) or more.

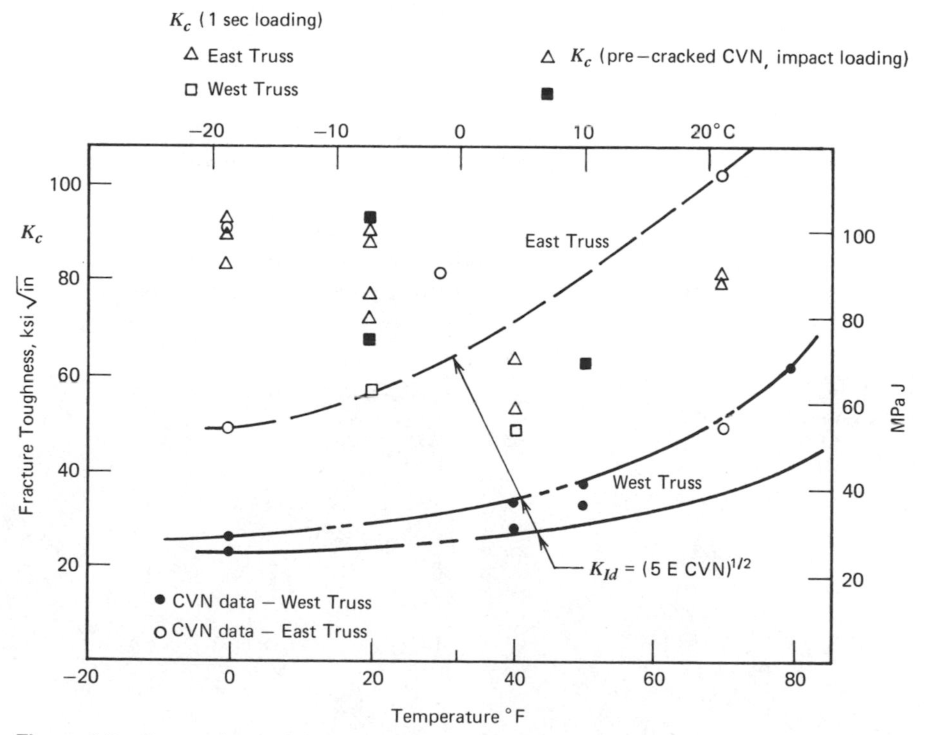

Figure 8.8 Comparison of dynamic fracture toughness based on CVN tests with results of compact tension (CT) and precracked CVN slow bend tests.

Visual and Fractographic Analysis

Several of the transverse weld cracks were removed by coring so that metallographic and fractographic studies could be carried out on the cracks. When the crack shown in Figure 8.5*b* was exposed by breaking open the segment, the exposed crack was found to extend into the heat-affected zones of both the flange and web plate. This can be seen in Figure 8.9 where the exposed crack surface and a polished and etched segment just behind it are shown. The etched section also reveals that multiple weld passes were made over the top of the original submerged arc weld pass. Fractographic studies were also carried out on the crack surface near the tip of the crack. Figure 8.10 shows a fractograph for the crack tip region. Extensive corrosion product was found on all surfaces. After preparing numerous replicas, a better indication of the crack surface was obtained. The fractograph shown in Figure 8.10 suggests that a possibility of fatigue crack extension existed. Striationlike features were found on the surface. On the whole the fractographic and metallographic examination suggested that the cracks were produced during or shortly after welding. Similar transverse weld cracking was reported in groove welds and attributed to hydrogen and residual stresses [8.5].

Analysis of Cracks

The cracks that formed in the box corner welds are variable in shape and size. The largest crack was found to be an elliptical-shaped crack with a minor axis diameter of $\frac{1}{4}$ in. (6.3 mm). Studies on the behavior of such defects have indicated that they can be evaluated by considering a circumscribed elliptical- or circular-shaped crack [8.6, 8.7]. The same type of fatigue crack growth was observed in fillet welds and partial penetration box corner welds.

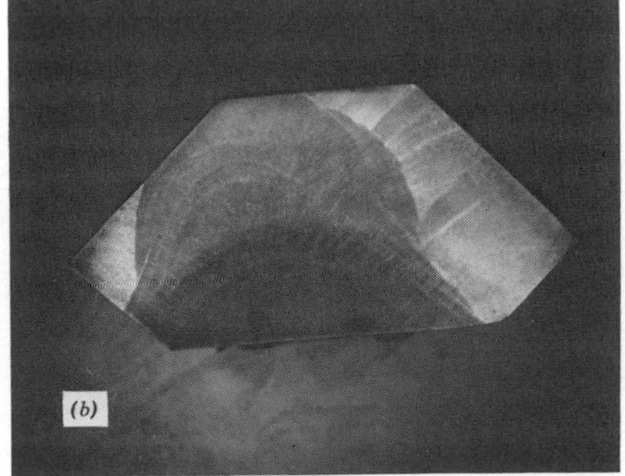

Figure 8.9 Photographs showing crack extends into the heat-affected zone (HAZ). (*a*) Exposed crack surface; (*b*) polished and etched section behind crack surface.

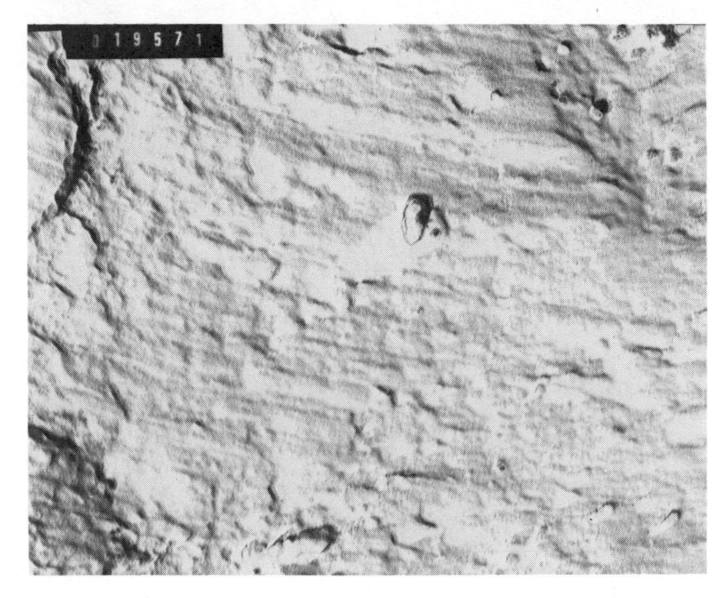

Figure 8.10 Fractograph of crack surface showing possible striations near crack tip, 45,000×.

The stress intensity factor for the elliptical shaped crack idealized in Figure 8.11 is given by

$$K = F_e F_w \sigma \sqrt{\pi a} \tag{8.2}$$

where

$$F_e = \frac{[\sin^2 \theta + (a/c)^2 \cos^2 \theta]^{1/4}}{E(k)} \tag{8.2a}$$

$$E(k) = \int_0^{\pi/2} \{1 - k^2 \sin^2 \theta\}^{1/2} d\theta \qquad k^2 = \frac{c^2 - a^2}{c^2}$$

$$\simeq \frac{3\pi}{8} + \frac{\pi}{8}\left(\frac{a}{c}\right)^2, \quad c < 2a$$

$$F_w = \sqrt{\sec \frac{\pi a}{2b}} \tag{8.2b}$$

For fatigue crack growth the crack growth threshold ΔK_{th} must be exceeded. The largest crack provided a minor semi-diameter of 0.25 in. (6.3 mm) and a major semidiameter of 0.31 in. (7.8 mm). The distance from the center of the ellipse to the back surface was estimated to be 0.4 in. (10 mm) so $2b = 0.8$ in. (20 mm). With the crack growth threshold equal to 3

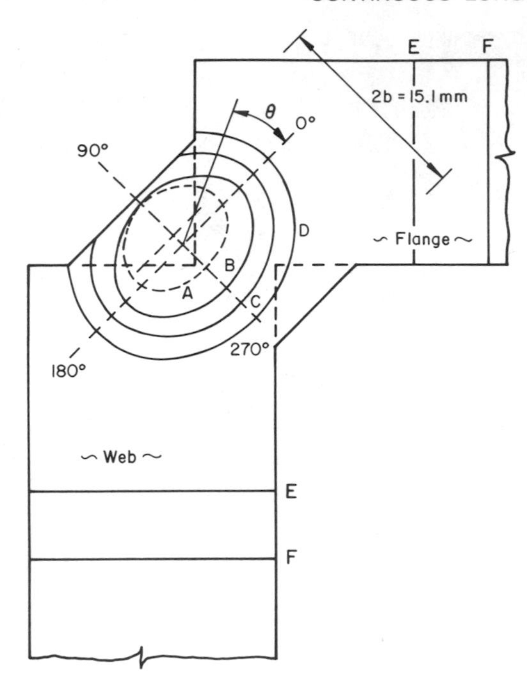

Figure 8.11 Idealized circumscribed cracks in the corner welds.

ksi $\sqrt{\text{in.}}$ (3.3 MPa $\sqrt{\text{m}}$), Eq. 8.2 indicates that the stress range needed to equal or exceed the crack growth threshold is 3.6 ksi (25 MPa). Therefore crack propagation is likely to have occurred in the largest cracks. The number of cycles needed to enlarge the initial flaw 0.005 in. (0.13 mm) can be estimated from Eq. 8.3:

$$N = \int_{0.25}^{0.255} \frac{da}{3.6 \times 10^{-10} \Delta K^3} \tag{8.3}$$

The stress intensity range is provided by Eq. 8.2, and the effective stress range was taken as 1.7 ksi (12 MPa). About 4.8 million stress cycles would be required to extend the crack tip 0.005 in. (0.13 mm). Hence not much crack enlargement was probable unless a local irregularity existed within the ellipse.

The maximum stress intensity was estimated for the crack geometries shown in Figure 8.11. Figures 8.12 and 8.13 show the variation in the stress intensity factor between 180° and 360° along the elliptical-shaped crack boundary for crack shapes A, B, C, and D. Equation 8.2 was used to evaluate the effect of the dead and live load stress. The residual stress field shown in Figure 8.6 was taken into account by using the point load model given in [8.8]. This considers the residual stress as point loads

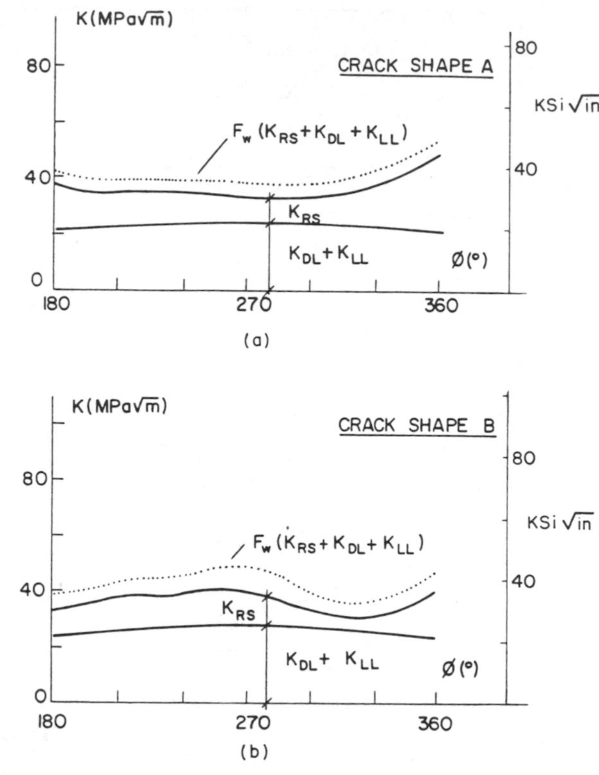

Figure 8.12 Stress intensity factor for cracks in corner weld, box girder. (a) Variation in *K* for idealized crack A; (b) variation in *K* for idealized crack B.

acting on a mesh that approximates the crack surface. The stress intensity factor contribution for each "point load" $\sigma_{rs}dA$ acting on a segment B (see Figure 8.14) is given by

$$K = \frac{P_{rs}\sqrt{a}}{\pi^{3/2}l^2}\sqrt{\frac{r}{R}}\frac{\sqrt{(1/\alpha^2) - 1}}{\{1 - [1 - (c^2/a^2)]\cos^2\Phi\}} \tag{8.4}$$

The geometric factors α, l, and Φ are shown in Figure 8.14, and $P_{rs} = \sigma_{rs}dA$.

The results summarized in Figures 8.12 and 8.13 indicate that the largest crack size (shape C) was approaching the fracture resistance of the plate. The maximum stress intensity factor occurred near the back free surface in the interior weld passes. It would be more likely for the weld metal to have much higher fracture toughness at the back surface. Hence the maximum stress intensity in the web plate is about 40 ksi $\sqrt{\text{in.}}$ (44 MPa $\sqrt{\text{m}}$) and occurs near 180°. Therefore the margin of safety at the time the cracks were discovered was about 1.4.

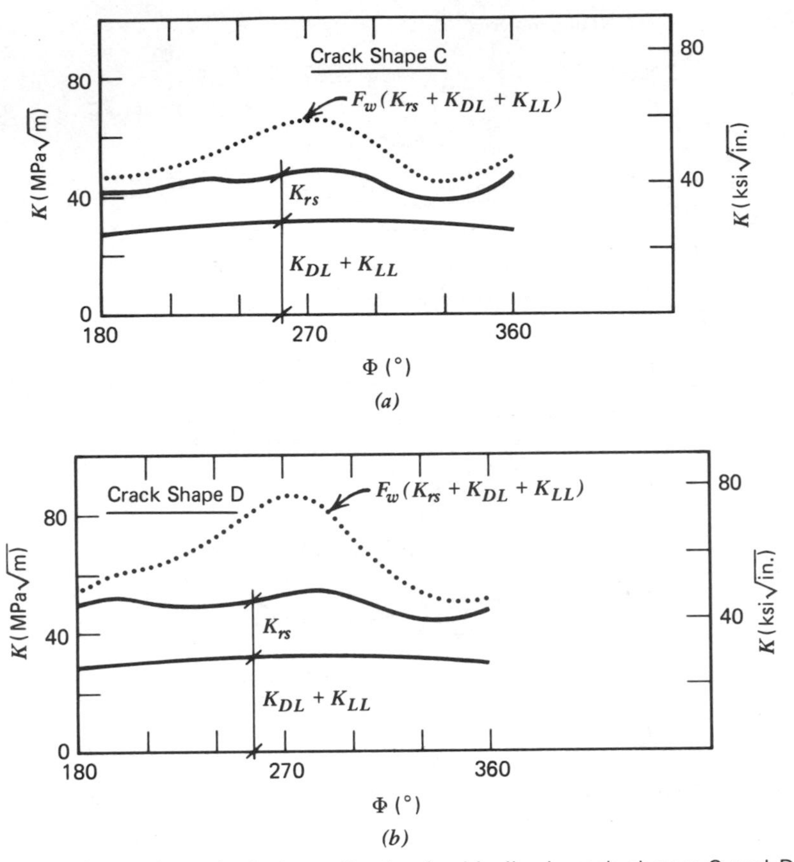

Figure 8.13 Stress intensity factor estimates for idealized crack shapes C and D. (*a*) Variation in *K* for idealized crack C; (*b*) variation in *K* for idealized crack D.

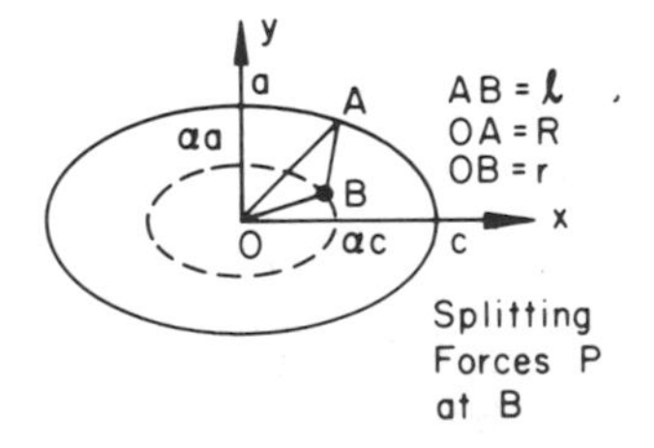

$$K = \frac{P\sqrt{a}}{\pi^{3/2}\ell^2}\sqrt{\frac{r}{R}}\;\frac{\sqrt{\frac{1}{\alpha^2} - 1}}{\left\{1 - k^2 \cos^2 \Phi\right\}^{1/4}}$$

$$x = \alpha c \cos \Phi_1, \quad y = \alpha a \sin \Phi_1$$

Figure 8.14 Residual stress point load model for elliptical cracks with the point splitting force acting at B and the stress intensity evaluated at A.

8.1.3 Conclusions

The transverse weld cracks in the longitudinal box corner fillet welds were all caused at the time of fabrication. They were located in regions where manually made weld passes were added to the submerged arc weld in order to increase its size. The fractographic and metallographic studies all suggested that the cracks were hydrogen related. All cracks removed for examination were found to originate at a porosity or entrapped piece of slag in a region of high residual tensile stress from weld shrinkage.

The largest embedded cracks in the web-flange connections were found to be susceptible to fatigue crack growth under the most severe service load conditions. Some evidence of propagation was detected.

The criticality of these cracks was assessed using a fracture mechanics model and by considering the contribution of dead load, live load, and residual stresses. This analysis indicated that it was desirable to remove the cracks in an expeditious manner.

8.1.4 Repair and Fracture Control

The examination of the cores with cracks and the field observation with nondestructive techniques indicated that it was desirable to grind the weld surface of all manually made welds and inspect them with dye-

Figure 8.15 Photograph of the ground and polished retrofit region after removal of crack.

penetrant. These locations were all suspect, as it was likely that inadequate levels of preheat were used when the manual weld passes were made. An examination of several submerged arc welds did not reveal any difficulties with those locations. Except for the pinned connection in the tie girder, all manual weld locations were confined to the bolted splice between the end of a tie girder section and the first internal diaphragm.

All cracks in the longitudinal box corner welds were removed by grinding, drilling, or coring. Figure 8.15 shows a ground and polished retrofit region after the removal of a crack. The ground area was also checked with dye-penetrant to ensure that no crack tip remained.

REFERENCES

8.1 Engineering News Record, Welding Flaws Close Interstate Tied Arch Bridge, *Engineering News Record,* August 16, 1979.

8.2 Fisher, J. W., Pense, A. W., and Hausammann, H., Fatigue and Fracture Analysis of Defects in a Tied Arch Bridge, Proc. IABSE Colloquium on Fatigue of Steel and Concrete Structures, Lausanne, 1982.

8.3 Guerrera, U., Private communication to J. W. Fisher, 1982.

8.4 Doi, D. A., and Guell, D. L., A Discrete Method of Thermal Stress Analysis Applied to Simple Beams, Appendix 4, Vol. 1, NCHRP Summary 1974, Transportation Research Board, October 1974.

8.5 Takahashi, E., and Iwai, K., Prevention of the Transverse Cracks in Heavy Weldments of $2\frac{1}{4}$ CR–1 Mo Steel Through Low Temperature Postweld Heat Treatment, *Trans. Japan Welding Soc.* 10 (April 1979).

8.6 Hirt, M. A., and Fisher, J. W., Fatigue Crack Growth in Welded Beams, *Engrng. Fracture Mech.* 5 (1973):405–429.

8.7 Tajima, J., Asama, T., Miki, C., and Takenouchi, H., Fatigue of Nodal Joints and Box-Section Members in a Bridge Truss, Proc. IABSE Colloquium Fatigue of Steel and Concrete Structures, Lausanne, 1982.

8.8 Roberts, R., Fisher, J. W., Irwin, G. R., Boyer, K. D., Hausammann, H., Krishna, G., Morf, U., and Slockbower, R. E., Determination of Tolerable Flaw Sizes in Full Size Welded Bridge Details, Report FHWA-RD-77-710, Federal Highway Administration, Office of Research and Development, Washington, D.C., 1977.

CHAPTER 9

Lamellar Tearing

Highly restrained welded connections are not frequently used in bridge structures. Generally, welded built-up girders and other welded structural members are fabricated so that the primary stresses from loads and the residual stresses from welding are acting in a direction parallel to the direction of rolling. Rigid frames are not frequently used in bridges in a manner that permits large flange plates to be welded perpendicular to the surface of other plates.

However, at least three bridge structures have developed cracks because of the highly restrained joints that occur when heavy flange plates are welded to opposite sides of a plate. In one case the I-275 bridge in Kenton County, Kentucky, the flanges of longitudinal girders were framed into the web of a box pier cap. Significant cracking was observed during fabrication because of the highly restrained joint. This led to lamellar tearing in the web plate which extended to the web plate surface near the weld toe [9.4]. A similar type of cracking was found to occur in the hanger brackets of the I-24 tied arch structures across the Ohio River near Paducah, Kentucky [9.5]. The third structure, the Ft. Duquesne Bridge in Pittsburgh, was found to have cracks at the approach ramps. Whereas the other two structures had the cracks detected at the time of fabrication, the Ft. Duquesne rigid frame bents were found to have cracks while in service. The cracks were found in the flange plates of the rigid frame connections of the steel support bents. A description of the cracking and retrofit of the bents is given in Section 9.1. Figures 9.5 and 9.6 show the cracks that formed at beam-column connections.

9.1 FATIGUE-FRACTURE ANALYSIS OF FT. DUQUESNE BRIDGE RIGID FRAME SUPPORTS

9.1.1 Description and History of the Bridge

Description of Structure

The northern approach ramp to the Ft. Duquesne Bridge in Pittsburgh, Pennsylvania, was constructed between 1966 and 1968. It carries southbound traffic onto the Ft. Duquesne Bridge over the Allegheny River [9.1, 9.2].

The approach ramp is a double deck roadway structure, consisting of several two-span and three-span continuous composite steel box girders supported by rigid frame steel box bents. A general plan of the approaches with an elevation view is given in Figure 9.1. Rectangular steel box girders support a composite reinforced concrete deck roadway. Each span has a centerline arc length of 100 ft (30.5 m). The radius of curvature for each span between 7 to 10 is constant. However, the radius varies between the spans from 853 ft (260 m) to 863 ft (263 m). The roadway box girders and bents were shop welded individually and were then field connected by welding. The cross girder center segments of the bents were field bolted. The roadway girders and bents act together structurally to form a series of space frames. Typical details of bent SB3 are shown in Figure 9.2. Figure 9.3 shows a typical view of the bent and box girder structure.

The cross section of the box girders which supports the roadway between bents varies along the length of each span. A typical cross section at an intermediate diaphragm is shown in Figure 9.4. The diaphragms are spaced at 10 ft (3.05 m) intervals. The box girders contain transverse and longitudinal stiffeners that are welded to the inner surfaces of the box.

Small rectangular backup bars were used to fabricate the web–bottom flange connection. The backup bars were tack welded to the web and flange plates along the length of the box girder. Discontinuous backup bars were used in the fabrication. This creates a cracklike discontinuity in the web–flange connection at each point that it occurs [9.1].

History of the Cracking

On September 20, 1978, a crack was discovered in steel box bent SB3. The crack was found in the north edge of the lower level cross girder top flange at the west column flange connection. The crack was spotted by an engineer employed by Richardson, Gordon and Associates as he was climbing the stairway adjacent to bent SB3, which provides access to a parking lot below the bridge (see Figure 9.1). A follow-up inspection by

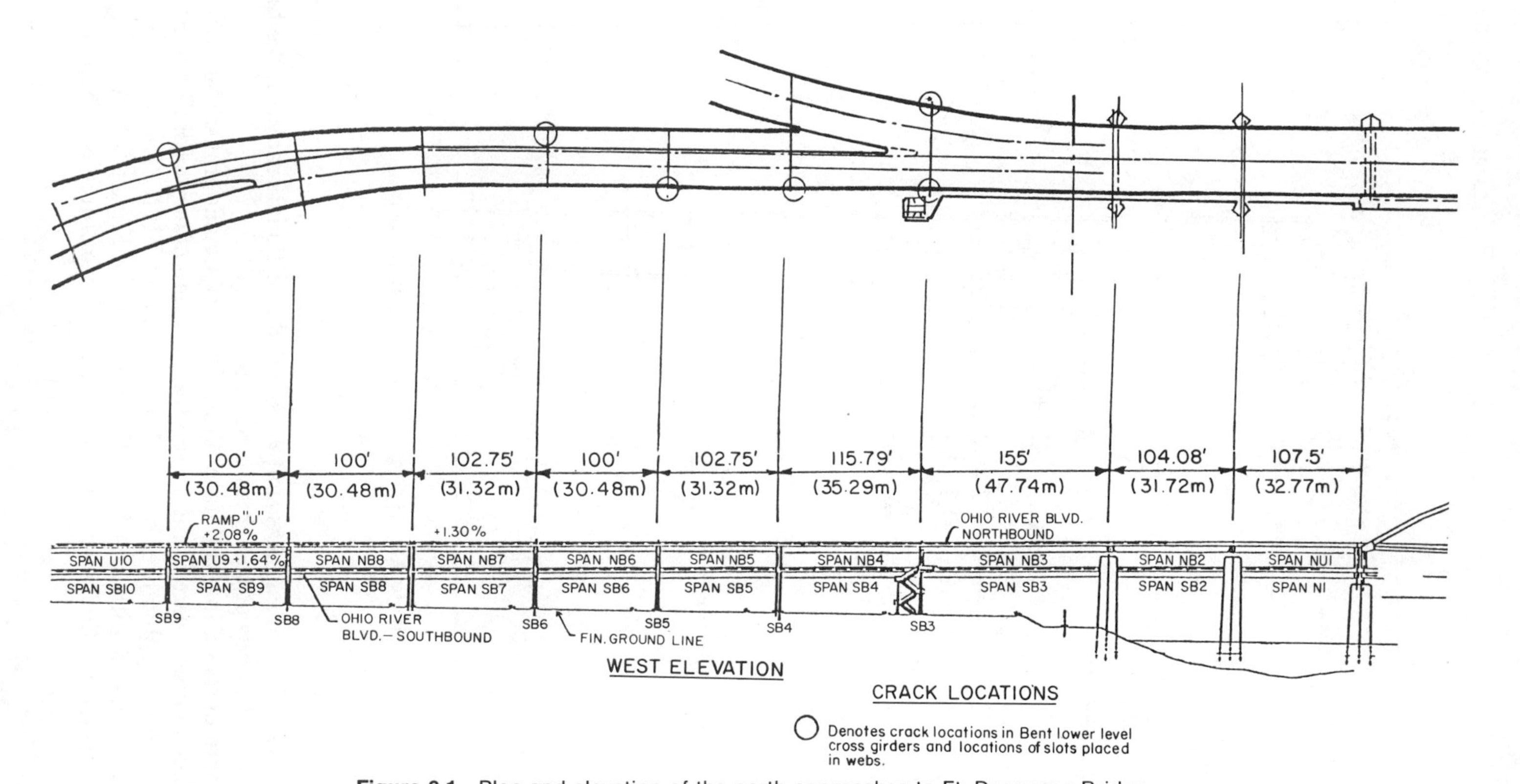

Figure 9.1 Plan and elevation of the north approaches to Ft. Duquesne Bridge.

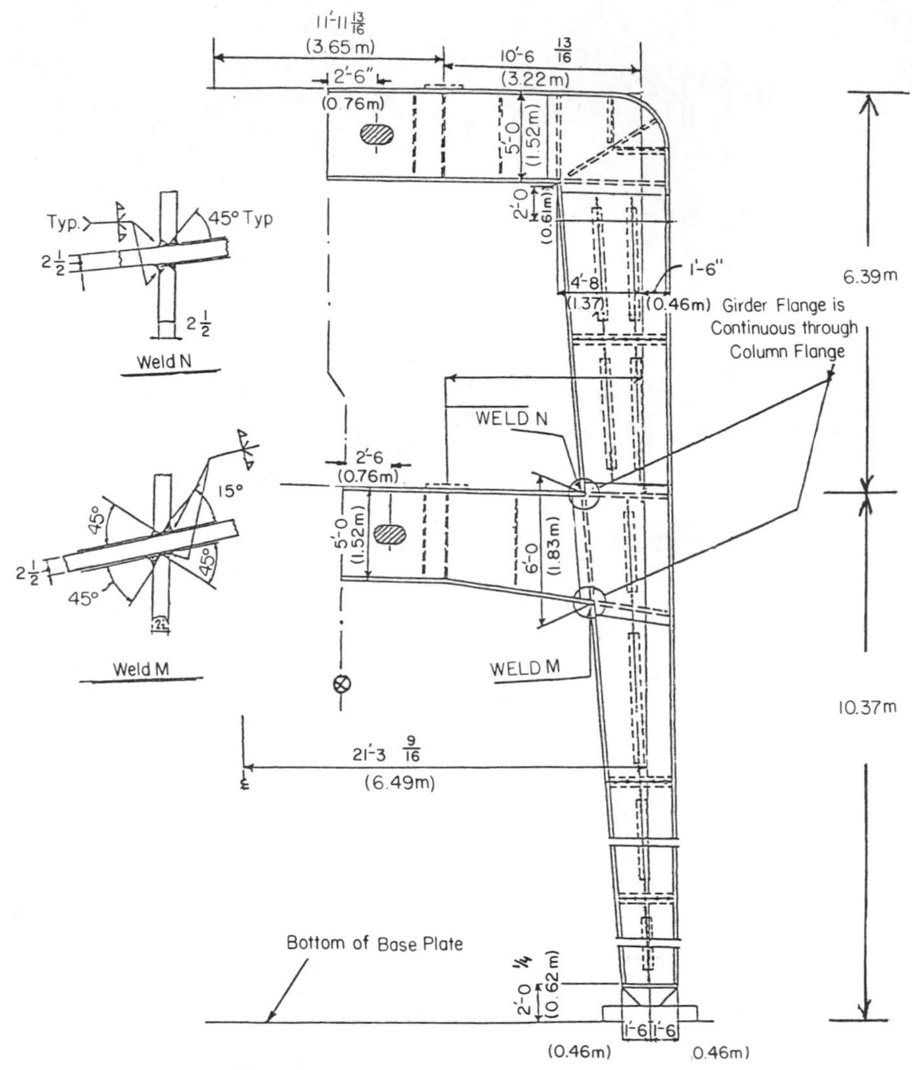

Figure 9.2 Typical details of rigid frame bent.

Richardson, Gordon and Associates revealed additional cracks on the top flange at both ends of the cross girder [9.2].

A more detailed inspection by Penn-DOT was made in October 1978. Additional cracks were found in the west column of bents SB4S and SB5 and in the east column of bents SB6, SB9, and SB10N. The crack locations are marked with an "X" in the plan view of Figure 9.1. All cracks were detected at the lower roadway cross girder–column connections.

Figure 9.3 Typical rigid frame bent and box girders.

Figure 9.5 shows the cross girder tension flange crack as seen in the paint film and after the paint was removed and the surface polished by grinding. Two distinctly different cracks can be seen at the corner connection. One is a lamellarlike tearing in the through flange plate near its mid-thickness. The second crack is a crack originating at the groove weld toe of the column flange–girder flange connection. Crack extension can be seen at the end of the lamellarlike tear which turns and becomes perpendicular to the girder flange.

Cracks were also detected in a number of the cross girder compression flanges (see Figure 9.6). This crack could be seen in the paint film and lies near the mid-thickness of the through flange plate. The crack was very tight and had obviously been pushed closed by the high compression closure forces in the box corner.

Several weld toe cracks were also detected at the cross girder tension flange connection (see Figure 9.7). These cracks occurred at either the beam flange or in the column flange depending on the reentrant corner weld angle.

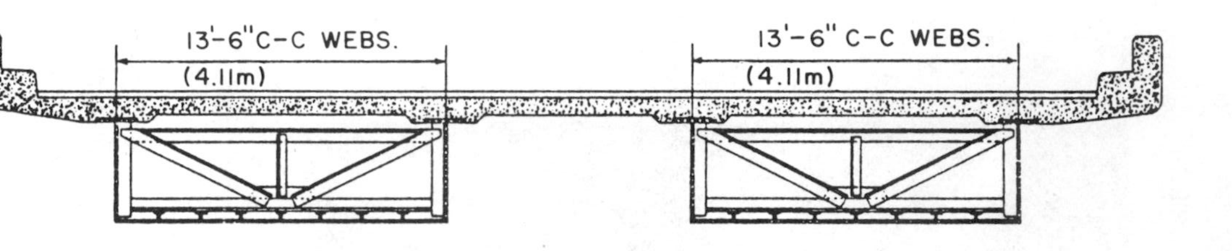

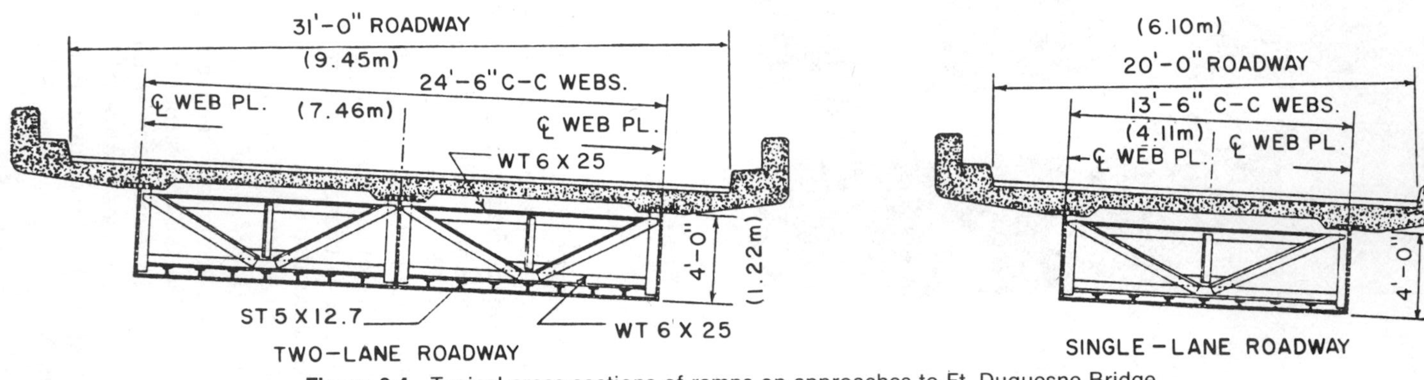

Figure 9.4 Typical cross sections of ramps on approaches to Ft. Duquesne Bridge.

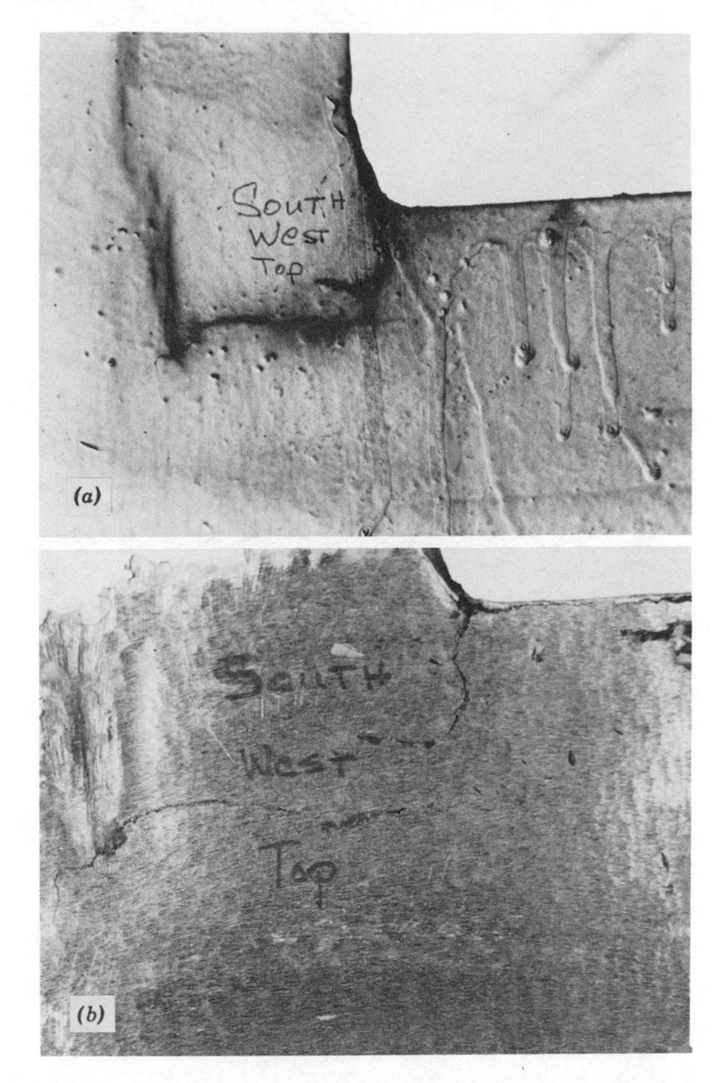

Figure 9.5 Cracks in cross girder–column connection at the top flange of bent SB3. (*a*) Crack in paint film at tension corner of cross girder–column connection of bent SB3; (*b*) cracks after removal of paint film.

9.1.2 Failure Modes and Analysis

Location of Cracks

The cracks were found in six of the rigid steel box frame bents that support the double deck roadway structures at the north end of approaches at the locations marked in Figure 9.1.

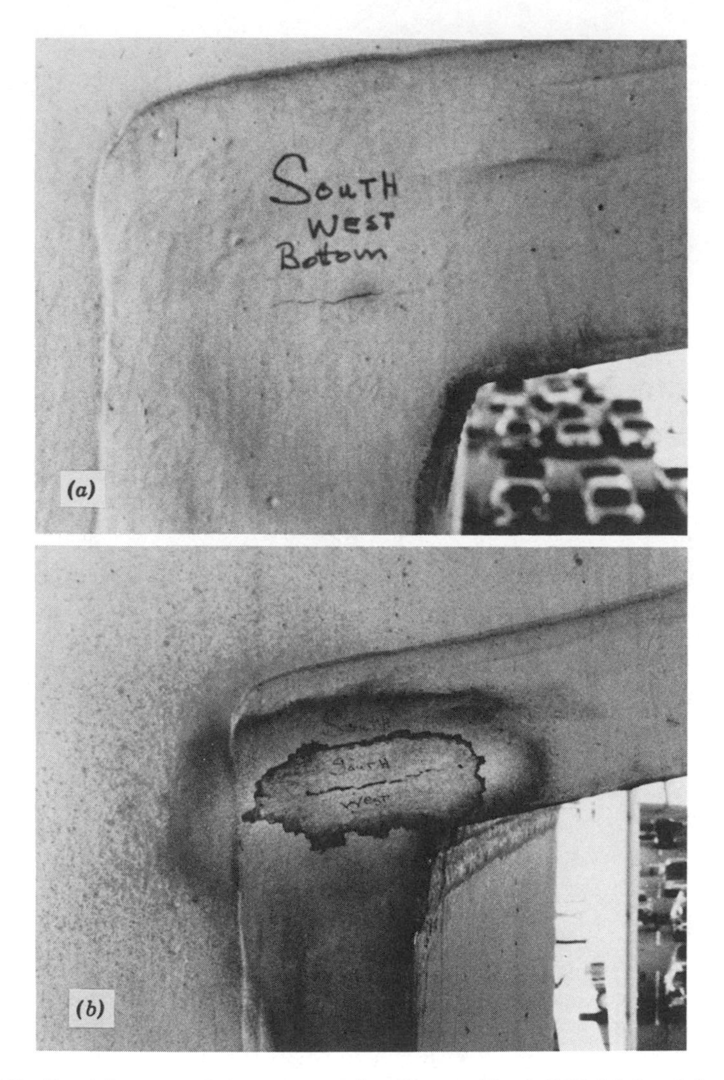

Figure 9.6 Crack in cross girder compression flange of bent SB3. (*a*) Crack in paint film at the girder compression flange; (*b*) crack in girder flange enhanced with dye-penetrant.

Most cracks were located in the top corner of the lower level cross girders–column connections. The girder flanges were continuous and extended into the column, as illustrated in Figure 9.2. Groove welds were used to attach the interior column flanges to the cross girder flanges at the rigid frame connections. The girder and column flanges of bent SB3 (see Figure 9.2) were $2\frac{1}{2} \times 48$ in. (64×1219 mm) A517 steel plates. Similar plate thicknesses were used at other bents. When the cracks exhibited lamellar tearing characteristics such as shown in Figure 9.5,

Figure 9.7 Small crack at one of the weld toes.

the cracks were visible on both edges of the girder flanges. This indicated that the lamellar tearing likely extended completely across the flange width.

Weld toe cracking was also detected at three top flange edges at the beam flange–column flange connection, as illustrated in Figure 9.7. These cracks all appeared to occur at one of the groove weld toes and extended into the base metal heat affected zone. Most of the weld toe cracks and the mid-thickness lamellar tears showed evidence of fatigue crack extension.

Only a few locations were found to have lamellar tears in the compression flange, as shown in Figure 9.6. No significant crack extension was observed at these cracks.

Cyclic Loads and Stresses

Traffic surveys in 1975 indicated that about 425 trucks used the inbound lanes which were located on the lower roadway each day. Available traffic counts suggested this was a reasonable daily volume for the 10 years the bridge was in service between 1968 and 1978. This would result in about 1.5 million variable amplitude stress cycles, assuming each truck produces one stress cycle.

A stress history study had been carried out in 1975 with primary focus on the rectangular box girder spans. A few gauges had been installed on one of the rigid frame bents supporting span 10. The gauge nearest the rigid frame connection yielded the stress range spectrum shown in Figure 9.8. This indicated that the maximum stress range was less than 4 ksi (28 MPa). The effective stress range $S_{r\,\mathrm{Miner}}$ was estimated to be 2 ksi (14 MPa) [9.1].

Low temperature and other environmental effects did not appear to have any influence on the behavior of the cracked bents. No evidence of unstable crack growth was observed.

The estimated dead load stress in the girder flange was about 32 ksi (220 MPa).

Mechanical, Chemical, and Fracture Properties of Material

The mill test reports indicated that the material satisfied the specification requirements of A517 steel. Samples have been removed from the structure, but no tests have been performed to date. The structure has been exposed to minimum temperatures as low as −20°F (−29°C) without any adverse effect. This suggests that the fracture toughness of the A517 steel is at least 100 ksi $\sqrt{\text{in.}}$ (110 MPa $\sqrt{\text{m}}$).

The structure was built prior to the mandatory fracture toughness requirements for steel used in tension components.

Most of the bents were fabricated from A36 and A441 steel. Only the longer spans were fabricated with A517 steel. The lamellar tearing was mainly confined to the A517 steel components. However, the cracks observed at the weld toe were found in both A517 and A441 steel.

Failure Analysis

The cracks in the steel welded cross box girder flange-to-box column flange connection regions of the rigid frame bents were found to be two types: (1) lamellar tearing at the mid-thickness of the beam flange and (2) groove weld toe cracks. The weld toe cracks exhibited lamellar tearing characteristics when they approached or linked up with the existing lamellar tears.

Fatigue crack growth was also observed at the ends of the coalesced cracks. The effective crack size appeared to vary between 1.5 and 2 in. (38 and 50 mm). Between 0.1 and 0.2 in. (2.5 and 5 mm) of crack extension was observed at the crack tip (see Figure 9.5*b*).

The crack extension can be roughly modeled by considering the deep corner cracks as an edge crack. The stress intensity factor is given by

$$K = 1.12\sigma\sqrt{\pi a}\left(\sqrt{\sec\frac{\pi a}{2t_f}}\right) \tag{9.1}$$

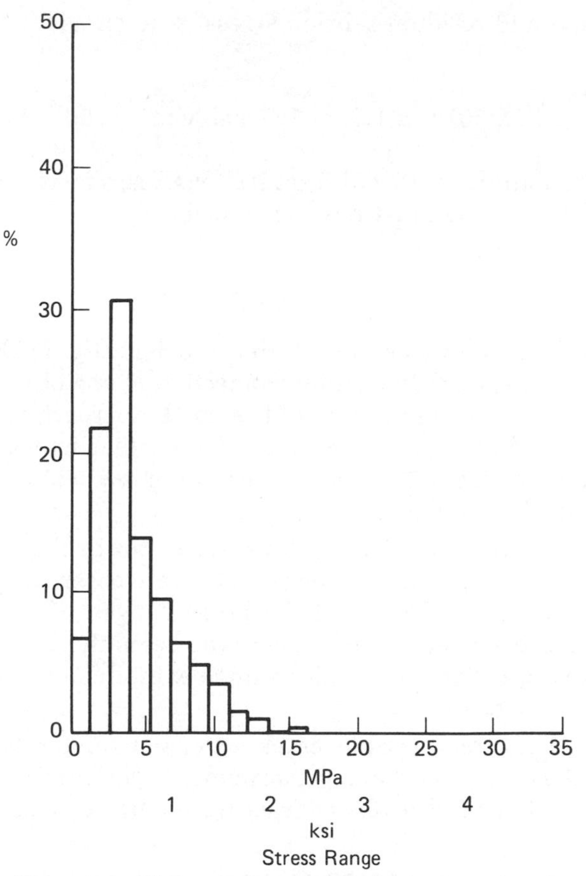

Figure 9.8 Histogram for stresses in the box section.

where a is the effective crack size and t_f is the flange thickness. For stable crack extension, the relationship

$$N = \int_{a_i}^{a_f} \frac{da}{3.6 \times 10^{-10}\,\Delta K^3} \tag{9.2}$$

can be used to estimate the random variable stress cycles. For an initial crack size of 1.8 in., and an effective stress range of 2 ksi (13.8 MPa), Eq. 9.2 yields 2.5×10^6 cycles. This is in reasonable agreement with the observed behavior of the cracks during the period the bridge was in service.

The maximum stress intensity at the crack tip is dependent on the dead and live load stresses and the residual stress from welding. Since the lamellar tears appear to have occurred at the time of fabrication, some relaxation of the residual stresses was likely. If the combined effect of applied load and residual tension stress is taken as 50 ksi (345 MPa) ($\sim$0.5 σ_y),

$$K_{\max} = 1.12(50)\sqrt{\pi(1.9)} = 137 \text{ ksi} \sqrt{\text{in.}} \quad (150 \text{ MPa} \sqrt{\text{m}}) \qquad (9.3)$$

No information on the material fracture toughness is available, but it is apparent that its value must have exceeded $K_{\max}$.

9.1.3 Conclusions

1. Analysis of the cracks discovered accidentally on September 20, 1978, at the beam to column connection of one of the bents of the two-story rigid frame bents of the north approach ramp of the Ft. Duquesne Bridge indicated that lamellar tearing was the primary cause of cracking. Comparable cracking was discovered in other bents in the complex.
2. All cracks were found at the lower-level box girder–column flange connection. The lamellar tear cracks were observed in the tension and compression flanges of the beam flanges. The column flange was groove welded to each side of the beam flanges. The cracks in the cross girder beam tension flange exhibited evidence of fatigue crack growth.
3. No evidence of unstable crack extension under traffic was observed. The behavior and performance of the cracked flange plates suggest that good levels of fracture toughness exist in the A517 steel.
4. All lamellar tears were produced at the time of fabrication. Crack growth from the weld toe developed while in service.

9.1.4 Repair and Fracture Control

In order to minimize the effects of unstable crack extension, slots were installed in the 1 in. (25 mm) web plates at the box girder tension flange–column connection, as shown schematically in Figure 9.9. This isolated the rigid frame webs from their connection with the intersecting flange plates and the cracks. With the large bending capacity of the web plates, no significant deformation or damage would develop at the box corner should a flange fracture before a final retrofit. These slots were installed in January 1979.

The final repair incorporated bolted splice plates on the box web and top flange. Figure 9.10 shows the general scheme used to reinforce a typical connection. Access holes were cut into the column webs and internal diaphragms for access to the joint and to provide material for further studies on the steel plate. Partial width flange plates were groove welded

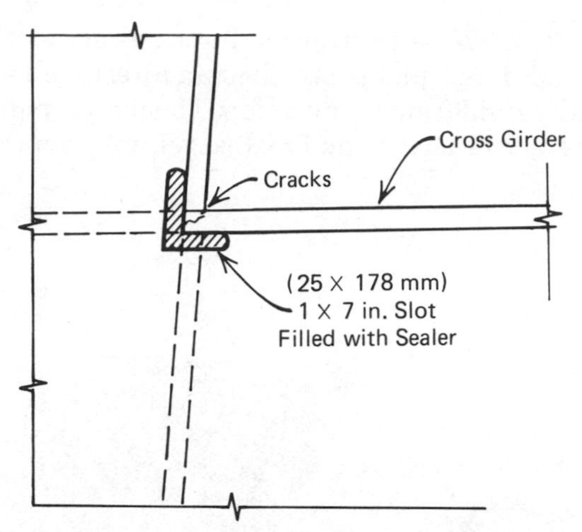

Figure 9.9 Schematic showing location of retrofit slots in girder webs.

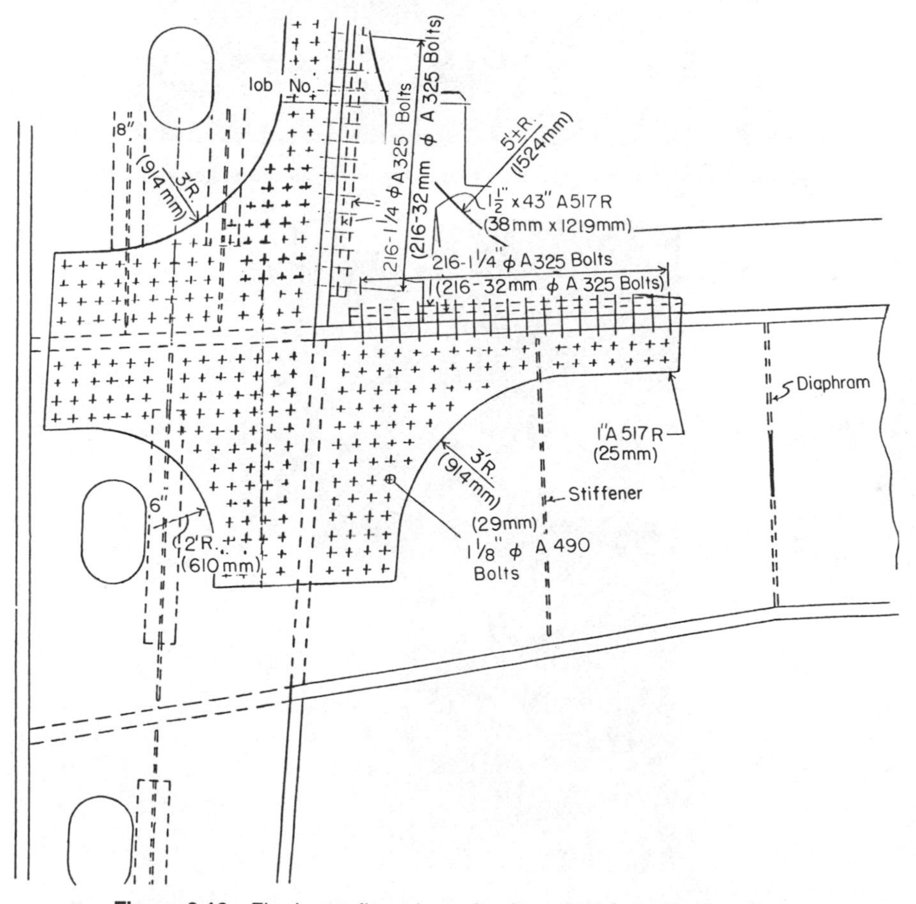

Figure 9.10 Final retrofit and repair of cracked frame connections.

to the web plates to assist with transferring loads across the rigid frame joints. The web and flange plates provided an alternate load path for the forces acting on the rigid frame connection. The final repairs were carried out in 1981. Figure 9.11 shows the finished reinforced joint at one of the details.

Figure 9.11*a* Fabricated splice plates for retrofitting cracked detail.

Figure 9.11*b* Splice plates installed on rigid frame joint.

None of the compression flange lamellar tears were retrofitted. They were all positioned in a plate parallel to the applied bending stresses, and none exhibited any evidence of fatigue crack growth.

REFERENCES

9.1 Inukai, G. J., Yen, B. T., and Fisher, J. W., Stress History of a Curved Box Bridge, Fritz Engineering Lab. Report 386-8 (78), Lehigh University, June 1978.

9.2 Richardson, Gordon & Associates, Inspection Report LR1039, Section IF, North Approach to Ft. Duquesne Bridge, Allegheny County, Pa., Richardson, Gordon & Associates, Pittsburgh, Pa., October 1979.

9.3 Fisher, J. W., Personal Communication to B. F. Kotalik, Pa. Department of Transportation, Bridge Engineering, October 10, 1978.

9.4 Aycock, J. N., Personal Communication to J. W. Fisher, October 1973.

9.5 Munse, W. H., Investigation of Hanger Connections on the I-24 Bridge over the Ohio River, Kentucky Department of Transportation, December 1974.

PART II

Secondary Stresses and Distortion-Induced Stress

The most common types of fatigue cracking that have developed in bridge structures have been the result of secondary and/or displacement-induced cyclic stresses. These problems have developed because of unforeseen interaction between the longitudinal and transverse members. This interaction does not alter the in-plane behavior of the structure, and hence the design for in-plane loading and deflection is adequate when proportioning the individual components. Generally, the effects of the secondary and displacement-induced cyclic stresses are seen at connections. Often short gaps in a girder web or greater than expected restraint results in a geometric amplification of the cyclic stress in the gap region, and this has resulted in cracking.

This type of cracking has occurred in many types of bridge structures. Stringer webs have cracked in suspension bridges at the stringer–floor-beam connections. Floor-beam webs have cracked in tied arch bridges. The longitudinal girders of girder–floor-beam bridges have experienced cracking in the girder web. Multiple beam bridges have experienced cracking in the girder webs at cross-frames and diaphragms, and at least one box girder structure has developed cracks in the girder web at interior cross-frames.

The cracking has been most extensive in welded structures where a weld toe has commonly existed in the high cyclic stress region. However, the problem has also occurred in riveted structures where similar conditions have existed. The riveted members have generally taken longer to develop cracks because the initial flaw condition and geometry of the base metal were not as critical.

The next series of examples shows the conditions that existed to promote this kind of fatigue cracking. Included are structural members that have cracked during handling or shipping of the components to the bridge site as well as structures that have cracked in service. The cracking that occurred during handling or shipping is discussed in one chapter where it was clear that the condition had developed prior to service. Such cracking prior to service may have contributed to some of the cases that are discussed in other chapters. However, reliance on prior cracking was not needed to explain the observed behavior in these cases of web gap cracking, even though it may have contributed to it.

The following table summarizes the cases of secondary and distortion-induced cracking that are discussed in Chapters 10 and 16. They include riveted as well as welded structures.

SUMMARY OF CASES OF SECONDARY AND DISTORTION-INDUCED CRACKS

Case	Type Condition	Bridge
10.1	In-plane bending cantilever bracket tie plates	Lehigh River and Canal Bridges, Bethlehem, Pa.
10.2	In-plane bending cantilever bracket tie plates	Allegheny River Bridge, Pennsylvania Turnpike
11.1	Transverse stiffener web gap distortion during transport	I-90 over Conrail Tracks in Cleveland
11.2	Transverse stiffener web gap distortion during transport	I-480 Cuyahoga River Bridge near Cleveland
12.1	Floor-beam connection plate web gap	Poplar Street Approaches East St. Louis, Ill.
12.2	Floor-beam connection plate web gap	Polk County Bridge, Des Moines, Ia.
13.1	Diaphragm connection plate web gap	Belle Forche River Bridge, S.D.
13.2	Diaphragm connection plate web gap	Missouri River Bridge, Chamberlain, S.D.
14.1	Floor-beam web gap at tie girder connection	Mississippi River Bridge, Praire du Chien, Wisc.
15.1	Stringer web gap	Walt Whitman Bridge, Philadelphia, Pa.
16.1	Coped stringers and restraint	Windermere Subdivision, Canadian Pacific Railroad, Ontario

CHAPTER 10

Cantilever Floor-Beam Brackets

One of the earlier types of cracking from unaccounted displacements that was found to develop in highway bridges was discovered in the tie plates connecting transverse floor beams and cantilever brackets across the main girders.

Such cracking was first found in the Allegheny River Bridge near Pittsburgh during the fall of 1971 and was observed to originate from the rivet holes used to attach the floor-beam–bracket tie plate to the main longitudinal girders. A short time later in the spring of 1972 cracks were found in the Lehigh River and Canal Bridges during the inspection of these details. The cracks in the Lehigh River and Canal Bridges were found to originate from tack welds used to connect the tie plate to the bracket prior to shop installation of the rivets.

A summary and review of the cracking and the resulting developments in the Allegheny River and the Lehigh River and Canal Bridges are provided in this chapter. Detailed measurements and an assessment of the crack development was made on the Lehigh Canal Bridge. This study demonstrated that unforseen in-plane displacements of the tie plate caused by a rotation of the longitudinal girders at a cross section were the reason for the cracking.

At least one other bridge has experienced distress at the girder-bracket joint. The South Bridge near Harrisburg, Pennsylvania was found to have a number of rivet heads cracked off. The in-plane movement of the connection plate distorted the rivets, prying off the rivet heads and thus creating separation at the joint.

10.1 CRACKING OF THE TIE PLATES OF THE LEHIGH RIVER AND CANAL BRIDGES

10.1.1 Description and History of the Bridge

Description of the Structure

The Lehigh Canal and River Bridges consist of two adjacent twin structures which carry the eastbound and westbound lanes of U.S. Route 22 over the Lehigh Canal and River near Bethlehem, Pennsylvania. Each

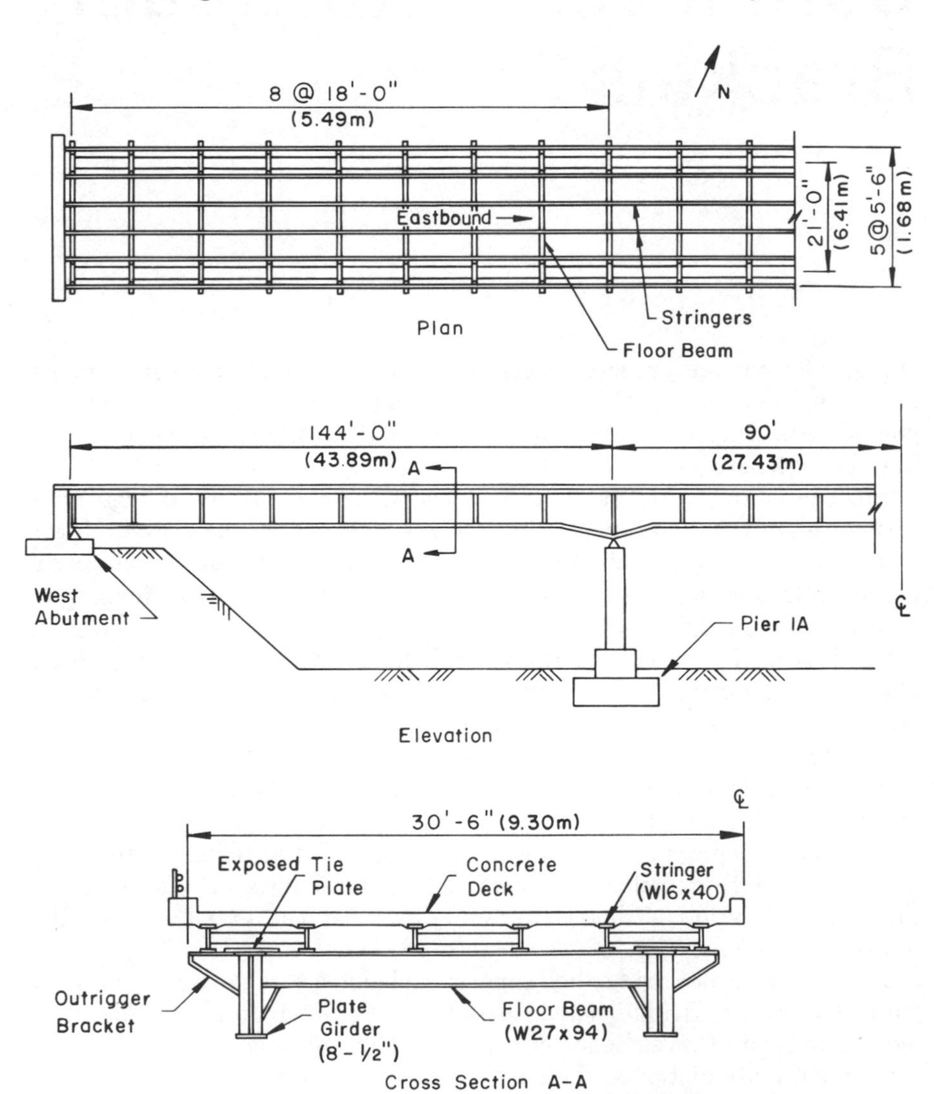

Figure 10.1 Plan, elevation, and cross section of the Lehigh Canal Bridge.

bridge is a two-girder continuous structure extending over three spans with small haunches at the interior piers. The plan and elevation of the eastbound Lehigh Canal Bridge is shown in Figure 10.1. The other three structures were nearly identical. A side view of the Lehigh Canal Bridge is given in Figure 10.2. The bridge cross section and the floor-beam–bracket tie plate detail are shown in Figures 10.1 and 10.3.

History of Structure and Cracking

The twin bridges (one for the westbound lanes and one for the eastbound lanes) were constructed in 1951 to 1953 and opened to traffic in November 1953.

During the spring of 1972 inspection by Pennsylvania Department of Transportation personnel revealed several fatigue cracks in the tie plates connecting the transverse floor beams and cantilever brackets across the main girders, as shown in Figures 10.4 and 10.5. A more detailed description and the location of these fatigue cracks in the tie plates is given in [10.1, 10.2, 10.3]. The cracks started at the edges of the tie plates at a tack weld which apparently connected the tie plates to the outrigger brackets during fabrication.

The cracks were discovered after the bridges were subjected to 19 years of traffic. U.S. Route 22 carries substantial amounts of heavy truck traffic. Annual inspections of the tie plates on the eastbound bridge were carried out by Fritz Engineering Laboratory between 1972 and 1976. The results of those inspections are given in [10.3]. Twenty of the 54 plates in that structure were found to have cracks of varying sizes.

Figure 10.2 Lehigh Canal Bridge (side view).

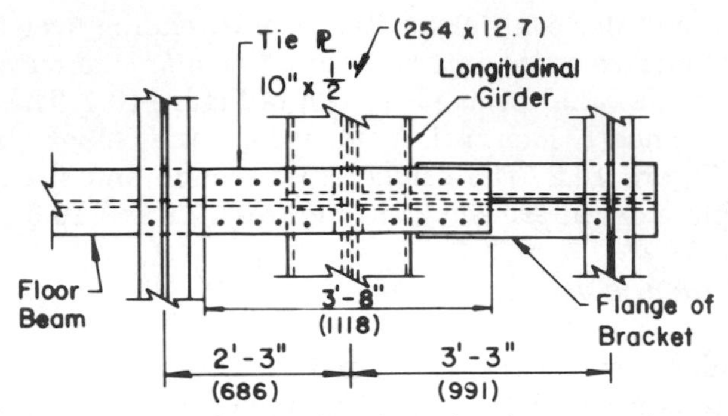

Figure 10.3 Tie plate detail at floor-beam bracket connection to girder.

10.1.2 Failure Modes and Analysis

Location of Cracks

The approximate location and length of the fatigue cracks and stresses in the tie plates of one side span are shown in Figure 10.5. Over the entire length of the Lehigh Canal and River Bridges most of the tie plate cracks were at or near the outside edge of longitudinal or main girders [10.1] near a pier or abutment. Several of the plates had cracked completely crosswise. All observed cracks appeared to be through the thickness of

Figure 10.4 Crack in tie plate originating at tack weld.

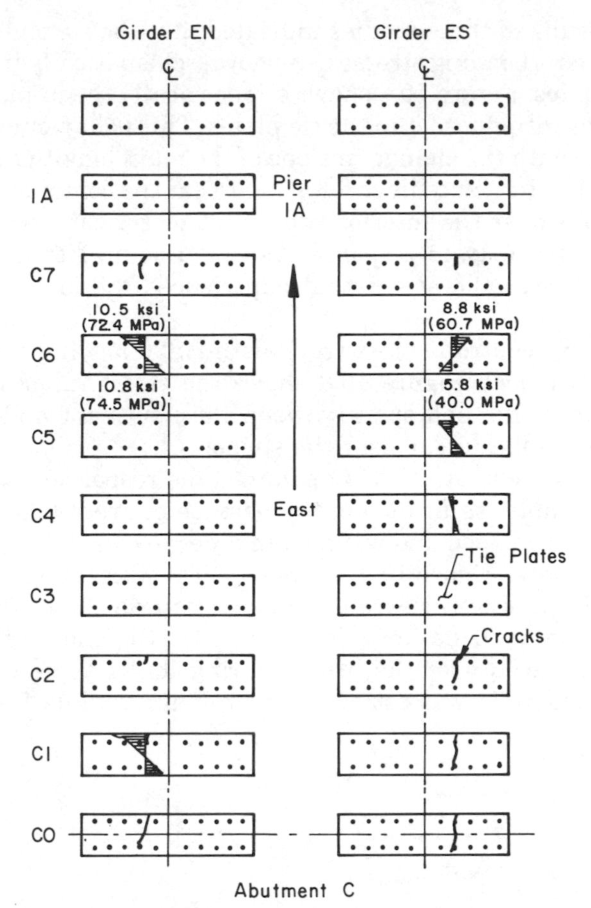

Figure 10.5 Schematic showing location of cracks and measured stresses in the tie plates of the west end of the eastbound span.

the tie plates and were fatigue cracks. All cracks initiated from the edge of the tie plates where a tack weld was used to connect the plates to the outrigger or cantilever bracket during fabrication. A typical tack weld can be seen in Figure 10.4.

A detailed study of the actual traffic conditions and the stress history measurements indicated that the fatigue cracks developed and propagated slowly from the ends of the tack welds due to out-of-plane displacements at the top flange of the main girders.

Cyclic Loads and Stresses

The behavior of the Lehigh Canal Bridges was established by strain measurements under controlled loading and random traffic in 1972 and

1974. The results of those studies indicated that the tie plates were subjected to in-plane bending stresses, as shown schematically in Figure 10.5 for several plates. Figure 10.6 shows schematically the in-plane displacement that was introduced into the tie plates. This behavior can be seen to be compatible with the change in slope of the main longitudinal girder in Figure 10.7. The influence lines for a point near the center of the end span and for a point near the interior support at pier 1 can be seen to agree with the measured strain response. As would be predicted, the primary cracking occurred at locations that experienced the largest changes in rotation.

Tests under a controlled test truck with HS20 loading also confirmed the analytical model. Figure 10.8 shows the stress range measured in each tie plate in the end span between the abutment and pier 1. The highest stress ranges occur near the supports.

The study carried out in 1974 examined the response of the structure to random variable traffic during the passage of over 8000 trucks. Over 200 trucks were stopped and weighed in order to correlate measured and computed response. Figure 10.9 shows a typical stress range spectrum for a tie plate. The gross vehicle weight distribution was found to be reasonably close to the 1970 nationwide average [10.3]. The total number of trucks that traveled over the bridge during its 21 years of service between 1953 and 1974 was estimated from Pennsylvania Department of

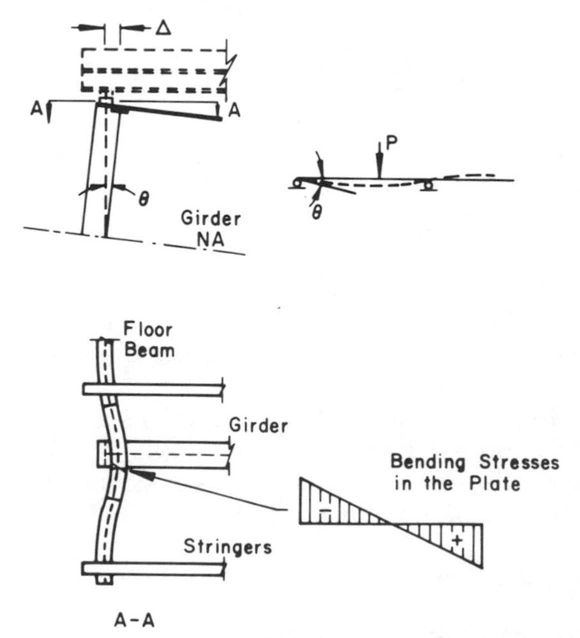

Figure 10.6 Longitudinal displacement at top flange of girder and horizontal bending of tie plate.

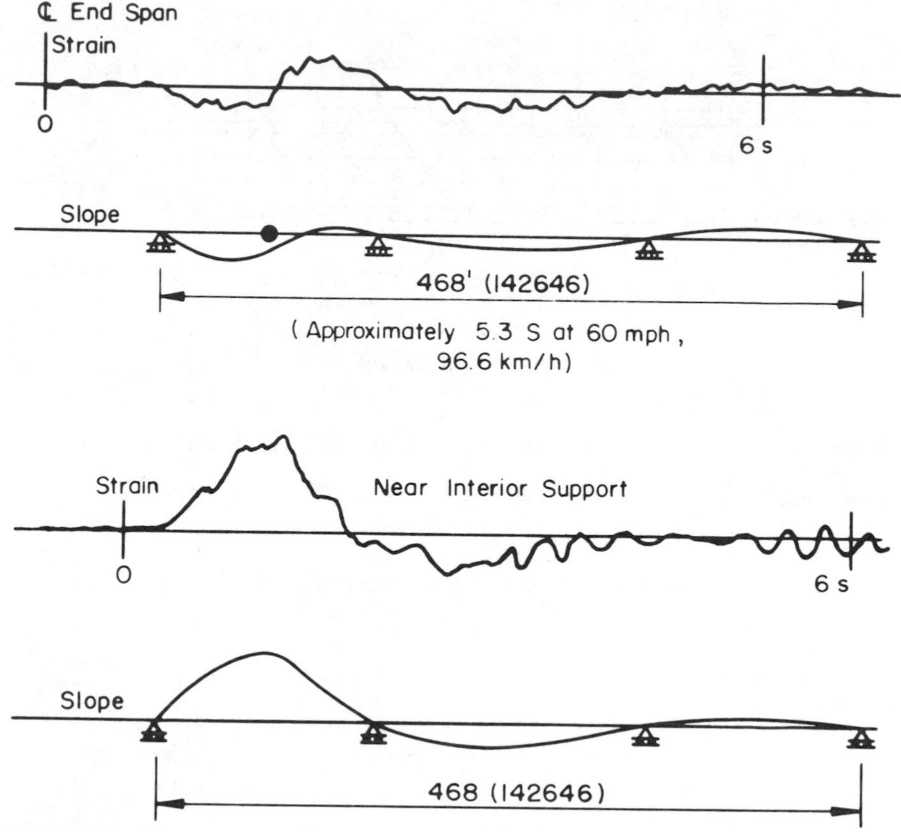

Figure 10.7 Comparison of measured history in tie plates with influence lines for girder slopes.

Transportation traffic counts. The estimated ADTT is plotted in Figure 10.10. The total volume between 1953 and 1974 was estimated to be 22 million trucks [10.3].

The measured stress range occurrences at the gauges on the tie plates were used to establish stress ranges at the plate edges where the crack formed and grew at the tack weld. Miner's effective stress range was established for each plate, using the relationship:

$$S_{r\text{Miner}} = (\Sigma \alpha_i S_{ri}^3)^{1/3} \tag{10.1}$$

For the distribution shown in Figure 10.9, this resulted in $S_{r\text{Miner}} = 11.6$ ksi (80 MPa).

The effective stress ranges for the eastbound bridge are plotted in Figure 10.11 based on the frequency of stress cycles for the measured stress sample. Many of the details experienced less than one stress cycle

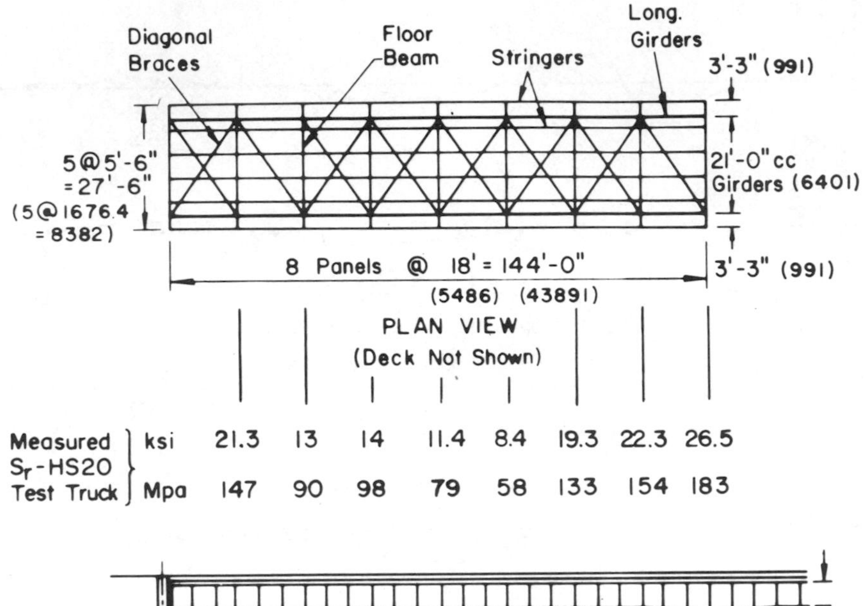

Figure 10.8 Plan and elevation of the side span.

per truck passage. The cracked details were in reasonable agreement with laboratory tests that simulated the tack welded condition [10.4]. The scatter bands for the test data are plotted in Figure 10.11. The cracked details in general were in or exceeded the upper limit which suggested that many had cracked after six or seven years of service. The uncracked details often plotted well below the lower confidence limit. However, many effective stress range points fell within the boundaries of the test data.

Material Properties and Crack Surfaces

None of the tie plates experienced crack instability. All cracks that were observed were propagated in a stable fatigue crack mode of growth. Sometimes the plate was nearly cracked in two, as can be seen in Figure

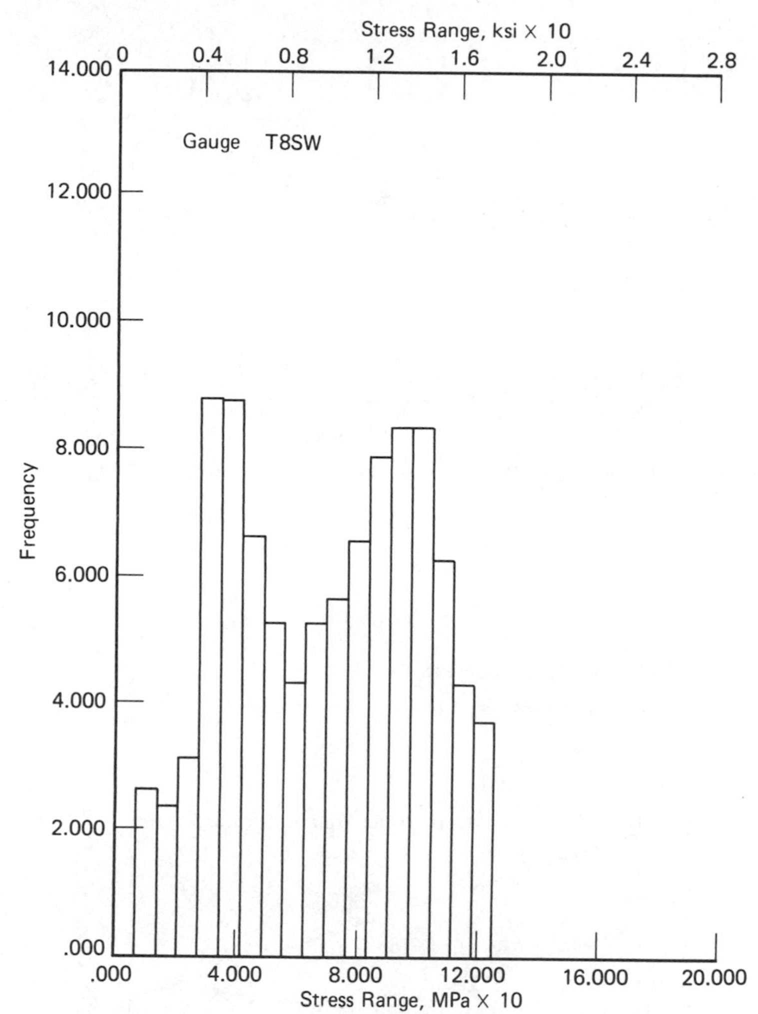

Figure 10.9 Stress range spectrum for gauge T8SW.

10.5. None of the cracked plates was removed and tested to establish their tensile properties and fracture toughness.

The material was supplied to the ASTM A7 specification. Since the plates carried very little dead load, the cyclic stress was the only significant stress that the plates were subjected to.

Analysis of Cracks

The tack weld on the tie plate edge ensured that edge cracks formed in the rectangular plate subjected to in-plane bending. The predicted behav-

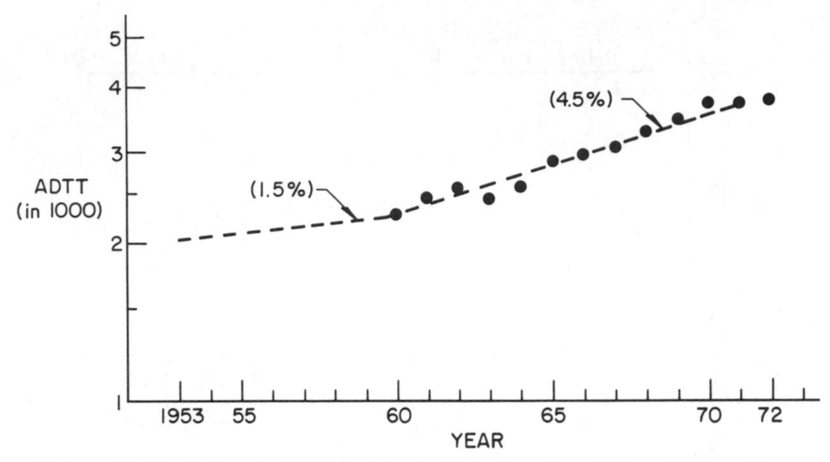

Figure 10.10 Estimated ADTT at the Lehigh Canal and River Bridge site.

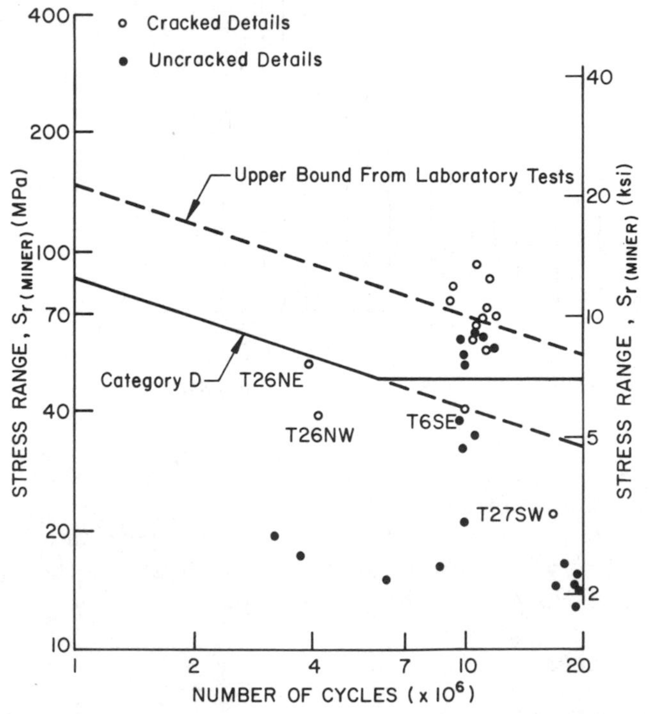

Figure 10.11 S–N Values using peak to peak method without multiple presence separation.

ior of these plates can be estimated by considering the stress intensity factor for an edge cracked plate [1.2]. This is given by

$$K = \left[\frac{0.923 + 0.199(1 - \sin \pi a/2b)^4}{\cos \pi a/2b}\right] \sigma\sqrt{\pi a} \left(\frac{2b}{\pi a} \tan \frac{\pi a}{2b}\right)^{1/2} \quad (10.2)$$

The plate width $b = 10$ in. (254 mm), and the effective stress range $S_{r\text{Miner}}$ varies between 6 and 12 ksi (41 and 83 MPa) for the cracked details plotted in Figure 10.11. Since tack welds are not high quality, an initial crack size 0.05 in. (1.3 mm) was selected as the fabricated condition. The number of cycles to grow a 4 in. (100 mm) crack was estimated for both the upper and lower levels of effective stress range. The cycles were estimated as

$$N = \int_{0.05}^{4.0} \frac{da}{3.6 \times 10^{10}\, \Delta K^3} \quad (10.3)$$

where ΔK was defined by Eq. 10.2. This resulted in 1.9 million cycles at the 11.6 ksi (80 MPa) stress range and 14 million cycles at the 6 ksi (41.4 MPa) level. It can be seen that these values are near the lower confidence limit of the test data shown in Figure 10.11.

10.1.3 Conclusions

The fatigue cracks that formed in the tie plates originated at tack weld ends where high stress ranges were introduced as a result of out-of-plane displacements. The cantilever load for which the bracket and tie plate were designed did not cause a significant stress cycle from traffic. The connection of the plate to the longitudinal girders, however, experienced a large secondary unaccounted-for cyclic stress that caused the fatigue cracks. Stress measurements indicated that the most severely stressed plates were near the reactions and that those particular plates had likely cracked after a few years of service.

Cracking of the tie plates did not seriously impair the performance of the structure, since the web connection of the bracket was able to resist the applied loads.

The web brackets had sufficient gap between the web gap and the connection angles so that, even though the tie plates were cracked, it did not adversely impair the performance of the bracket. Very little load was placed on the cantilever in this structure because of its size and the location of the curb. Most traffic did not load the outside stringer a significant amount.

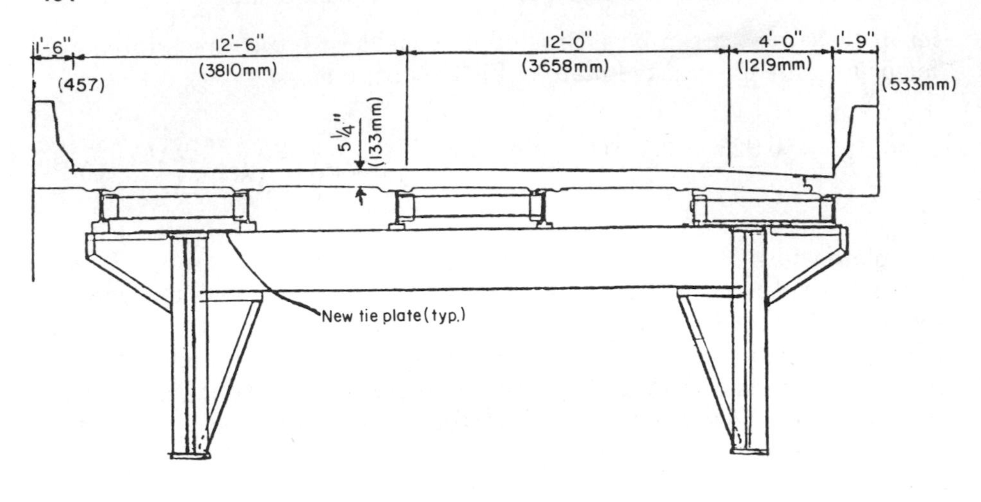

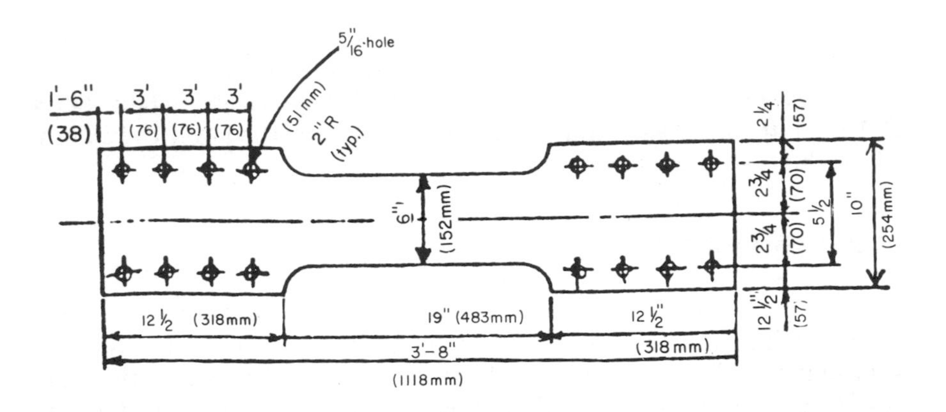

Figure 10.12 Retrofit detail for Lehigh Canal and River Bridge (note the change in tie plate detail and configuration).

10.1.4 Repair and Fracture Control

The tie plates on the Lehigh Canal and River Bridges were replaced with new tie plates during the bridge deck replacement undertaken during 1977 and 1978. This involved removal of all existing cracked and uncracked plates and replacing them with the type of plate shown in Figure 10.12. Since the bolted connection between the tie plate and girder was eliminated, the out-of-plane displacement has been minimized. Field studies confirmed the feasibility of the corrective action [10.1].

10.2 CRACKING OF TIE PLATES OF THE ALLEGHENY RIVER BRIDGE

10.2.1 Description and History of the Bridge

Description of Structure

The Allegheny River Bridge is a four-span continuous beam-girder bridge and a five-span truss bridge over the Allegheny River outside Pittsburgh, Pennsylvania, on the Pennsylvania Turnpike. The bridge deck carries both eastbound and westbound traffic. A view of the steel structure is shown in Figure 10.13, which also shows the cantilever brackets.

The plan and elevation of the end span and the second span of the four-span continuous girder bridge are shown in Figure 10.14. A typical cross section and the original tie plate details for the floor-beam bracket con-

Figure 10.13 Side view of Allegheny River Bridge.

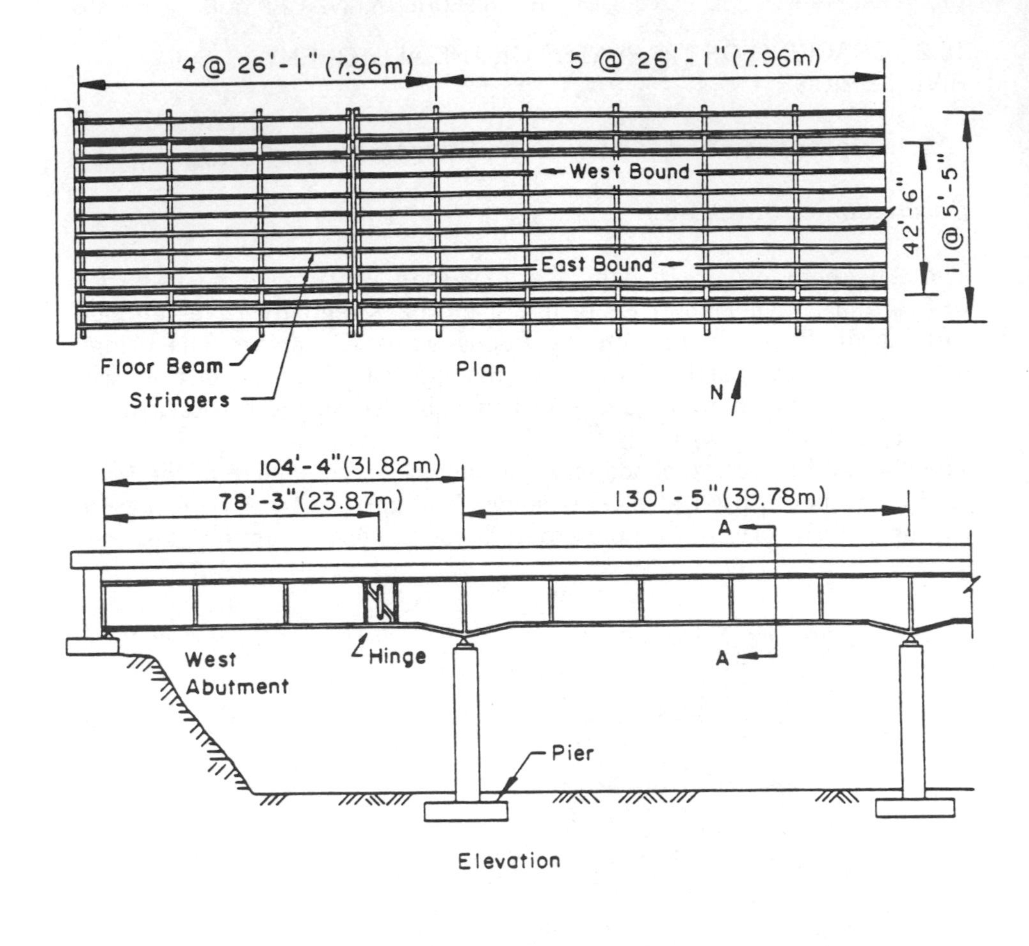

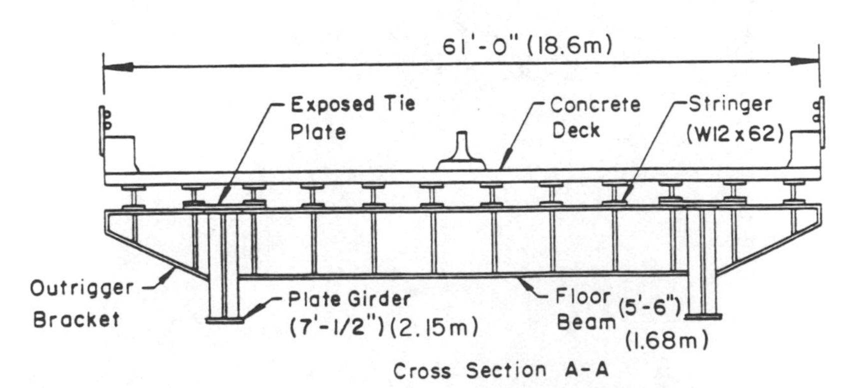

Figure 10.14 Allegheny River Bridge plan, elevation and cross section of end and second span.

nection are shown in Figure 10.15. The end span is 104 ft 4 in. (31.82 m) and the second span is 130 ft 5 in. (39.78 m) long. The end span contains a hinge 78 ft 3 in. (23.87 m) from the west abutment, as shown in Figure 10.14. The longitudinal girders are 7 ft 0.5 in. (2.15 m) deep with a haunch at the piers of 9 ft 1.5 in. (2.78 m).

The reinforced concrete slab in the end span is supported by 12 longitudinal W21 × 62 stringers, resting on the floor beams and outrigger brackets. The floor beams are built-up I-beams (the web being 66 × 0.37 in., or 1676 × 10 mm, and the flanges 6 × 6 × 0.5 in., or 150 × 150 × 12.7 mm) angles with outrigger brackets.

The original tie plate (14 in. × 0.5 in. × 4 ft 8 in., or 355 mm × 127 mm × 1421 mm) which connected the top flange of the floor beam to the outrigger bracket flange is shown in Figure 10.15.

History of Structure and Cracking

The Allegheny River Bridge was constructed in 1952. During the fall of 1971 inspections by the Pennsylvania Turnpike Commission personnel revealed several fatigue cracks in the riveted tie plates [10.4]. In January 1972 all tie plates were again checked for fatigue cracks. The approximate locations and length of cracks in the end and second spans are given in Figure 10.16. It was also observed that some of the rivets connecting the tie plates to the first inboard or to the first outboard stringer failed. These rivets are indicated by darkened circles in Figure 10.16. A typical cracked tie plate is shown in Figure 10.17.

In the spring of 1972, the original cracked tie plates were repaired with groove welds and reinforcement tie plates were added over the top of the repaired plate and bolted into place with high strength bolts.

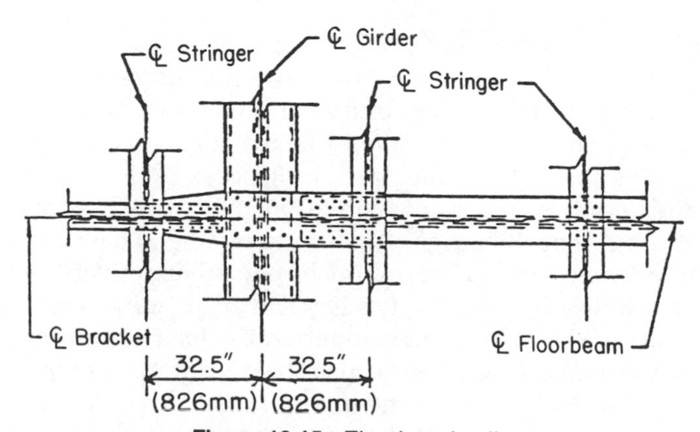

Figure 10.15 Tie plate detail.

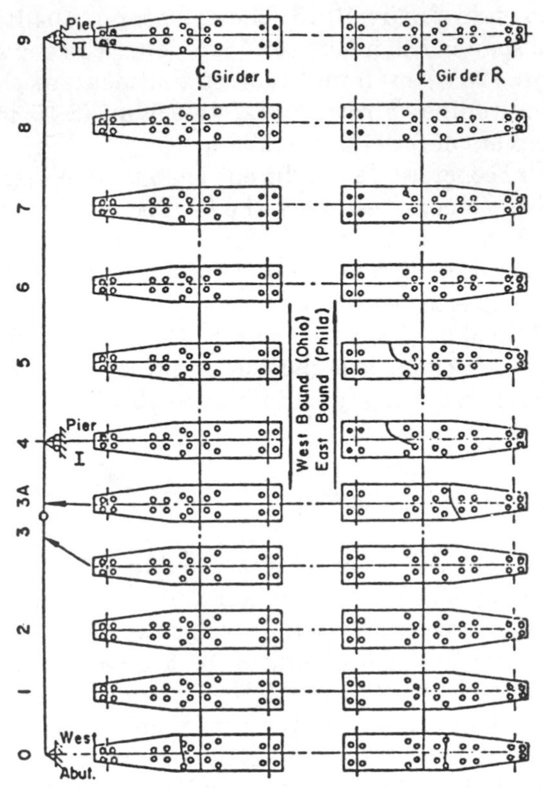

Figure 10.16 Cracked tie plates and loose rivets in test spans; ● loose rivet.

10.2.2 Failure Modes and Analysis

Location of Cracks

The location and approximate lengths of the fatigue cracks in the tie plates in the end and second spans are shown in Figure 10.16. Most of the cracks were at or near the piers and abutments (i.e., near the supports). All cracks originated from rivet holes in the region where the tie plates were connected to the main longitudinal girders.

A detailed study of the traffic conditions and stress history measurements indicated that the fatigue cracks initiated and developed from the tie plate rivet holes due to horizontal in-plane bending of the tie plates. This horizontal bending was induced by the longitudinal displacement of the top flange of the main girder under traffic loads. Similar cracks due to out-of-plane displacements were also observed in the Lehigh Canal and River Bridges on U.S. Route 22, near Allentown, Pennsylvania [10.1]. As shown in Figure 10.16, some of the tie plate rivets (indicated by the solid circles) became ineffective or failed completely.

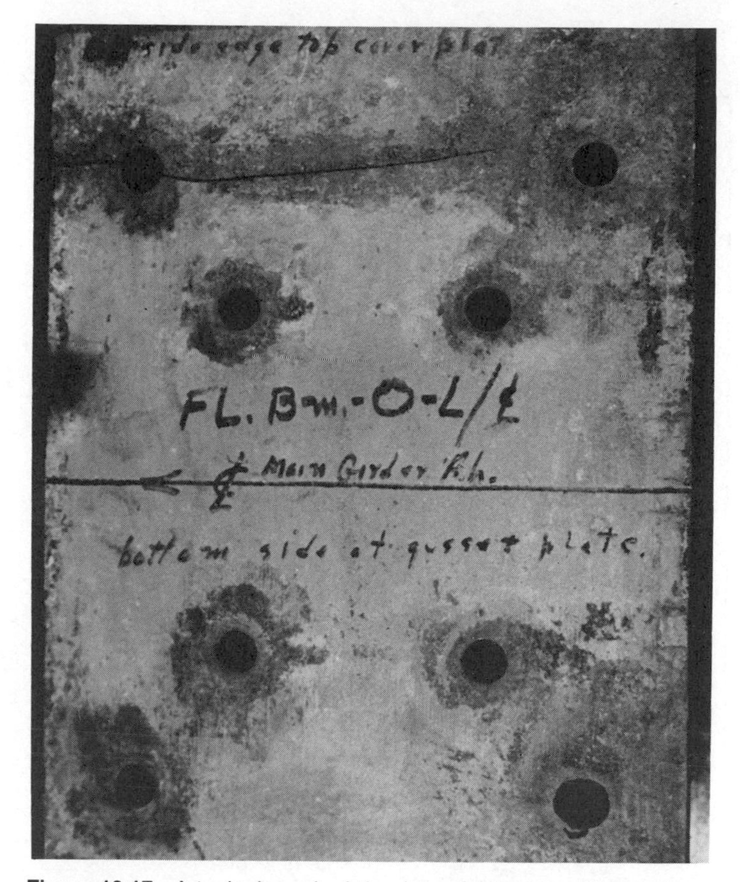

Figure 10.17 A typical cracked tie plate of Allegheny River Bridge.

Cyclic Loads and Stresses

Strain gauges were used to record the strains in the tie plates of the Allegheny River Bridge during the period November 10, 1972, to November 17, 1972 [10.4]. Several stress range histograms were developed from the measurements under traffic. A typical stress range histogram for a tie plate adjacent to pier 1 is shown in Figure 10.18. All tie plates near supports provided maximum stress ranges between 12 and 18 ksi (83 and 124 MPa).

The measured stress distributions in the reinforced and repaired tie plates indicated that the tie plates were subjected to high bending stresses in a horizontal plane, as shown in Figure 10.19. The measured stresses suggested that the horizontal bending was induced by a relative movement between the girder and each end of the tie plate. Bending stresses in tie plates were higher near the abutment and pier which agreed with the cracked tie plate pattern shown in Figure 10.16.

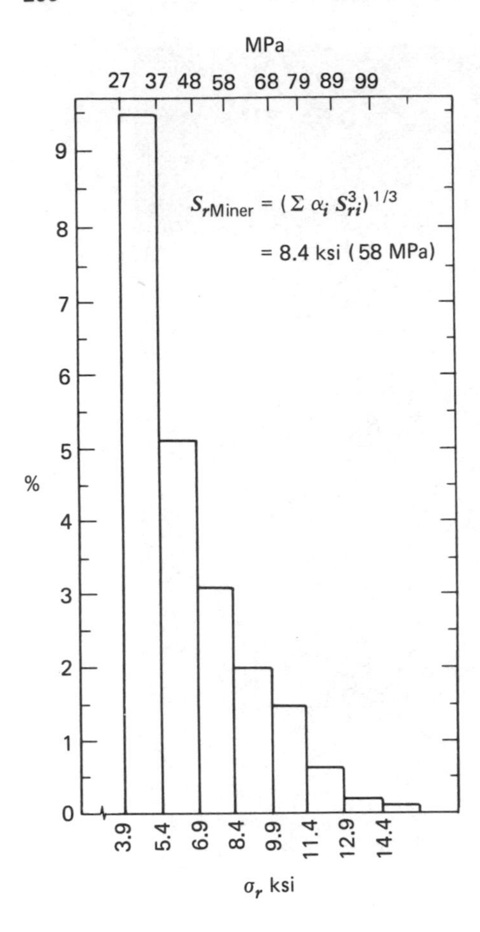

Figure 10.18 Stress histogram for tie plate near pier 1.

Very low live load stresses were observed in the longitudinal girders and the floor beams. The stresses on the main girder caused by a HS20-44 test truck in the outside lane of the bridge provided a maximum live load stress of 2.4 ksi (16.5 MPa) and a stress range of 4.65 ksi (32.04 MPa). Both of these measurements were well below the calculated design live load stress of 9.6 ksi (66.2 MPa).

Traffic counts and gross vehicle weight were also taken into consideration during the in-service testing in 1972. The survey provided a gross vehicle weight distribution comparable to the 1970 FHWA nationwide survey [4.3]. The total volume of truck traffic with weight exceeding 20 k (89 kN) that crossed the bridge during its 20 years of service between 1952 and 1972 was 20.7 million vehicles. Hence each tie plate was subjected to at least 10 million variable stress cycles, assuming an equal number of trucks traveling in each direction. This also assumed that the tie plates on the opposite side of the bridge were not significantly loaded by vehicles on one side of the bridge.

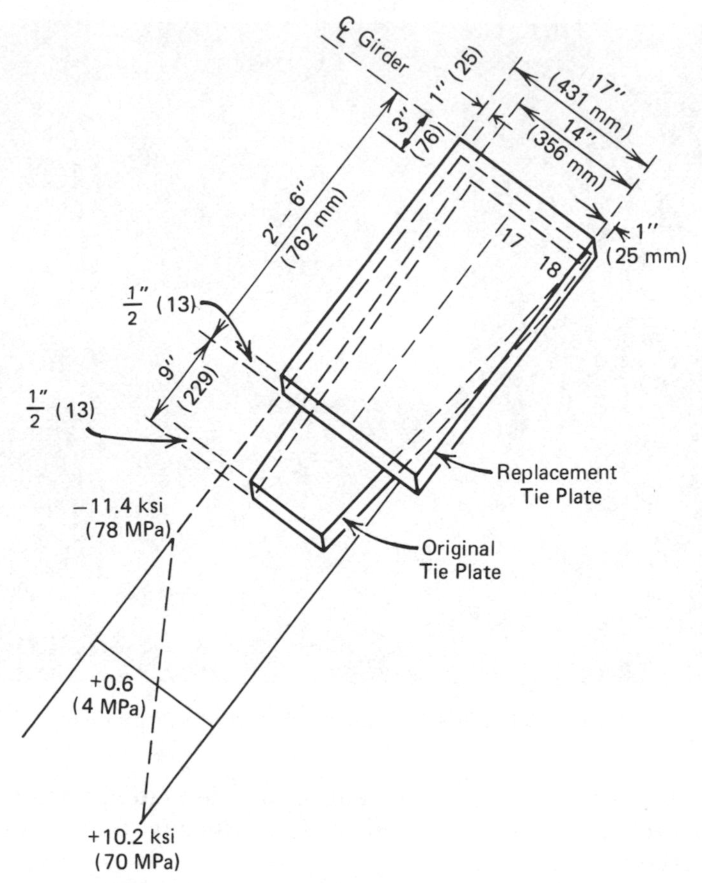

Figure 10.19 Stress distribution in tie plate R-5.

Material Properties and Crack Surfaces

None of the tie plates appeared to exhibit crack instability. The cracks frequently turned and followed a principal tension field path that was not perpendicular to the plate. This can be seen in Figure 10.16. It primarily occurred near pier 1.

One of the tie plates was removed from the structure, and a photograph of it is shown in Figure 10.17. This tie plate was located at the north side of the west abutment. Each of the rivet holes was further examined, and smaller cracks were detected in several of these holes. Figure 10.20 shows one such crack surface after the plate was sawed off and the crack surface exposed. The fatigue crack formed as a semicircular corner crack from one edge of the rivet hole. All of the crack surfaces showed clear evidence of crack growth.

Figure 10.20 Small fatigue crack at edge of rivet hole.

No material tests were carried out on the tie plate. It was fabricated from A7 steel, and the cracking did not appear to have any relationship to the material properties.

The stress distribution in the tie plates suggested the cracks originated at the rivet holes nearest the outside edges of the plates.

Several rivets in the tie plates of span 2 (see Figure 10.16) were cracked because of the in-plane bending of the plate. None of these rivets was available for examination.

Analysis of Cracks

The stress range spectrum at the rivet holes was determined from strain measurements for each of the four tie plate locations. They included the north and south plates at the west abutment and the south plates on each side of pier 1. The effective stress range for these four locations was evaluated as

$$S_{r\text{Miner}} = (\Sigma\ \alpha_i S_{ri}^3)^{1/3} \tag{10.4}$$

The effective stress range varied between 6.6 ksi (45 MPa) and 8.6 ksi (59 MPa). The effective stress range values are all plotted in Figure 10.21 and

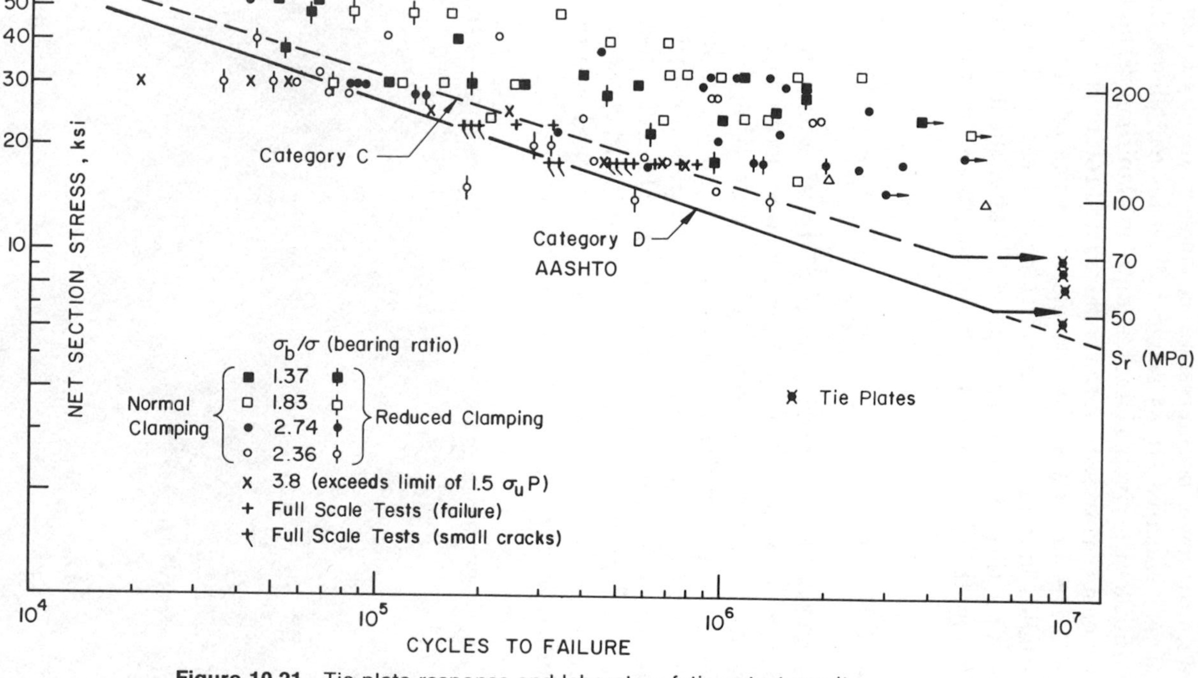

Figure 10.21 Tie-plate response and laboratory fatigue test results.

compared with the lower bound resistance curve (Category D) for riveted connections. All effective stress range values appear above the lower confidence limit for the riveted joint test data at 10 million variable stress cycles. The measured stresses in the repaired and reinforced system may be slightly lower than those that developed in the original structure. However, studies on the Lehigh Canal Bridge demonstrated that increasing the size of the tie plates did not appreciably alter the magnitude of the stress cycle from the displacement-induced conditions that exist at the tie plates.

10.2.3 Conclusions

All fatigue cracks found in the tie plates of the Allegheny River Bridge originated from rivet holes; most of the cracks were located near the piers or supports and were formed where tie plates, connecting the brackets to the floor beams, were riveted to the main longitudinal girders. The fatigue cracks developed due to the out-of-plane displacements which introduced high horizontal bending stresses in the tie plates. The connection of the plate to the longitudinal main girders resulted in cyclic stresses.

Stress relief in several tie plates was accomplished naturally under service conditions when some of the rivets cracked.

Figure 10.22 Reinforcement and original tie plates (bottom view).

10.2.4 Repair and Fracture Control

During the spring of 1972 reinforcement tie plates (17 in. × 0.5 in. × 4 ft 7.94 in., or 431.8 mm × 12.7 mm × 1420.8 mm) were added to the structure. The cracked tie plates were groove weld repaired on site, with reinforcement plates placed over them. The bolt holes in the reinforcement plates were matched to those of original tie plates for ease of installment. The reinforcement plates were high-strength bolted to the original tie plates, girders, floor beams, outrigger brackets and the neighboring stringers, as shown in Figure 10.22.

The stress measurements obtained on the repaired and reinforced tie plates have revealed that displacement-induced cyclic stresses have continued to develop in the tie plates. Hence the basic cause of the cracking has not changed, and it is very likely that the new plates will crack in another 10 or 20 years. Although high-strength preloaded bolts would normally increase the fatigue resistance, this increase will likely be offset because the nuts were tack welded to the plates at the time of installation. This will result in crack initiation sites comparable to the tack welds on the Lehigh River and Canal Bridges tie plates.

REFERENCES

10.1 Fisher, J. W., Yen, B. T., and Daniels, J. H., Fatigue Damage in the Lehigh Canal Bridge from Displacement Induced Secondary Stresses, Transportation Research Record 607, Transportation Research Board, 1977.

10.2 Fisher, J. W., Fatigue Cracking in Bridges from Out-of-Plane Displacements, *Can. J. Civ. Eng.* 5 (1978):542–556.

10.3 Woodward, H. T., and Fisher, J. W., Predictions of Fatigue Failure in Steel Bridges, Fritz Engineering Laboratory Report 386-12(80), Lehigh University, August 1980.

10.4 Marchica, N. V., Stress History Study of the Allegheny River Bridge–Pennsylvania Turnpike, M.S. Thesis, submitted to Lehigh University, Bethlehem, Pa., May 1974.

CHAPTER 11

Transverse Stiffener Web Gaps

Cyclic out-of-plane bending stresses have been introduced into several bridge girders during handling or shipping. This resulted from the relative rotation and displacement in the web gap between the end of the stiffener on the web and the girder flange. The extensiveness of such cracks has depended on the girder size, how the girder was handled in the shop and in transit, what method of transportation was used, the length of the trip, and the degree of cyclic swaying motion in transit.

In at least one case cracking was observed in the fabricating shop as the girders were handled and turned. A similar type of cracking occurred in a large stiffened web plate. Web gaps existed at the intersection of vertical and transverse stiffeners that were not connected to each other. Cracking was also detected as the stiffened web plate was moved and turned in the fabrication shop.

In all of these cases handling and/or shipping created high cyclic stresses in the girder webs at a short gap. Cracks formed at the weld termination of the stiffener in nearly every case. In addition uneven rail elevations and joints are capable of contributing to fatigue cracking because dynamic loads are transmitted to girder support points that are near web gap details.

11.1 FATIGUE ANALYSIS OF CRACKING IN I-90 BRIDGE OVER CONRAIL TRACKS IN CLEVELAND

11.1.1 Description and History of the Bridge

Description of the Structure

The I-90 Bridge carries traffic over the Conrail yard tracks in Cleveland, Ohio. Figure 11.1 shows the elevation of the five-span structure. The east- and westbound structures are not connected and are composed of nine continuous welded built-up steel girders with a noncomposite $8\frac{1}{2}$ in. (215 mm) concrete slab. The steel framing is shown in Figure 11.2 for both bridge structures. A concrete guard wall is placed along the centerline of the two roadways.

The five spans have lengths between 62 ft (19 m) and 140 ft (43 m). All of the longitudinal girders have transverse stiffeners which are cut short of the tension flange by 1 in. (25 mm). The skewed structures have transverse diaphragms placed perpendicular to the main girders which is also shown in Figure 11.2. Lateral bracing is installed between six of the girders.

The structure was designed in accord with the 1965 AASHO specifications and Ohio-DOT supplements.

History of Structure and Cracking

Between May 1 and 3, 1973, during the erection of the steel girders of the eastbound structure, an inspection by Ohio Department of Transportation personnel led to the discovery of cracks in some of the erected girder webs on the eastbound structure near the west abutment [11.1]. All of the steel superstructure for the eastbound bridge was in place, as was the first span of the westbound bridge. Erection was interrupted so that magnetic-particle and dye-penetrant inspections could be carried out [11.2, 11.3]. Two types of cracks were found to exist and are shown schematically in Figure 11.3. Figure 11.4 shows a photograph of the web gap area and the crack detected by dye-penetrant is visibie.

The field examination indicated that cracks had formed at the location identified by solid dots in Figure 11.2.

11.1.2 Failure Modes and Analysis

Location of Cracks

The cracks were only observed in the positive moment region of the girders where the transverse stiffeners were cut short of the tension flange, as illustrated in Figure 11.4. All cracks were located near the

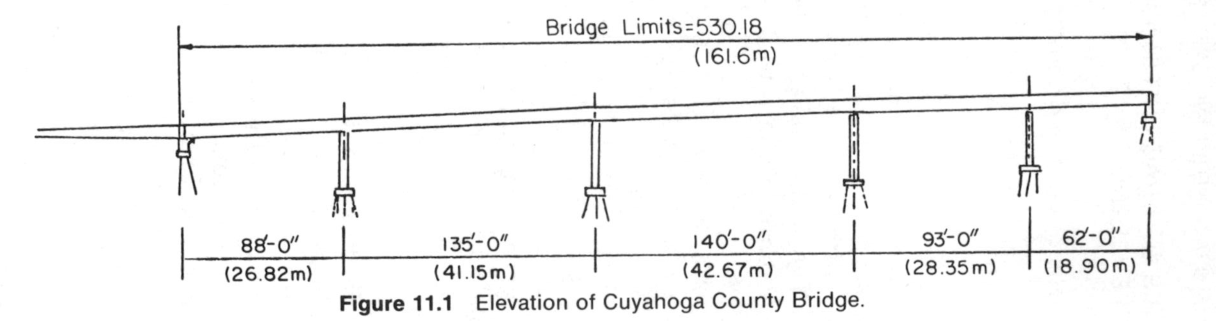

Figure 11.1 Elevation of Cuyahoga County Bridge.

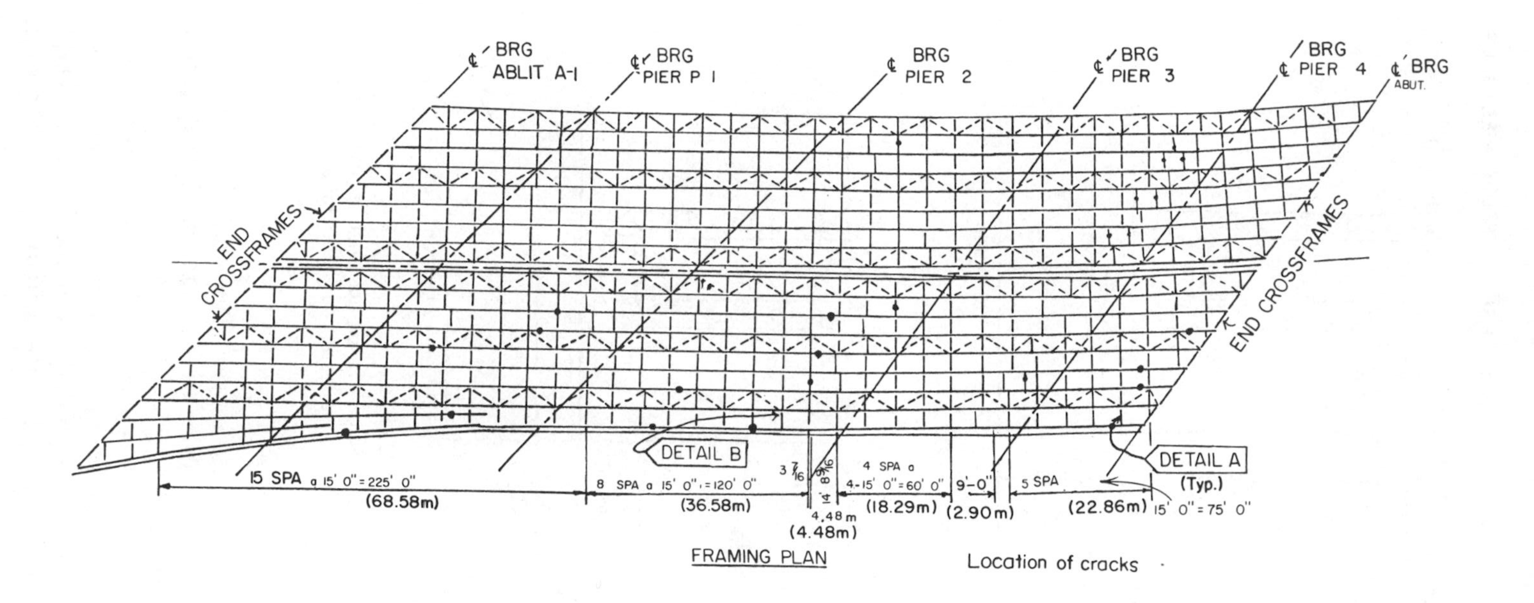

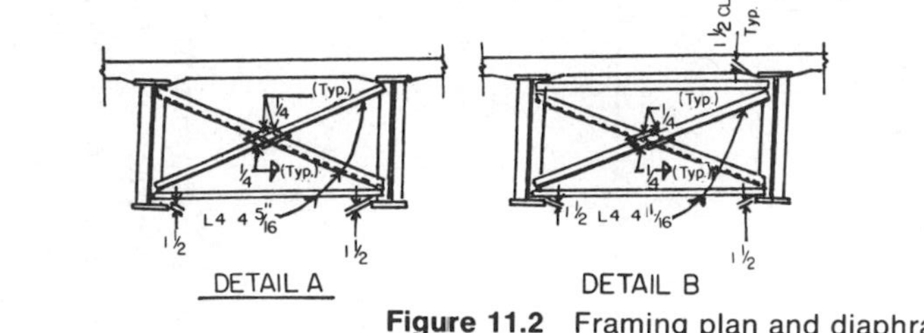

Figure 11.2 Framing plan and diaphragm detail of Cuyahoga County Bridge.

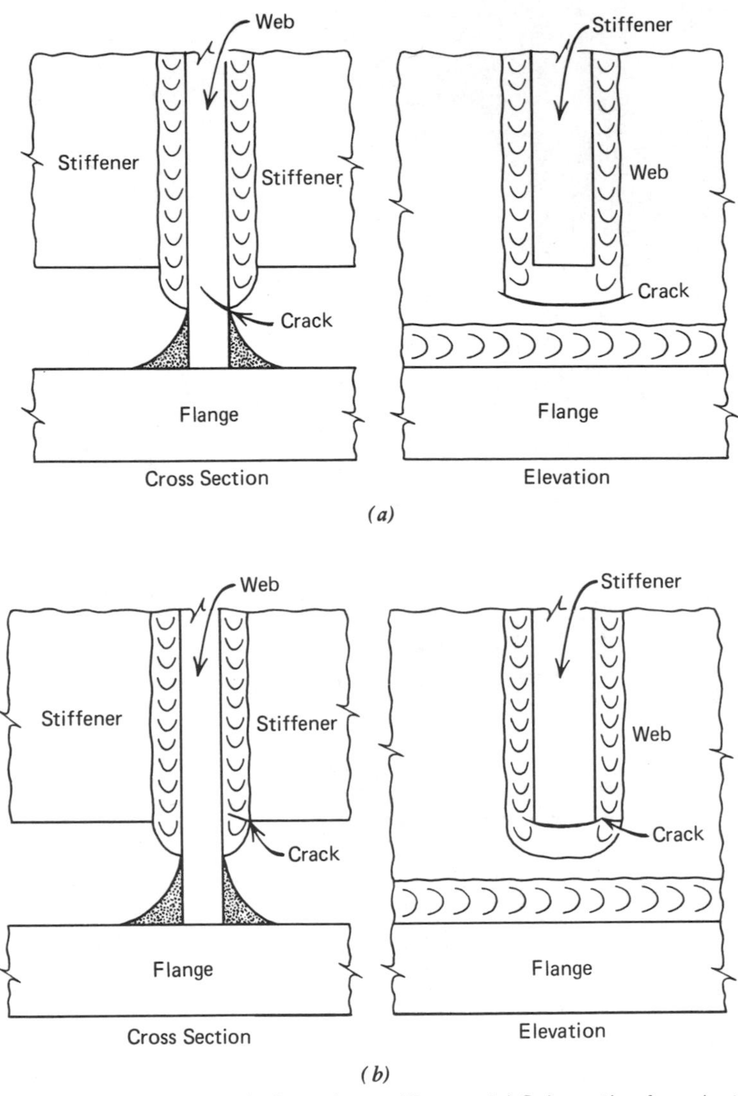

Figure 11.3 Cracks found at end of cut short stiffeners. (*a*) Schematic of crack at end of stiffener welds; (*b*) schematic of crack at end of stiffener into weld (and web).

girder support points during shipping by rail. The transverse stiffeners were attached to both sides of the girder web, and cracks were observed on each side in many instances.

Two types of cracks were detected as was shown schematically in Figure 11.3. One type formed at the termination of the transverse stiffener weld and propagated into the girder web at the weld toe. The second type of crack formed in the web at the end of the stiffener when the weld

Figure 11.4 Photograph of stiffener end crack prior to removing sample core.

extended some distance beyond the end of the stiffener. Sometimes the crack in the weld extended into the web plate.

Cyclic Loads and Stresses

The girders were all fabricated and shipped from Columbus, Ohio, to the bridge site by rail and truck. All girders shipped by rail were arranged on a flatbed rail car in the configuration shown in Figure 11.5. Each girder was supported on the bed of the flatbed rail car at two locations using wooden blocks and tie down rods.

The cracks were only observed in the girders that were shipped to the construction site by rail. There were no cracks found on girders shipped by truck. Furthermore no cracks were found when the web gap was at the top flange, shown as case 2 in Figure 11.5. All cracks were located at stiffeners that were at or adjacent to the wooden support blocks.

Although no stresses were measured in any of the girders during transport, any sway movement would of course be transmitted into the girder web at the web gap shown in Figure 11.6. In the process the girder weight would deform the web in the gap if the flange rests on a block. At the top flange there would be enough slack so that no relative twist of any significant magnitude would be introduced into the girder web.

The vertical tie down rods used to support the girders offer little restraint to lateral movement of the girder sections. Consequently, local

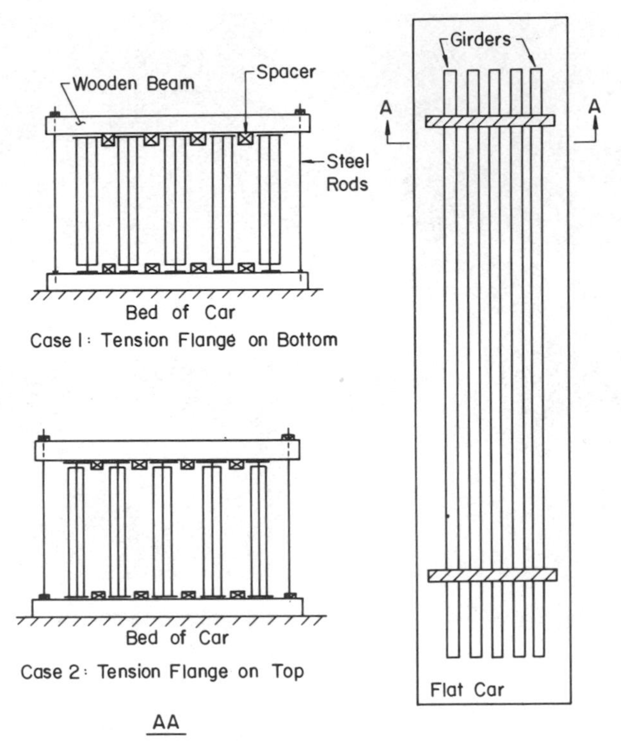

Figure 11.5 Tie down arrangement for moderate-size girders during rail shipment.

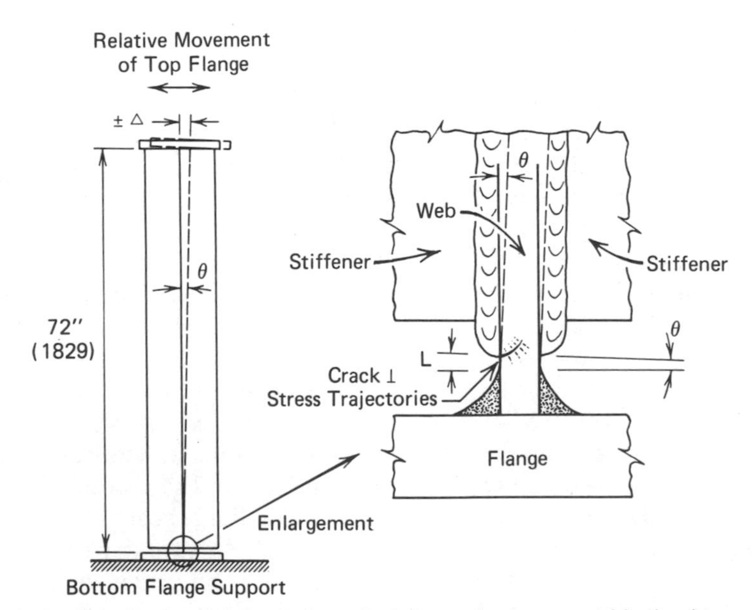

Figure 11.6 Schematic of deformation of girder web at support block when cut short stiffeners are adjacent to block.

twisting of the girder flanges resting on the wooden support blocks could occur from the lateral sway of the girder and tie down system.

Mechanical, Chemical, and Fracture Properties of Material

The girder webs and stiffener were fabricated from A36 steel. A review of the mill reports indicated that all of the material satisfied the mechanical and chemistry requirements of the specifications.

A small section of the web plate was removed from a girder near the abutment in order to make standard Charpy V-notch tests [11.4]. The test results are summarized in Figure 11.7. The average absorbed energy at 40°F (4°C) was 16.8 ft-lb (23 J). Hence the web plate provided a good level of fracture toughness. The web plate satisfied the current impact energy requirement for AASHTO zone 2.

Visual and Fractographic Examination of the Crack Surfaces

The external surface examinations showed that the cracks all formed in the web gap between the cut short end of the transverse stiffeners and the girder flange. Figure 11.4 showed one such crack.

In order to evaluate the causes of the cracking, several core specimens were removed from the stiffener end [11.4]. Figure 11.8 shows the sample removed from the cracked detail shown in Figure 11.4. A crack can be seen in the weld segment that extended beyond the end of the stiffener. A saw cut was made through the center of the stiffener, and the section was polished and etched. Figure 11.9 shows the polished and etched surface of

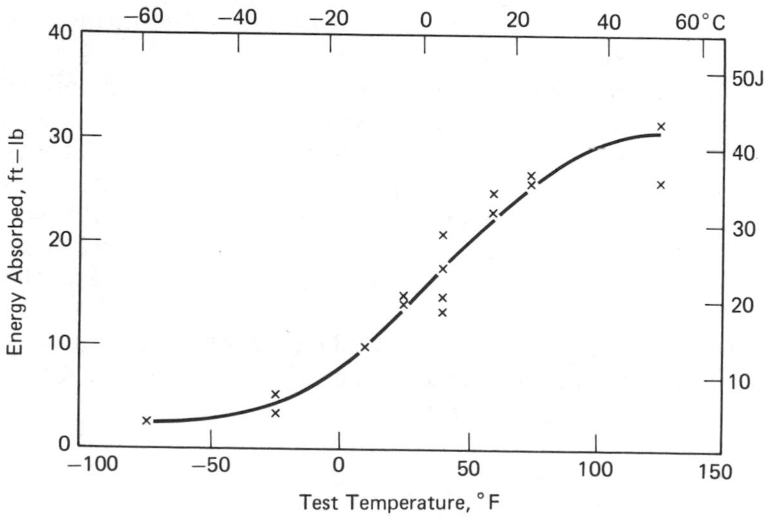

Figure 11.7 Transverse Charpy V-notch data for A36 steel web plate.

Figure 11.8 Sample core showing crack in weld at end of stiffener. Near side.

the web plate, stiffeners, and welds. Fatigue cracks can be seen to exist on each side of the girder web in the web gap. The crack on the left is shown in Figure 11.8, the crack on the right in Figure 11.4. Each crack can be seen to turn after it propagated a short distance into the web. The crack follows the cyclic principal tensile stress direction that results when the web gap is twisted out of plane.

The crack surface was exposed and examined in some detail with the transmission electron microscope. Figure 11.10 shows the resulting fractograph of the crack surface in the web plate at the weld toe. The striation spacing was observed to vary between 5 and 10 microinches. The fatigue crack extended about 0.2 in. (5 mm) into the girder web plate. The striation spacing of 5×10^{-6} to 10^{-5} in. per cycle indicates that relatively high web bending stresses of variable amplitude were responsible for the crack propagation.

The fatigue crack surface in the weld on the other side of the web was covered with much heavier oxides and corrosion product. No fatigue striations were detected on the crack surface.

Failure Analysis

The cyclic stress from relative rotational movement in the web gap is the cause of crack propagation. The rotation introduced in the gap depends on

Figure 11.9 Macro-etched section of saw cut showing cracks into web and weld.

the magnitude of the sway ± Δ shown in Figure 11.6. The cyclic stress range is given by

$$\Delta\sigma = \frac{4EI}{g}\,\Theta_r = \frac{4EI}{g}\left(\frac{2\Delta}{h}\right) \tag{11.1}$$

where g is the web gap $\sim \frac{1}{2}$ in. (12.7 mm), h the girder height = 72 in. (1829 mm), and $I \sim t_w^3/12 = (\frac{7}{16})^3/12 = 0.00698$ in^4. Therefore

$$\Delta\sigma = 46.5\Delta \quad \text{(ksi)} \tag{11.2}$$

The striation spacing can be used to estimate the stress intensity range using the empirical correlation proposed to Bates and Clark [11.5]:

$$\text{Striation spacing} \sim 6\left(\frac{\Delta K}{E}\right)^2 \tag{11.3}$$

The crack surface examination indicated that the striation spacing varied between 5 and 10 microinches (0.13 to 0.26 micrometers). Hence Eq.

Figure 11.10 Fatigue crack striations in web plate, 7500×.

11.3 suggests that ΔK varies between 27 ksi $\sqrt{\text{in.}}$ (30 MPa $\sqrt{\text{m}}$) and 39 ksi $\sqrt{\text{in.}}$ (43 MPa $\sqrt{\text{m}}$). Since the stress gradient effect, F_g, will be small at the crack tip: $a \sim 0.5$ in. (3.8 mm), the stress intensity factor can be estimated as

$$K = \left[\frac{0.923 + 0.199 \left(1 - \sin \dfrac{\pi a}{2t_w}\right)^4}{\cos \dfrac{\pi a}{2t_w}} \right] F_e F_s F_w F_g \sigma \sqrt{\pi a} \tag{11.4}$$

where

$$F_g = \frac{K_{tm}}{1 + 2.776(a/t_w)0.2487} \tag{11.4a}$$

$$K_{tm} = 1.621 \ln \left(\frac{Z}{t_w}\right) + 3.963 \tag{11.4b}$$

$$Z = \text{weld leg size} = \tfrac{1}{4} \text{ in. (6.4 mm)}$$

$$F_e = \frac{1}{E(k)}$$

$$E(k) \simeq \frac{3\pi}{8} + \frac{\pi}{8}\left(\frac{a}{c}\right)^2 = \frac{\pi}{2} \quad c \simeq a \tag{11.4c}$$

$$F_s = 1.211 - 0.186\sqrt{\frac{a}{c}} = 1.025 \tag{11.4d}$$

$$F_w = \left[\frac{2t_w}{\pi a} \tan \frac{\pi a}{2t_w}\right]^{1/2} \tag{11.4e}$$

For $a = 0.15$ in. (3.8 mm), Eq. 11.4 yields

$$K = 0.583\sigma \tag{11.5}$$

When equated to the ΔK values predicted from Eq. 11.3, the reversal bending stress range, $\Delta\sigma$, is between 46 ksi (371 MPa) and 67 ksi (460 MPa).

The number of stress cycles needed to grow the initial flaw $a_i = 0.03$ in. (0.75 mm) to the maximum depth of the crack (0.20 in. (5 mm) can be estimated from Eq. 11.4 and the relationship

$$N = \int_{0.03}^{0.20} \frac{da}{3.6 \times 10^{-10}\,\Delta K^3} \tag{11.6}$$

This results in 29,300 cycles if the effective stress range is taken as the average value equal to 55 ksi (380 MPa). If the stress range were 46 ksi (317 MPa), this would increase the estimated stress cycles to 50,000. Since the girders were shipped from Columbus to Cleveland and are relatively short, they would likely sway with the car frequency. This would result in about 500 cycles per mile. For the 100 mile trip, about 50,000 random load cycles would be expected.

It should also be noted that Eq. 11.4 assumes that the crack will propagate into the web plate perpendicular to the bending stress and plate surface. As is readily apparent in Figure 11.9, the crack tends to turn from this path and move up the web. Equations 11.4 and 11.6 will tend to underestimate the cyclic life because the cyclic stress intensity at the crack tip will be less severe than implied by Eq. 11.4.

11.1.3 Conclusions

The cracks found at the ends of transverse stiffeners cut short of the tension flange on continuous plate girders were fatigue cracks. No brittle fractures were observed. All cracks were caused by large cyclic web bending stresses in the short length of the girder web between the stiffener end and the flange-to-web fillet weld.

The cracks were only found in the positive moment regions of the girders. All cracking was confined to the stiffeners located near the wooden blocks used to support the girders during rail transit to the bridge site. Relatively small out-of-plane sway movements of the girders during transit introduced large bending strains at the ends of the short stiffeners in the girder web near the wooden supports, resulting in fatigue cracks.

The cracks were all parallel to the bending stresses that would occur in the webs during normal service.

11.1.4 Repair and Fracture Control

The following repair procedures were used in order to prevent future crack growth during the service life of the Cuyahoga County I-90 bridges over the Conrail tracks.

Cracks in Stiffener Welds

Cracks that propagated into the weld metal near the stiffener end were repaired by grinding out the crack and the weld metal around it. The ground-out region was checked with dye-penetrant in order to ensure the complete removal of crack tips.

Cracks Propagated into the Web Plate

Cracks that initiated from the weld toe at the end of the stiffener or propagated through the weld into the web were retrofitted by drilling $\frac{7}{16}$ in. (11 mm) holes at the end of each crack and grinding the cylindrical surface of the hole smooth, as shown schematically in Figure 11.11.

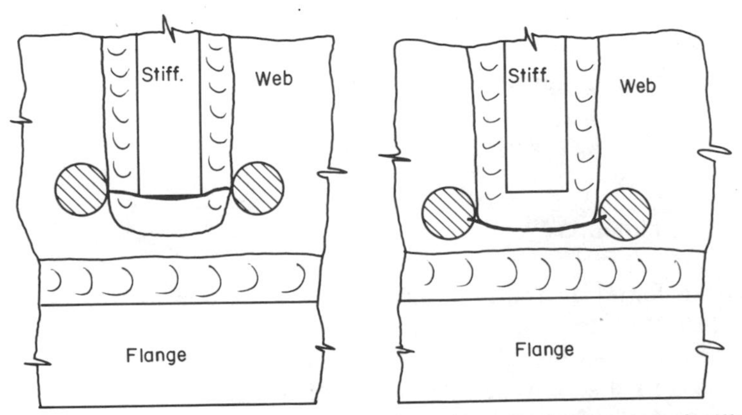

Figure 11.11 Retrofit detail for two types of cracks at the bottom end of stiffeners.

The holes that were cut into the girders to obtain cores for study were ground smooth, and any crack tip that extended beyond the core hole was removed as well.

High-strength bolts were inserted into the holes and tightened to a high preload level.

11.2 FATIGUE ANALYSIS OF I-480 CUYAHOGA RIVER BRIDGE

11.2.1 Description and History of the Bridge

Description of Structure

The Cuyahoga River Bridges are twin structures on Interstate I-480 carrying east- and westbound traffic on separate roadways over the Cuyahoga River, the Ohio Canal, the B&O Railroad, and the abandoned Penn Central Railroad tracks. The twin bridges are located near Independence, Ohio. The total length of each bridge is 4150 ft (1266 m) over 15 spans with a width of 71 ft (22 m). The spans adjacent to the west and east abutments are 220 ft (67 m) and 180 ft (55 m), respectively. Eleven interior spans are 300 ft (92 m) long, and the two remaining spans are 225 ft (69 m) long. The overall view of the bridges is given in Figure 11.12.

Figure 11.12 General view of I-480 Cuyahoga River Bridge (courtesy of Ohio Department of Transportation).

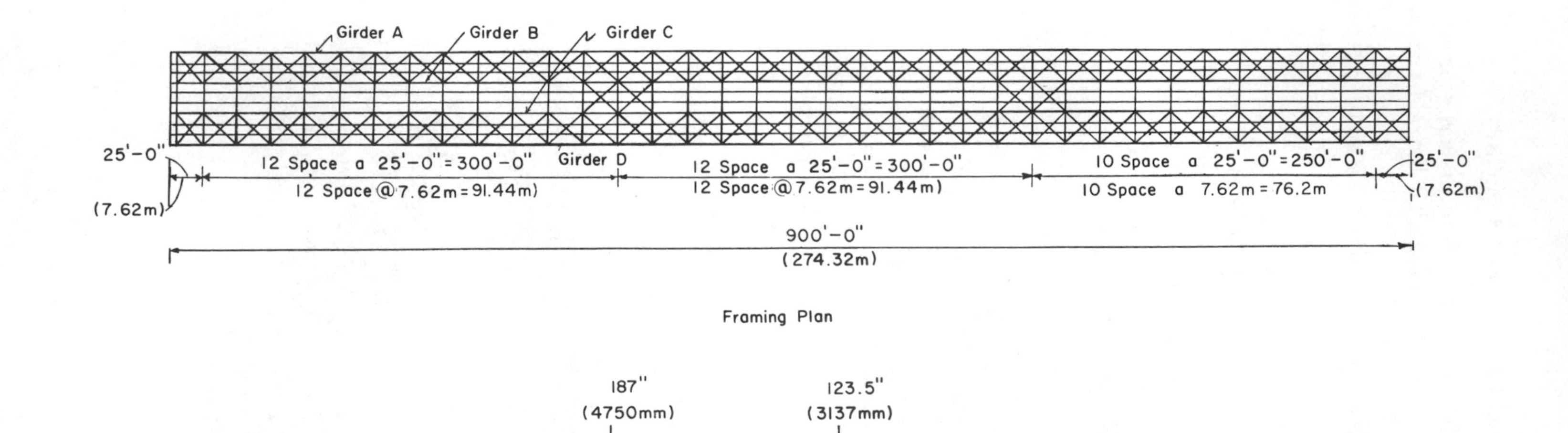

Figure 11.13 Typical main girders.

Each roadway is supported by four continuous main girders with two stringers resting on a transverse floor beam between girders as illustrated in Figures 11.13 and 11.14. The main girders are about 10 ft (3 m) deep and shopwelded. Segments are field assembled with bolted splices. The main girders are haunched at the supports between the west abutment and pier 11, which can be seen in Figure 11.12.

The structure was designed to satisfy the 1965 AASHO Specifications, the 1966–1967 Interim Specifications, and the Ohio "Supplement."

History of Structure and Cracking

In April 1973 while erection of the structural steel was progressing from west to east, inspection of a girder damaged when blown over by wind revealed a crack in the web at the open end of an intermediate stiffener. This led to an examination of other girders on the ground as well as erected on piers which revealed cracks near the bottom end of stiffeners cut short of the tension flange [11.6].

All cracks were found to be due to fatigue [11.7]. They were caused by the relative out-of-plane movement and bending of the short length of web between the end of the stiffeners and the web-to-flange fillet weld.

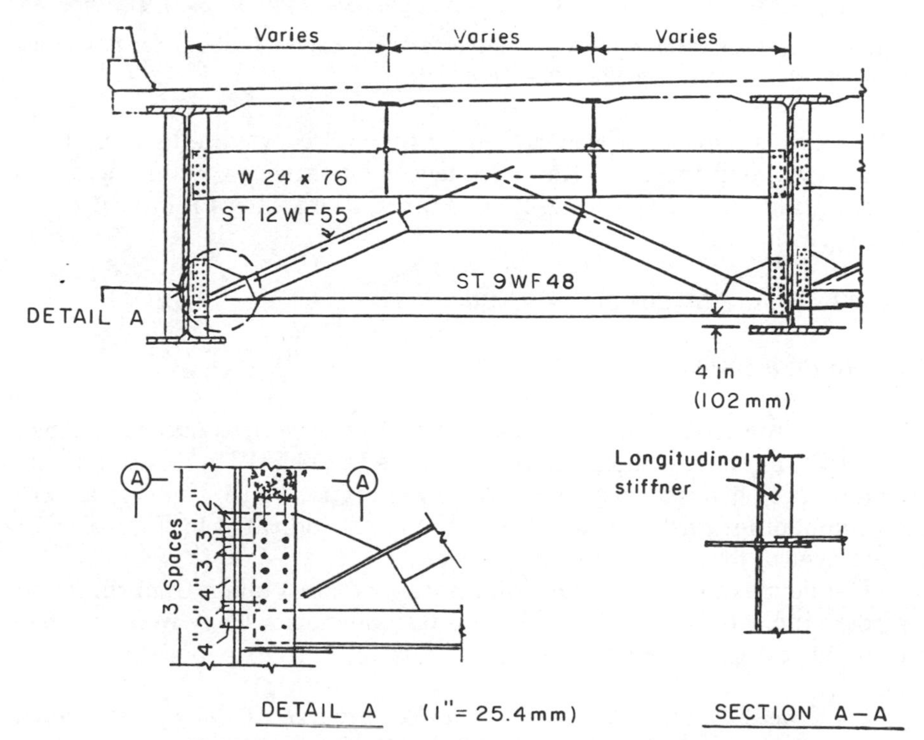

Figure 11.14 Typical cross sections and diaphragms.

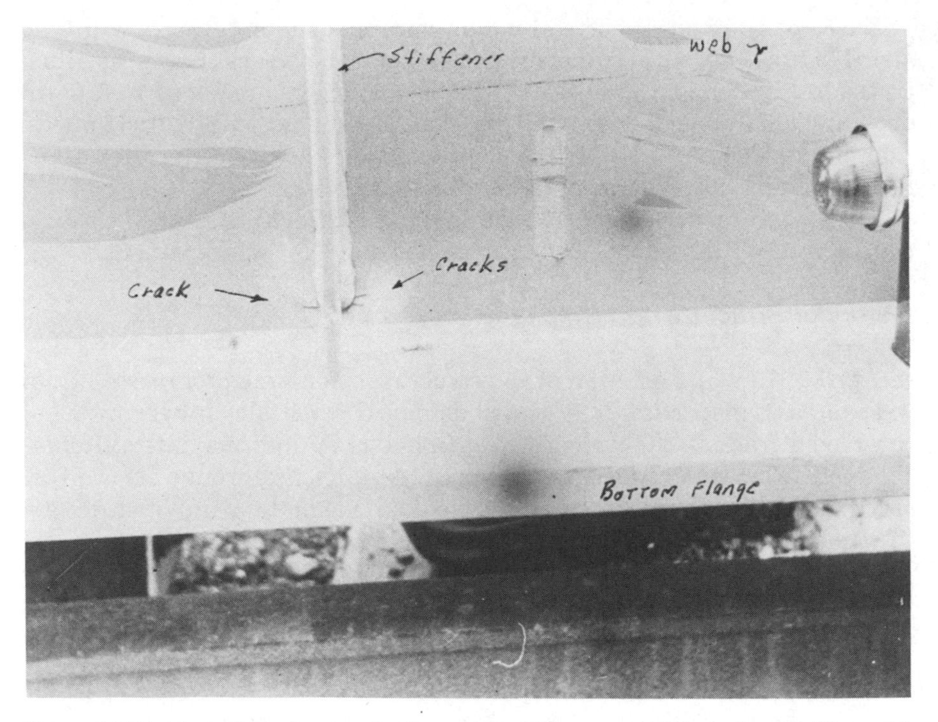

Figure 11.15 Typical cracks at end of transverse stiffener and stiffener-to-web fillet welds (courtesy of Ohio Department of Transportation).

The relative movement was caused by the cyclic swaying motion of the girders while in transit to the construction site and/or wind-induced motion during storage on the ground. Figure 11.15 shows a typical crack found in the web gap.

11.2.2 Failure Modes and Analysis

Location of Cracks

Several girders with the cracked stiffener details were examined on April 17, 1973, at the Cuyahoga River Bridge site [11.7]. The results of the examination of a typical girder is shown in Figure 11.16. The field examination of other girders showed that the crack indicated in Figure 11.16 were typical for all the cracked girders.

The field examination and examination of cores taken from the main girders indicated that all cracks were fatigue cracks. They were grouped into four categories as follows:

1. *Cracks between the stiffener weld and web.* This type is shown schematically in Figure 11.17. Cyclically applied loads caused fatigue cracking and peeled the weld away from the web surface.

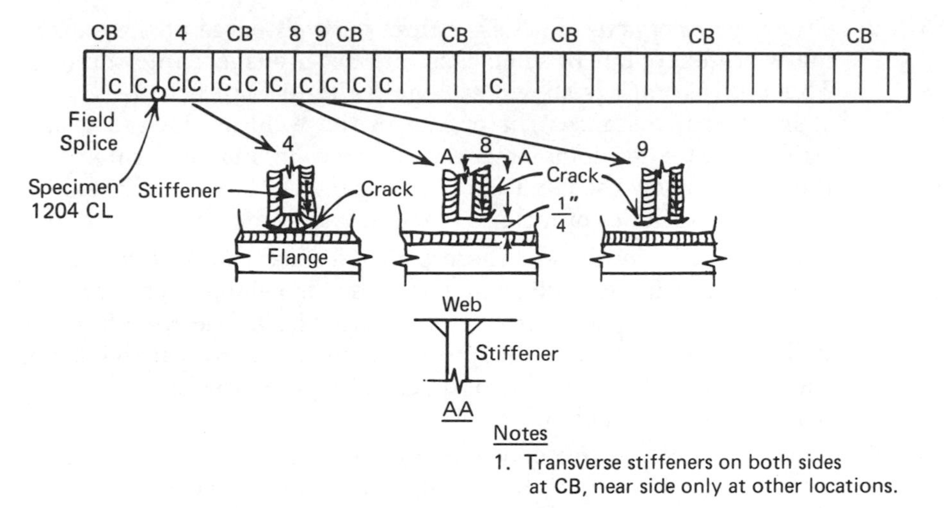

Figure 11.16 Schematic showing cracks in one of the main girders.

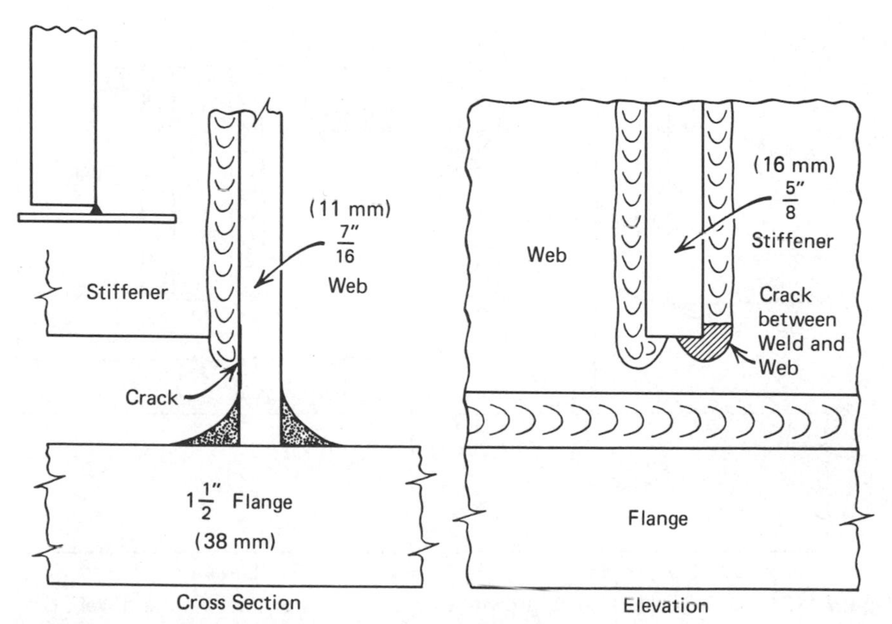

Figure 11.17 Schematic of crack growth between transverse stiffener weld and web. Crack type 1.

2. *Cracks across the weld.* These cracks were observed at the ends of many stiffeners but in some cases did not propagate into the web. This type of crack is shown schematically in Figure 11.18.

   In several instances the cracks in the weld at the end of the stiffener extended completely across the weld and penetrated into the web, as shown schematically in Figure 11.18. All cracks exhibited characteristics of fatigue crack propagation.

3. *Cracks at fillet weld toes.* These cracks formed at the weld toes at the end of stiffeners adjacent to the tension flange. This type of crack is shown schematically in Figure 11.19. The sample cores indicated that the crack had propagated into the web at the end of the fillet weld and turned and moved up the web after a short distance into the web.

4. *Cracks in web surface opposite stiffeners.* Several surface cracks were observed in the web opposite the stiffener, as illustrated in Figure 11.19. One core removed from a girder contained this type of crack. This crack originated on the web surface and did not join the crack propagating into the web from the other surface at the end of the transverse stiffener. The crack initiating from the web surface opposite the stiffener indicates cyclic strains on the web surface near the yield point.

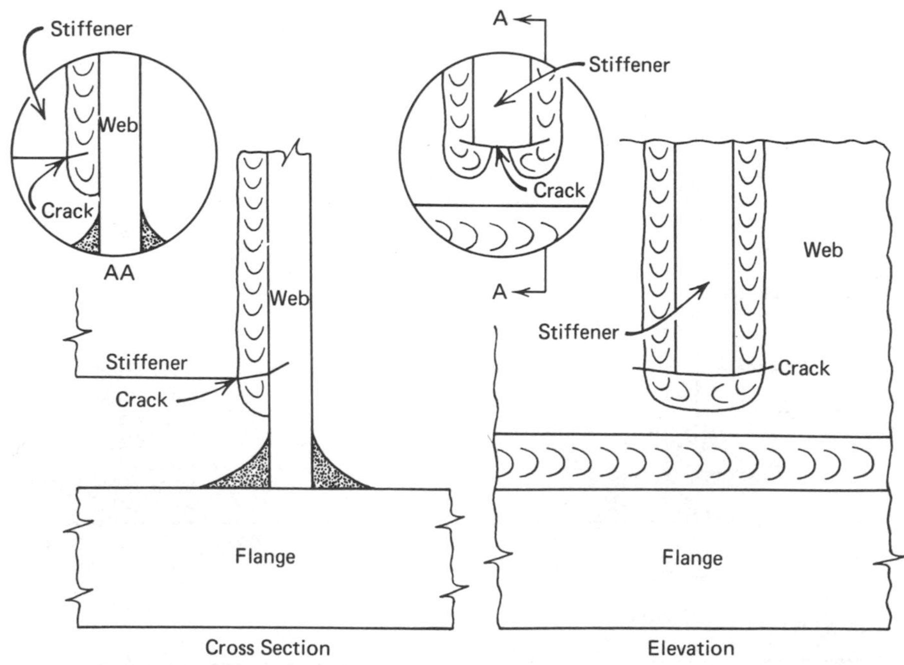

Figure 11.18 Schematic of crack growth at end of stiffener into weld and web. Crack type 2.

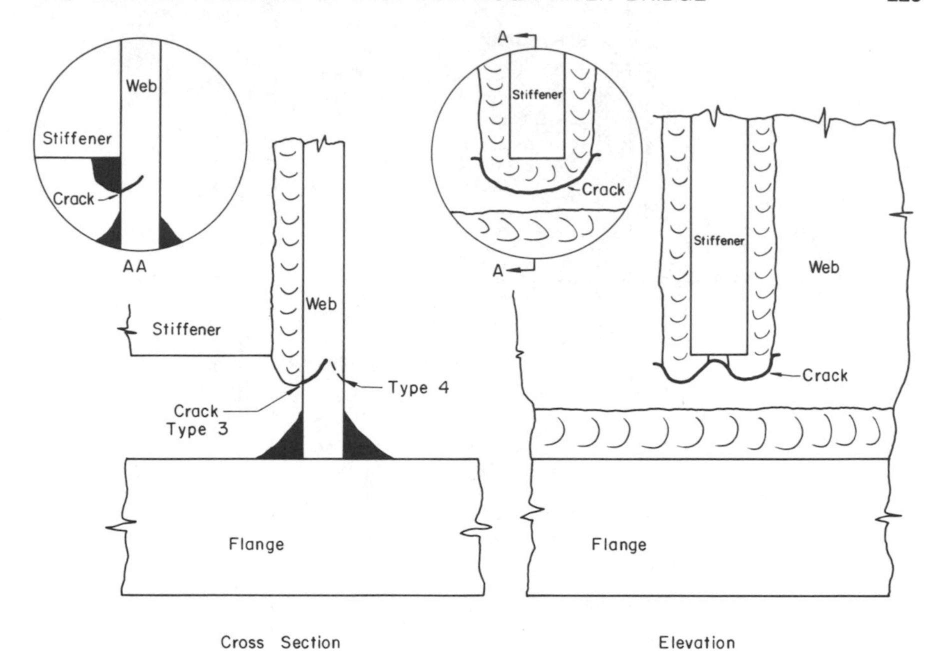

Figure 11.19 Schematic of crack growth into web at end of stiffener welds. Crack types 3 and 4.

Eighty to 90% of the cracks observed in the main girders were type 1. The remaining cracks were distributed among other types described.

All visual evidence from the field inspection and subsequent laboratory examination of the crack surfaces of the sample cores removed from the girders indicated that the cause of crack formation and propagation was cyclic loading.

Cyclic Loads and Stresses

The main girders of the Cuyahoga River Bridge were shipped from Chicago to the bridge site by rail transportation. Because of their size, a single girder was transported on three flatbed cars, as shown in Figure 11.20. All girders were supported at two locations using the tie down arrangement shown in section A–A. The two supports were either located on the first and third flat cars or both supports were placed on the center flat car over its axles. One in. steel rods were used to provide longitudinal restraint to movement. This restraint system permitted relative movement between the top and bottom flange during rail shipment. The relative movement was caused by swaying of the girder and resulted in large cyclic out-of-plane bending stresses in the short length of web between the end of the stiffener and the flange-to-web fillet welds. Similar cyclic stresses may have occurred due to wind-induced swaying motion during

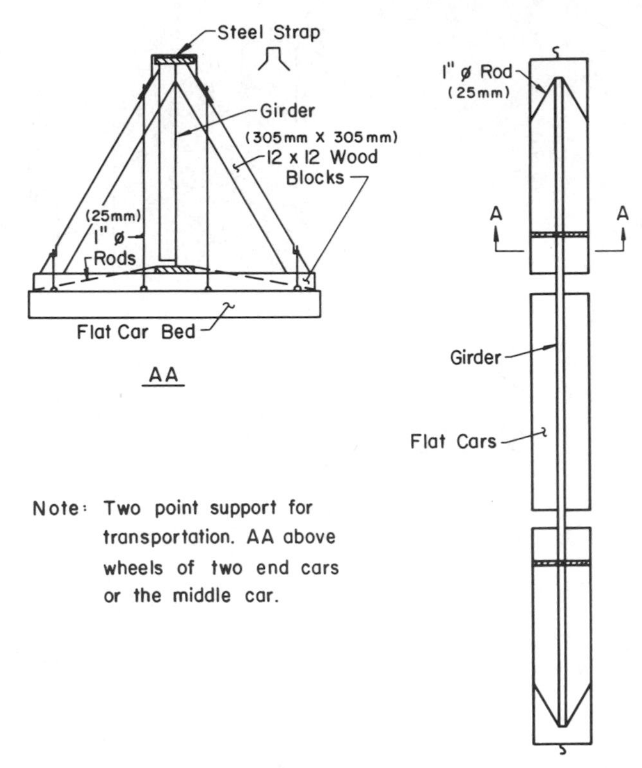

Figure 11.20 Tie down arrangement for girder during rail shipment.

storage. The cyclic motion of the girder cross section is shown schematically in Figure 11.21.

Thus a condition was created that produced large cyclic strains and fatigue crack propagation. The cyclic stress in the web gap from sway can be approximated as

$$\Delta\sigma = \frac{4EI}{g}\,\Theta_r + \frac{4EI}{g}\left(\frac{2\Delta}{h}\right) \tag{11.7}$$

Since $g \approx t_w = \frac{7}{16}$ in. (11 mm), and $h = 120$ in. (3.05 m), the stress range is

$$\Delta\sigma = 1900\Theta_r + 32\Delta$$

where Θ_r is the rotation due to twisting the flange and Δ is the relative out-of-plane movement between the top and bottom flange.

Mechanical, Chemical, and Fracture Properties of Material

The web plates were fabricated from ASTM A588 steel. Mill test reports indicated that the steel met the physical and chemical requirements [11.7].

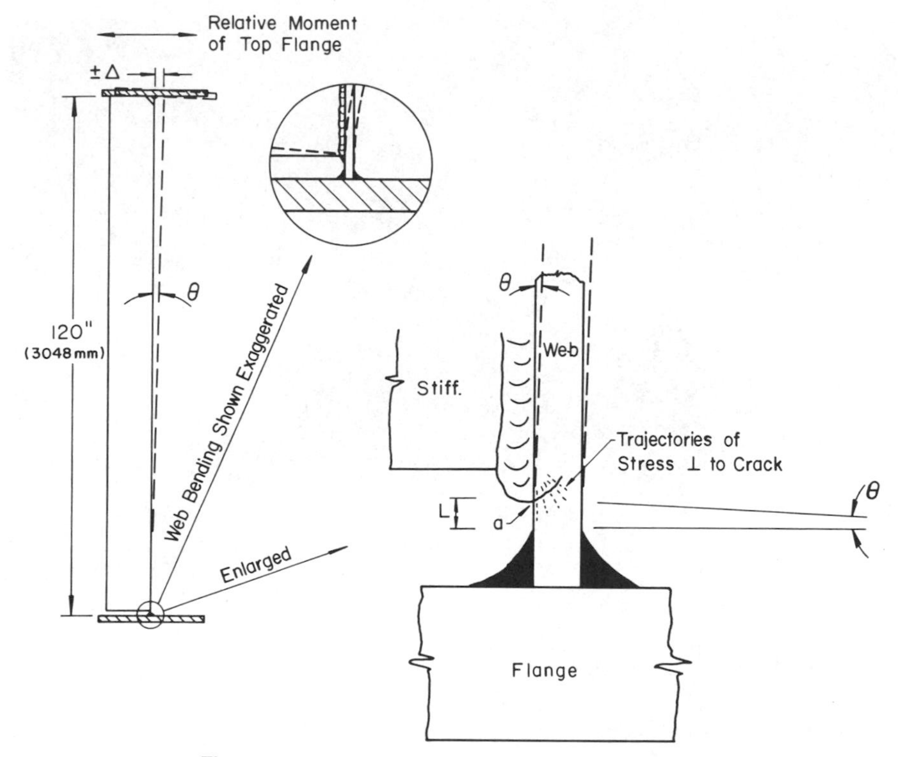

Figure 11.21 Out-of-plane motion of girder web.

Longitudinal Charpy V-notch (CVN) tests at 40°F (4°C) were reported for information from the thickest plate in a heat. The mill reports showed absorbed energy values that varied from a low of 24 ft-lb (33 J) to a high of 130 ft-lb (176 J). Since the web plates were 123 in. (3124 mm) wide, they had been cross-rolled, and hence the CVN values are indicative of the toughness in both the transverse and longitudinal directions.

The cracks in the girder webs did not show any evidence of cleavage or brittle fracture.

Visual and Fractographic Examination of Crack Surfaces

The cores removed from the ends of several cracked stiffeners were examined and verified the four types of cracking described earlier. Figure 11.22 shows a core containing the type 1 and type 2 cracks shown schematically in Figures 11.17 and 11.18. Figure 11.23 shows a polished and etched segment of the core where the path of the crack has extended between the weld and web.

The weld crack shown in Figure 11.22 was also exposed, and Figure 11.24 shows the crack surface. Notice in the figure that the weld crack intersected a smaller crack between the weld and web at the fusion line.

Figure 11.22 Sample core showing crack in weld and crack between weld and web.

A type 3 weld toe crack into the girder web is shown in Figure 11.25. The crack can be seen to originate at the weld toe and propagate into and up the web behind the stiffener.

None of the fracture surfaces exhibited any cleavage nor brittle fracture characteristics.

The fractographic examination of the crack surfaces revealed the presence of fatigue striations which verified the development of fatigue cracks. The transmission electron microscope fractograph in Figure 11.26 shows the fatigue striations from the surface of the cracked weld shown in Figure 11.22. The striations are in the small segment which is adjacent to the stiffener end near the point of crack initiation at a magnification of 19,000×. The striation spacing was between 10 and 15 microinches (0.25 and 0.4 micrometers), and height of the photograph is about 0.0001 in. (0.0025 mm).

The field inspection and visual and fractographic examination of the crack surfaces confirmed that the cause of cracking was cyclic loading. Fatigue cracks developed at the end of the cut short stiffeners as a result of large cyclic out-of-plane bending stresses in the short gap between the stiffener end and the flange-to-web fillet weld. The out-of-plane movements of the top flange relative to the bottom flange due to cyclic swaying

Figure 11.23 Macro-etched section of stiffener weld showing type 1 crack between weld and web.

Figure 11.24 Macro-etched section of weld and web showing crack surface in weld (see Figure 11.22).

Figure 11.25 Macro-etched section showing crack growth into and up web.

motion of the girder occurred during rail transportation to the bridge site and/or wind-induced motions while in storage. The effects of the relative lateral deformation between the two flanges were concentrated in the gaps between the end of the transverse stiffener and web-to-flange fillet weld.

The stiffener welds were terminated very close to the web-to-flange weld. The residual stresses due to stiffener-web welds create high tensile residual stresses between the stiffener end and the flange. The presence of residual tensile stresses in the gap makes each cycle of web bending stress effective in causing crack growth at the ends of the stiffeners. The cracks follow the direction of the cyclic bending stresses and propagate in planes parallel to the longitudinal direction of the girder.

Failure Analysis

The cyclic stresses from the rotation of the flange relative to the stiffened web was approximated from the relationship

$$\Delta\sigma = 1900\Theta_r + 32\Delta \tag{11.8}$$

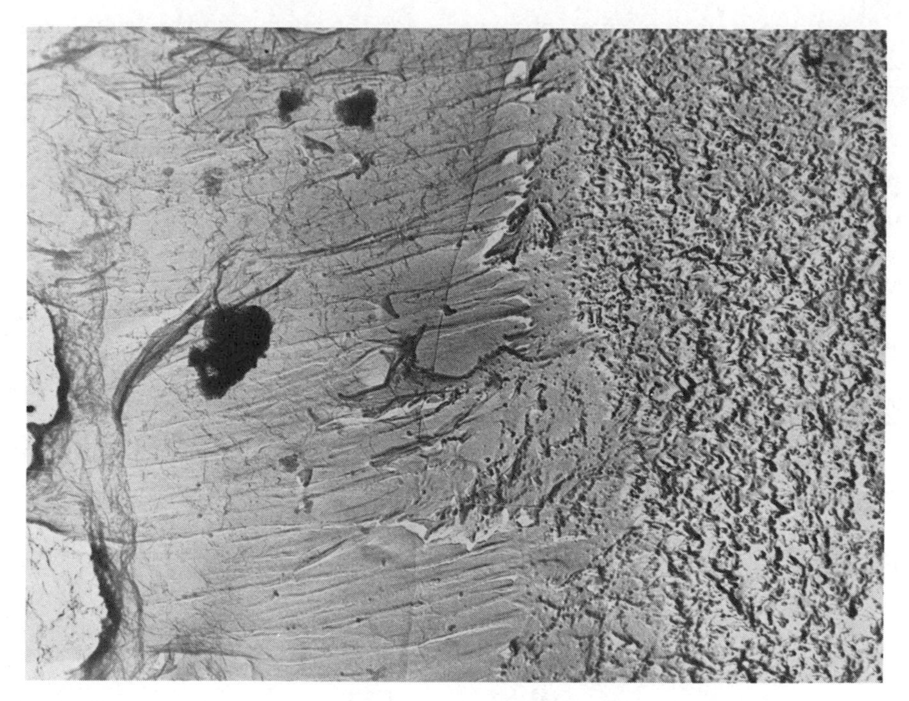

Figure 11.26 Fatigue crack striations at the weld toe crack shown in Figure 11.24. Height of the photo about 0.0001 in., 19,000×.

where Θ_r is the rotation of the flange due to twisting and Δ is the lateral movement between the top and bottom flanges of the girder. Since the girder probably swayed, the rotation and displacement resulted in reversal of the stresses. Therefore the maximum stress range appears to have equaled or exceeded $2\sigma_Y \approx 80$ ksi (552 MPa). Hence Eq. 11.8 indicates that a relative displacement of 2 or 3 in. (50 or 75 mm) would result in cyclic stress or a rotation of 0.04 radians or a combination of these two distortions.

The striation spacing can be used to estimate the stress intensity range, a la the relationship of Bates and Clark [11.5]. For the observed 10 to 15 microinches

$$\text{Striation spacing} = 6\left(\frac{\Delta K}{E}\right)^2 \tag{11.9}$$

This yields ΔK values between 39 ksi $\sqrt{\text{in.}}$ (43 MPa $\sqrt{\text{m}}$) and 47.4 ksi $\sqrt{\text{in.}}$ (52 MPa $\sqrt{\text{m}}$).

Since the striations were observed near the surface of the weld bead, the stress intensity factor for a small crack is about equal to

$$K \approx 1.12\sigma\sqrt{\pi a} \tag{11.10}$$

The crack depth was about 0.05 in. (1.3 mm), and this results in a stress range between 88 ksi (607 MPa) and 106 ksi (737 MPa) when Eq. 11.10 is equated to Eq. 11.9. At such high stress ranges crack propagation is rapid and in the low cycle fatigue region.

Crack propagation at the weld toe crack shown in Figure 11.25 can be evaluated by the relationship

$$K = \left[\frac{0.923 + 0.199\left(1 - \sin\dfrac{\pi a}{2t_w}\right)^4}{\cos\dfrac{\pi a}{2t_w}}\right] F_e F_s F_w F_g \sigma\sqrt{\pi a} \tag{11.11}$$

where F_e, F_s, F_w, and F_g are defined by Eqs. 11.4. Assuming the effective stress range for crack growth between the initial defect and the final crack size is 88 ksi (607 MPa), the number of cycles was estimated from

$$N = \int_{0.03}^{0.22=t_w/2} \frac{da}{3.6 \times 10^{-10}\,\Delta K^3} \tag{11.12}$$

This results in 7300 cycles of stress reversal. The period of sway for 160 ft (48.8 m) long girder segments supported on the rail cars would be much slower than the response of the smaller girders. Hence fewer stress cycles per mile would occur, although they were shipped over a greater distance (Chicago to Cleveland).

The predicted number of cycles is a conservative estimate because Eq. 11.11 assumes that a crack propagates normal to the plate surface. As the crack tip turns and moves up the web, the stress intensity factor will be less severe than implied by Eq. 11.11.

11.2.3 Conclusions

The cracks at the end of stiffener welds in the main girders of the Cuyahoga River Bridge were fatigue cracks created by cyclic stresses. No cleavage or brittle crack propagation was observed. All cracks were caused by large cyclic out-of-plane bending stresses in the gap between stiffener ends and the flange due to the swaying motion of the girder during rail transit to the site and wind-induced motions during storage.

The cracks are primarily parallel to the longitudinal direction of the girder and to the nominal bending stresses. The high toughness of web material and the orientation of fatigue cracks indicate that there is no

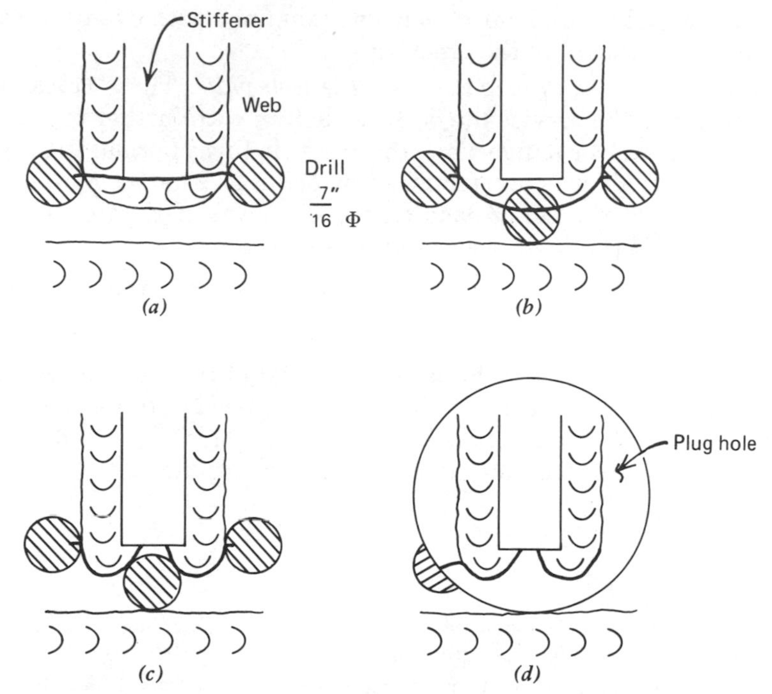

Figure 11.27 Schematic for suggested retrofit and repair procedures. (*a*) Cracks in weld; (*b* and *c*) cracks in web at weld toe; (*d*) cracks at locations where core was removed.

danger of sudden brittle fracture of the main girders. Construction and erection loads should have no significant effect on the cracks. Hence it was concluded that repairs could be made while erection and construction continued.

11.2.4 Repair and Fracture Control

In order to prevent possible subsequent crack growth during the service life of the Cuyahoga River Bridge, the following repair procedures were used.

1. *Cracks between stiffener welds and webs*. Since these cracks were parallel to the stress field, they were not repaired. The short length of the weld which cracked from the web at the end of the stiffener has no significant influence on the performance of the web stiffeners, and any crack there will not propagate under subsequent traffic loads.
2. *Cracks in the stiffener welds*. The cracks that propagated into the weld near the stiffener end were removed by grinding out the

crack. The ground-out region was examined with dye-penetrant to ensure removal of the crack tip.

3. and 4. *Cracks propagated into the web plate.* These cracks could influence the service life of the structure with further propagation into the web or flange when the crack tip turns normal to the stress field. Repair was accomplished by drilling $\frac{7}{16}$ in. (11 mm) holes at the end of the tip of each crack, as shown in Figure 11.27. The surface of the holes were ground smooth.

Sample Core Holes

The holes cut into the web to permit removal of the sample core eliminated most of the crack. Only the crack tips needed to be removed by grinding or drilling holes, as shown in Figure 11.27. These holes were ground smooth, painted, and filled with a rubber plug.

REFERENCES

11.1 Ohio-DOT Internal Memo dated May 7, 1973, from J. D. Jones, Shop Supervisory Engineer, to R. B. Pfeifer, Engineer of Bridges.

11.2 Ohio-DOT Interoffice Communication dated May 11, 1973, from R. Van Horn, Structural Steel Engineer, to R. B. Pfeifer, Engineer of Bridges.

11.3 Fort Pitt Bridge Co., Canonsburg, Pa., Memo dated June 13, 1973, from R. Osmond to W. H. Wallhauser (on "Field Inspection of Open-Ended Stiffeners").

11.4 Fisher, J. W., and Pense, A. W., Report on Cracked Stiffener Details on Girders for Cuyahoga County Project Bridge No. CUY-90-1365 L&R (Prepared for the C. E. Morris Co. and Fort Pitt Bridge Works Co.), Fisher, Fang and Associates, Engineering Consultants, Bethlehem, Pa. 18015, June 1973 (limited distribution).

11.5 Bates, R. C., and Clark, W. G., Fractography and Fracture Mechanics, *Trans. Amer. Soc. Metals* 62 (June 1967):380–390.

11.6 Ohio-DOT Interoffice Communication dated April 4, 1973, from J. D. Jones to B. Pfeifer.

11.7 Fisher, J. W., Yen, B. T., and Pense, A. W., Report on Structural and Fractographic Investigation of Cracked Stiffener Details on Girders for Bridge over Cuyahoga River, Fisher, Fang and Associates, Bethlehem, Pa., June 1973 (limited distribution).

CHAPTER 12

Floor-Beam Connection Plates

One of the most common sources of fatigue cracks during the past decade (1970s) is the cracking in the web gaps at the ends of floor-beam connection plates. These plates have been welded to the web, and often the compression flange of the main longitudinal girders, but have not been attached to the tension flange. At least 28 bridge structures have developed cracks at these connections.

Distortion of the transverse connection plate web gap in the longitudinal girders occurs as the floor beam deforms. The floor-beam end rotation caused by bending of the floor beam and the relative end deflection creates large cyclic out-of-plane stresses that lead to rapid development of fatigue cracks when the structure is on a high volume road. Even less heavily traveled bridges crack with time.

Two examples of this type of cracking are reviewed in this section. The Poplar Street Bridge was one of the first bridge structures where cracking was detected in the web gap at floor-beam connection plates. After discovery of similar web cracks in several floor-beam–girder bridges in 1978, Iowa-DOT conducted a detailed survey of 39 structures with comparable details. The Polk County Bridge near Des Moines is also reviewed and is typical of the type of structure used in Iowa. The cracking that developed was similar to the cracking in the Poplar Street structure; however, the retrofit procedures used to correct the condition were different. Twenty-one of the 39 structures in Iowa were found to have some evidence of cracking by 1982.

Comparable cracks have been found in at least six states with this type of bridge and general detail at the floor-beam girder connections.

12.1 FATIGUE CRACKING OF THE POPLAR STREET BRIDGE APPROACHES

12.1.1 Description and History of the Bridge

Description of Structure

The Poplar Street Approach Bridges are located on the east bank of the Mississippi River in East St. Louis, Illinois. The complex is one of the largest and busiest interchanges in the state of Illinois. It serves as a focal point for Interstate Highways I-55, I-70, and U.S. Route 40 which join and cross the Mississippi River. It starts at the Poplar Street Bridge over the Mississippi River on the west and ends northeast of Broadway in East St. Louis, as shown in Figure 12.1. It consists of several ramps and viaducts consisting of multispan continuous two girder–floor-beam bridges, as shown in Figure 12.2.

The majority of the two-girder I-beam bridges in the complex are on horizontal curves with approximately 1800 ft (550 m) radii. The torsional and side-sway (transverse) rigidity of the system are provided by transverse floor beams which are connected to the two main girders. The floor beams support longitudinal stringers and the reinforced concrete deck, as shown in Figure 12.3. The girder webs were generally $\frac{1}{2}$ in. (12.7 mm)

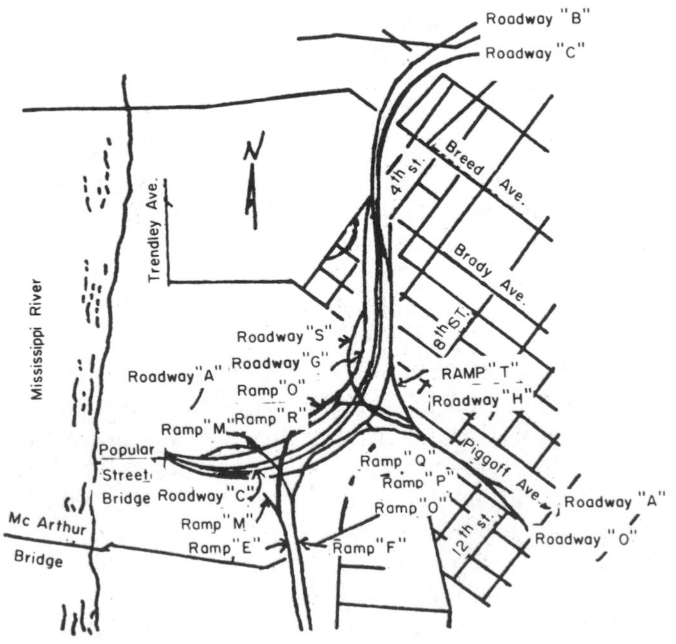

Figure 12.1 Layout of the Poplar Street Bridge approaches.

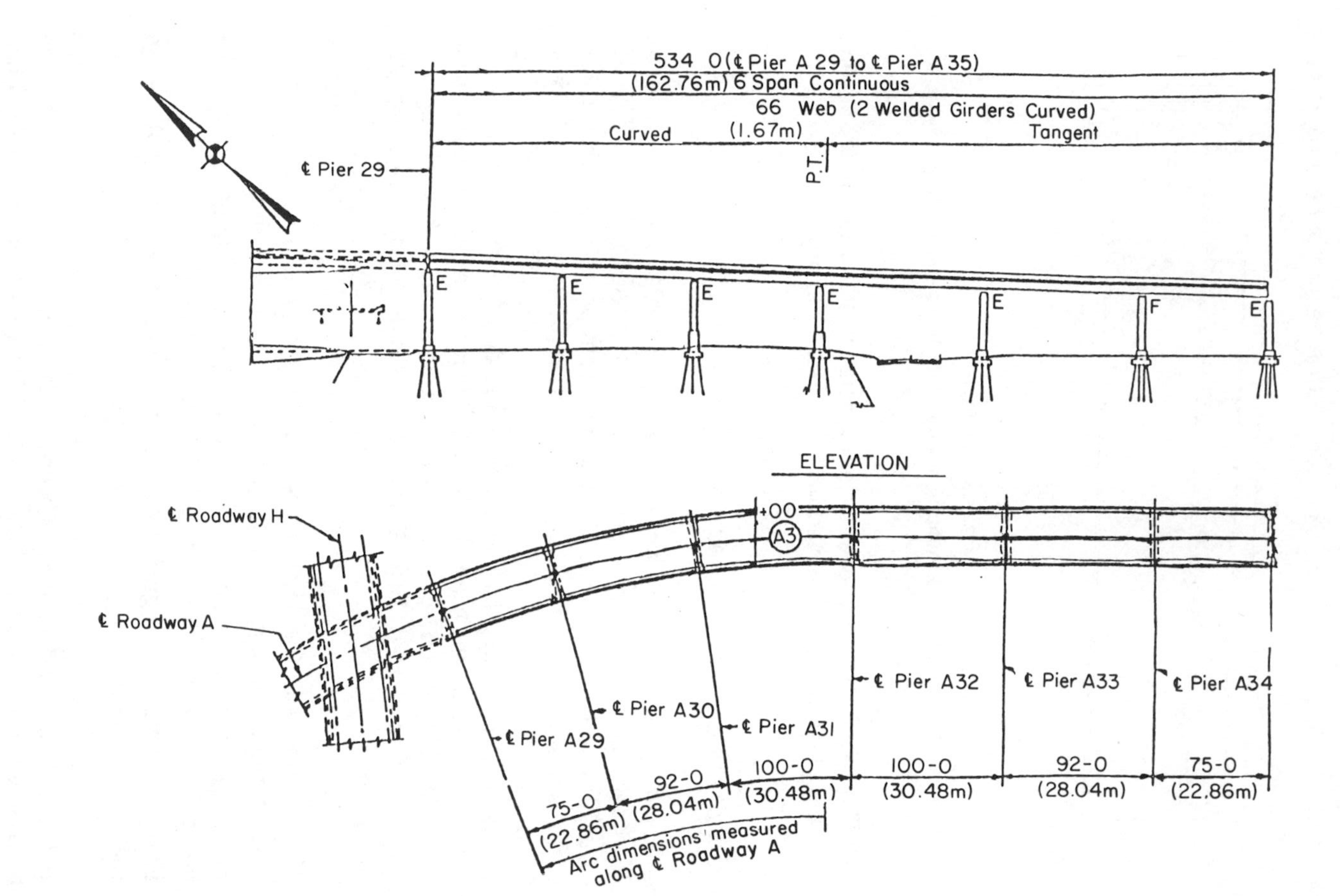

Figure 12.2 Typical plan and elevation of section of approach structure.

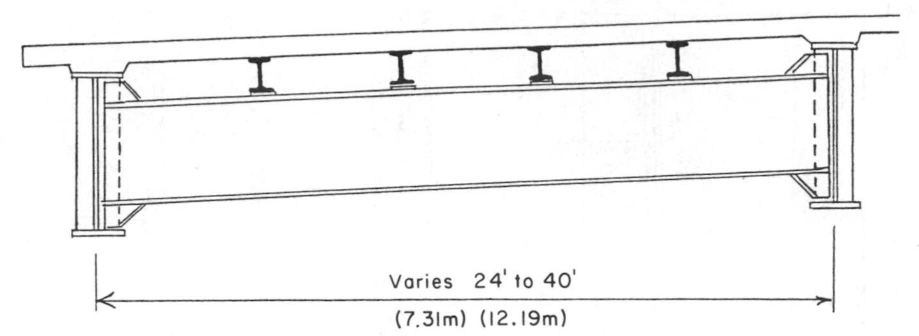

Figure 12.3 Typical cross section showing floor-beam–girder connection.

thick. The gap between the web-flange fillet welds and the transverse connection plate weld end varied between $\frac{1}{2}$ in. (12.7 mm) and 1 in. (25.4 mm).

History of the Bridge and Cracking

The Poplar Street Bridges were designed in 1964. The design of the welded details and transverse stiffeners conformed to the AASHO and state of Illinois design specifications and current practice in effect in 1964 and throughout the design period. The structures were built between 1967 and 1971.

In late 1973 the complex was inspected in depth for the first time by a bridge inspection team from the Illinois Department of Transportation. During this inspection several types of cracks and web buckling were discovered and reported [12.1]. The cracks consisted of the following types of conditions:

Fatigue cracks in the bottom "web gap" region of main girders at floor-beam connection plates,

Fatigue cracks in the upper "web gap" region of main girders at floor-beam connection plates adjacent to end-bearing stiffeners,

Fatigue cracks in the girder webs at the ends of bearing stiffeners of the main girders.

The latter two types of cracks were located at the ends of the continuous main girders. In addition to these cracks there were instances of web buckling at the girder end supports a few inches above the bottom flange.

Additional inspections and field measurements in 1975 showed that cracks also existed in the upper web gap of the main girders in the negative moment regions [12.2].

12.1.2 Failure Modes and Analysis

Location of Cracks

The fatigue cracks were located near the abutment end supports or adjacent to the interior piers of the bridges in the complex. These cracks can be grouped into three general types as follows [12.2, 12.3, 12.4].

1. Fatigue cracks in the girder web near the support in the gap between the lower end of the floor-beam–main girder connecting plate and the bottom flange of the main girder, as illustrated in Figure 12.4.
2. Fatigue cracks in the web in the region between the top end of the floor-beam–girder connecting plate and the top flange of the main girder, as illustrated in Figure 12.5.
3. Fatigue cracks at the ends of bearing stiffeners which were also used as floor-beam–girder connection plates.

The type 1 and 3 web cracks were at the end supports of the main girders. The type 2 cracks were only observed at the upper end of the floor-beam connecting plates which were located in the negative moment regions. There were no web cracks detected in the positive moment regions of any span.

The type 1 web cracks were first observed near the end of the girder under the end floor-beam connecting plate which was positioned 7 in.

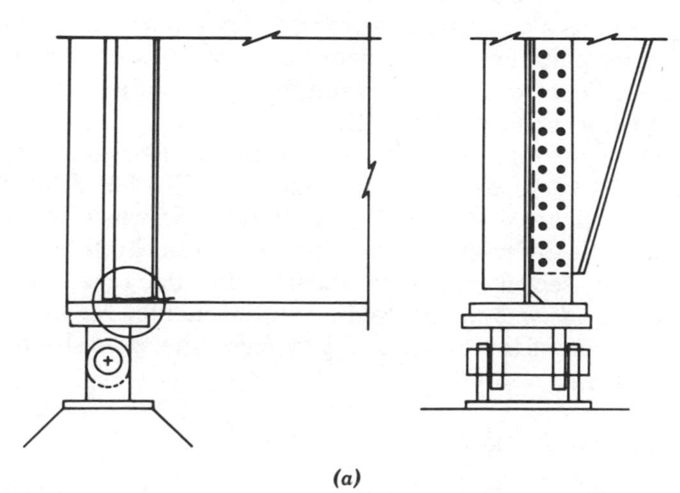

(*a*)

Figure 12.4 Cracks in web near end support. (*a*) Schematic of detail.

Figure 12.4 Cracks in web near end support. (*b*) Crack at end reaction.

(180 mm) toward the center of the span from the bearing stiffeners, as shown in Figure 12.4. Most of these cracks started near the lower end of the vertical connection plate and extended in both directions along the web-flange weld, as shown in Figure 12.4. The longest crack was 19 in. (48 cm) long. Occasionally two cracks were observed at the same location one immediately under the end of the connecting plate, and the other along the web-to-flange weld. The type 2 cracks in the upper gap regions were only found in the negative moment areas (see Figure 12.5) where the floor-beam connection plate was not welded to the top tension flange.

The type 3 cracks occurred at bearing stiffeners which were also used as a connecting plate for the end floor beam at a few end supports or at interior piers. These cracks were usually at the top end of the bearing stiffener which was not welded to the top flange.

In addition to the fatigue cracks, web buckling was also observed at the ends of several girders, as shown in Figure 12.6. The buckling of the web occurred a few inches above the bottom flange and was often associated with the separation of the top bearing plate from the bottom flange of the girder due to the seized expansion bearings. In some cases web cracking developed along the web-flange weld toe in the buckled region. This crack started at the end of the girder and progressed horizontally toward the bearing stiffeners.

Cyclic Loads and Stresses

The I-55 and I-70 east- and westbound structures were subjected to 5250 ADTT (average daily truck traffic) in 1980. Several of the ramps were

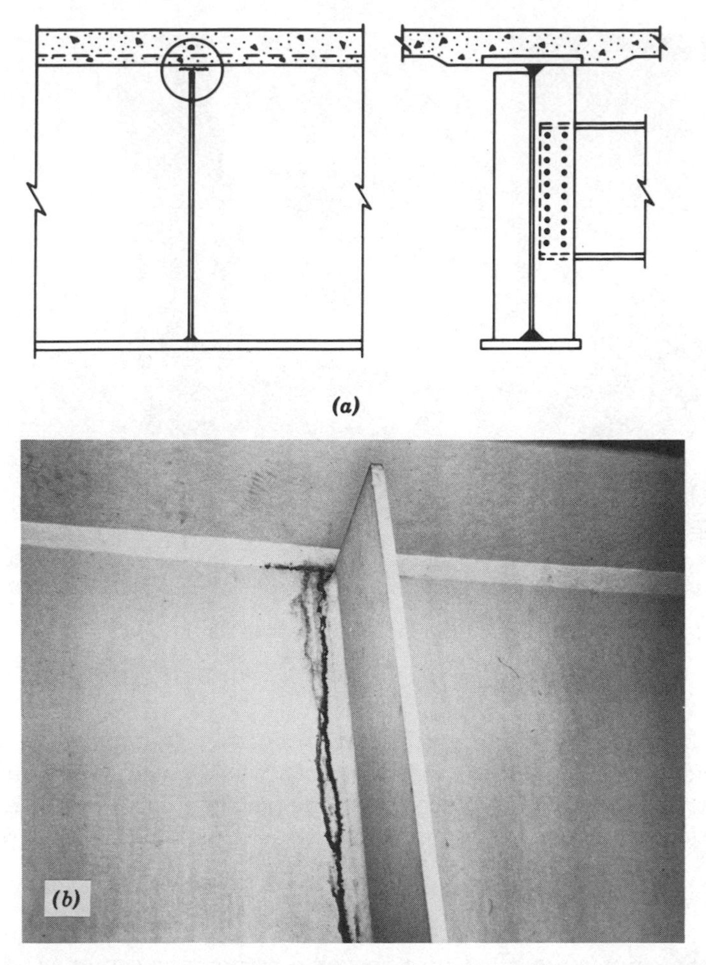

Figure 12.5 Crack in web of negative moment region. (*a*) Schematic of negative moment detail; (*b*) crack along outside web-flange weld toe.

only subjected to about 200 ADTT. The ADTT has varied between 4000 and 5250 since the structures were opened in 1967.

In order to determine which mode of displacement was causing the web cracking, field measurements were conducted in July 1975 [12.2]. The web gaps at several floor beams were selected for the strain measurements. These measurements were acquired at the following locations:

1. At the end of a girder where the floor beam was bolted to a connection plate 7 in. (180 mm) from the bearing stiffener.
2. At the end of a girder where the floor beam was bolted directly to the bearing stiffener.
3. At intermediate floor beams in both positive and negative moment regions.

Figure 12.6 Web buckling at end support (courtesy of Illinois Department of Transportation).

At the floor-beam–bearing stiffener connection, gauges were placed at each end of the stiffener, as neither end was welded to the flange. All other strain gauges were placed near the web gap between the end of the connection plate and the tension flange, as illustrated in Figure 12.7.

Behavior at End Supports

When the connection plate also served as a bearing stiffener, it was not welded to the flange at either end. No crack had formed at the lower gap where the end reaction forced the flange into bearing with the stiffener. However, cracks had developed at the upper end of the stiffener connection plate. The strain gradient and stress range at the lower web gap were relatively small; that is, the effective stress range was $S_{r\text{Miner}} = 7.5$ ksi (52 MPa), while at the upper web gap the effective stress range was $S_{r\text{Miner}} = 16.6$ ksi (114 Mpa). Figure 12.8 shows the time-stress response at the end support and at two interior floor-beam connections. At the bottom web gap at the end reaction the stress range was always less than 5 ksi (34 MPa) which indicated that cracking would not likely develop.

When the connection plate was adjacent to the end reaction-bearing stiffeners, large cracks developed at the bottom end, as illustrated in Figure 12.4. Since the connection plate adjacent to the end reaction was welded to the top flange, no movement could occur, and no damage was detected at the welded ends.

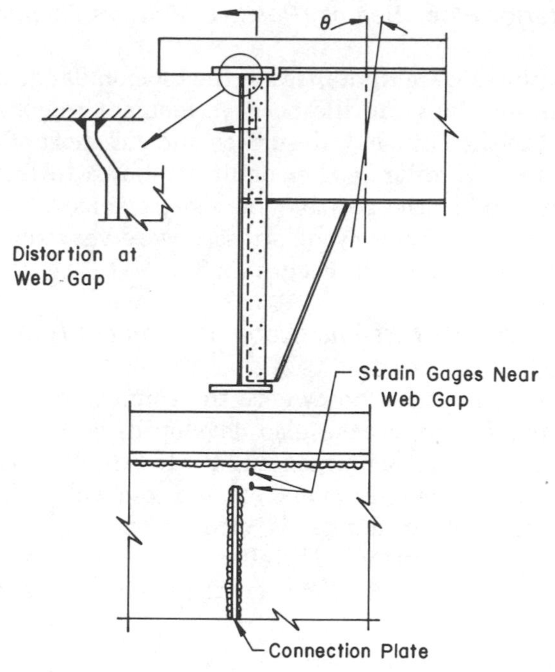

Figure 12.7 Schematic of distortion in web gap and location of strain gauges.

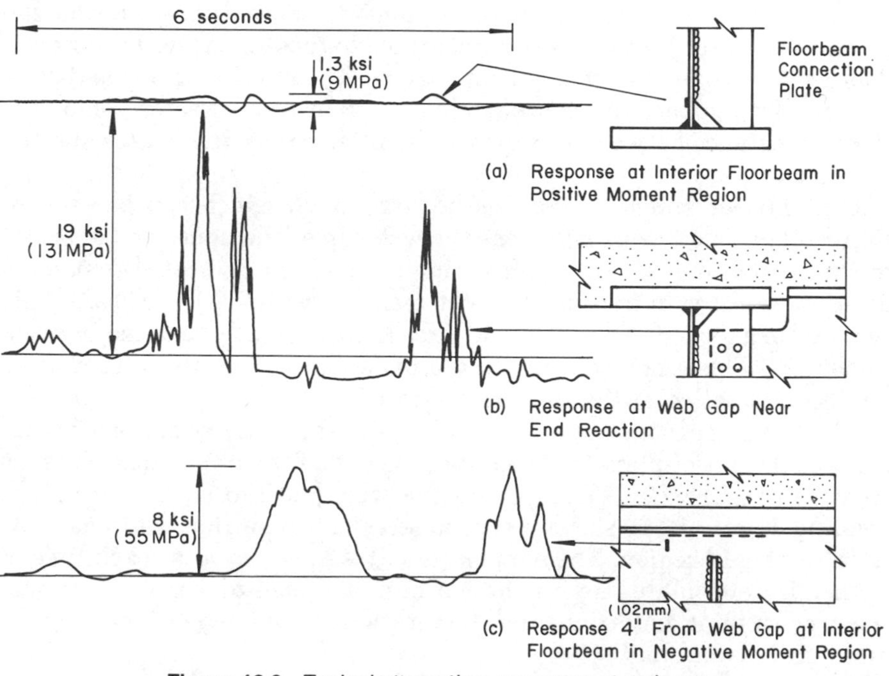

Figure 12.8 Typical stress-time response at web gaps.

Behavior at Interior Floor Beams (Positive Moment Region)

The connection plates were tight fitted to the tension flange and welded to the top compression flange in the positive moment regions. A stiffener was also welded to the outside web surface and cut short of the flange by about $\frac{5}{8}$ in. (16 mm), similar to the condition shown in Figure 12.5. No cracks were observed in the positive moment regions. The strain gauge measurements indicated that cyclic stresses were very small in the bottom web gap, as can be seen in Figure 12.8.

Behavior at Interior Floor Beams (Negative Moment Region)

In the negative moment regions where the connection plates were not welded to the top flange, cracks also developed, as was illustrated in Figure 12.5. The strain measurements indicated that the stress range at the weld toe was high even at the end of a cracked web, as can be seen in Figure 12.8. At the bottom flange the connection plates were welded to the flange, and no cracks formed. The effective stress range of the cracked detail was $S_{r\text{Miner}}$ = 12 ksi (83 MPa) for the measured random variable stress spectrum.

Behavior at Interior Supports

The strain measurements at interior supports showed that the web gap strains were usually small and comparable to those measured at interior floor beams in the positive moment regions. The reason for this is that at the interior support the concentrated reaction force creates some frictional resistance between the ends of the fitted connection plates and the top flanges.

One interior support developed a large crack which can be seen in Figure 12.9. The crack origin was the web gap at the upper end adjacent to the tension flange. The deformation out-of-plane indicated that some difference existed in the elevation between the two ends of the beam. This resulted in a locked in out-of-plane force and eventually resulted in crack instability. This crack was discovered in January 1978. No retrofit holes had been installed at the interior support.

The strain measurements all verified that the primary cause of cracking was the out-of-plane movement of the end of the connection plate, as shown in Figure 12.7. This placed the web gap into double curvature bending. Figure 12.10 shows a typical stress gradient that was measured in the web gap region. The same gradient was observed at each type of connection end plate. Hence the out-of-plane push-pull movement was the predominant distortion, and the rotation of the flange of the longitudinal girder was small.

The stresses in the gap can be approximated from the relationship

$$M = \frac{4EI\Theta}{L} + \frac{6EI\Delta}{L^2} \tag{12.1}$$

where the displacements Θ and Δ are the relative rotation of the flange relative to the web and out-of-plane movement of the web. The contribution due to the rotation Θ appears to be very small and can be neglected for most conditions. Hence the cyclic stress in the gap becomes

$$\Delta\sigma = \frac{M}{S} \approx \frac{6EI}{2I/t}\frac{\Delta}{L^2} = \frac{3Et\Delta}{L^2} \tag{12.2}$$

The magnitude of the deformation, Δ, depends on the amount of end rotation of the floor beam due to bending and relative vertical movement between each end of the floor beam.

Figure 12.9 Large crack at interior support (courtesy of Illinois Department of Transportation).

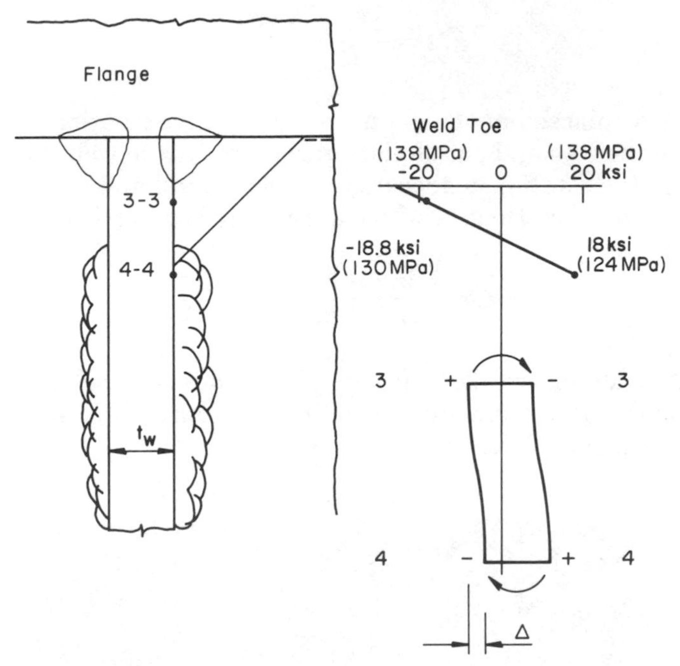

Figure 12.10 Typical stress gradient in the web gap.

Temperature and Environmental Effects

The temperature and environment did not appear to play a significant role in the fatigue cracks that developed in the main girder webs at the Poplar Street structures. The thermal movement of the girders caused the girder to bend because of the seized expansion bearing. This resulted in a separation between the bottom flange and the top bearing plate and likely contributed to web buckling, as illustrated in Figure 12.6. The full depth crack shown in Figure 12.9 did experience cleavage fracture over much of the web depth. The fracture likely occurred at a reduced temperature near the time of its discovery in January 1978.

Failure Analysis

Since stress range measurements were available from several web gap details, it was possible to estimate the stress intensity range and establish an estimated number of variable stress cycles. The stress intensity factor was modeled as

$$K = \left[\frac{0.923 + 0.199(1 - \sin \pi a/2t_w)^4}{\cos \pi a/2t_w}\right] F_e F_w F_g \sigma \sqrt{\pi a} \qquad (12.3)$$

where

$$F_e \approx 1.0 \tag{12.3a}$$

$$F_w = \left[\frac{t_w}{\pi a}\tan\frac{\pi a}{t_w}\right]^{1/2} \tag{12.3b}$$

$$F_g = \frac{K_{tm}}{1 + 2.776(2a/t_w)^{0.2487}} \tag{12.3c}$$

$$K_{tm} = 1.621\ \ln\left(\frac{Z}{t_w}\right) + 3.963 \tag{12.3d}$$

The web thickness $t_w = \frac{1}{2}$ in. (12.7 mm) and the weld leg $Z = \frac{5}{16}$ in. (8 mm) were used to estimate the stress concentration factor. Since cracks propagated from each side of the web surface at the weld toe of the connection plate and web-flange connection, the final crack size was taken as $t_w/2$.

The number of cycles for an assumed initial crack of 0.015 in. (0.38 mm) to penetrate to the mid-thickness of the web was estimated as

$$N = \int_{0.015\text{ in.}}^{0.25\text{ in.}} \frac{da}{3.6 \times 10^{-10}\,\Delta K^3} \tag{12.4}$$

The effective variable stress range $S_{r\text{Miner}}$ was measured to be between 16 ksi (110 MPa) and 20 ksi (138 MPa) at the most highly stressed web gap details. This would result in 260,000 to 510,000 cycles to achieve through web thickness cracking. Even the lower gap at an end support, where the effective stress range was 7.5 ksi (52 MPa), would be predicted to crack after 4.94 million cycles, since many of the stress cycles exceeded the crack growth threshold. These results indicate that cracks probably formed during the first year of service at the more highly stressed details.

It should be noted that a smaller initial crack size was assumed at the weld toe because the web-flange connection was produced by the automatic submerged arc process, and larger initial flaws were not as probable.

12.1.3 Conclusions

The cracks in the web gaps between the end of the floor-beam connection plates and the girder were all caused by cyclic stress. These fatigue cracks resulted from repeated out-of-plane web bending stresses due to out-of-plane displacement of the web in the gap region, as shown schematically in Figure 12.7. The severity of this displacement was greatest at the end bearings, the top flanges of intermediate floor beams in the negative moment regions of girders, and at the top flanges at interior support. This was confirmed by field test measurements in typical gap regions under actual traffic [12.2]. The end rotation of the floor beams, which were rigidly connected to the main girder, pushed the web out of plane and caused high cyclic stresses along the toe of the web-to-flange welds and at the stiffener connection plate ends.

The longitudinal orientation of the cracks in the girder web relative to the stresses from loading are such that they do not seriously affect the girder strength. However, corrective repairs were needed to prevent the cracks from turning and growing perpendicular to the in-plane tensile bending stresses in girder webs.

12.1.4 Repair and Fracture Control

Based on the observed behavior at the Poplar Street Complex, the following procedures for repair and corrective action were used [12.4].

At Girder Ends

1. The floor-beam connection plates were welded to the top and bottom flanges in order to prevent relative displacement between the ends of the connection plates and the girder flanges. Figure 12.11 shows retrofit procedures for the cracked and buckled ends.
2. One-half in. (13 mm) holes were drilled through the web at the ends of the existing web-to-flange connections.
3. Drilled holes were also placed at the ends of the web cracks at the ends of connection plates or stiffener connection plates.
4. The cracks were gouged out and welded with a full-penetration groove weld up to the hole.

At Negative Moment Regions

1. Holes were drilled at each end of the cracks, as shown schematically in Figure 12.12. Holes were drilled near the ends of the crack along the web-flange weld and on each side of the stiffener. This procedure permitted the crack to develop between the holes and thus softened the connection to accommodate the out-of-plane displacements.

The retrofit and repair work was carried out in 1975 to 1976. Holes were drilled in the girder webs at 751 locations in accord with the scheme shown in Figure 12.12. Ninety-seven of these locations were known to have cracks. At the girder ends, stiffeners were welded to the tension

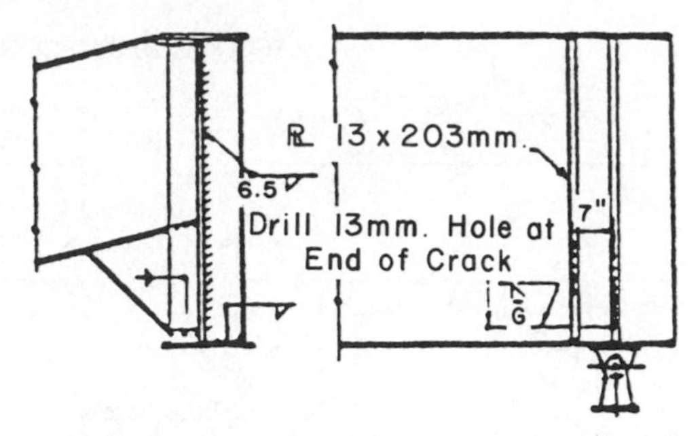

(a) Without intermediate stiffener on outside face

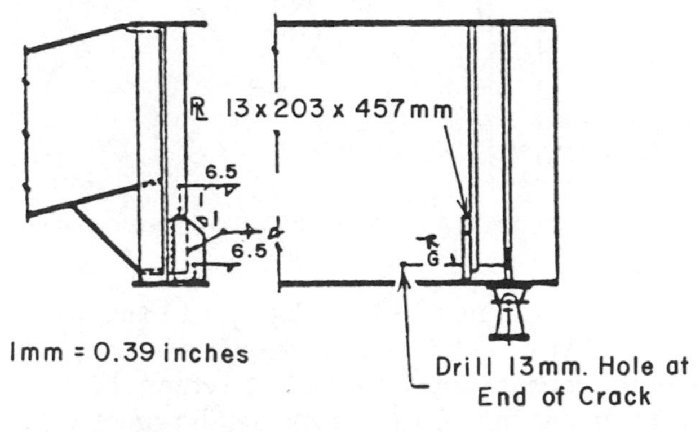

(b) With intermediate stiffener on outside face

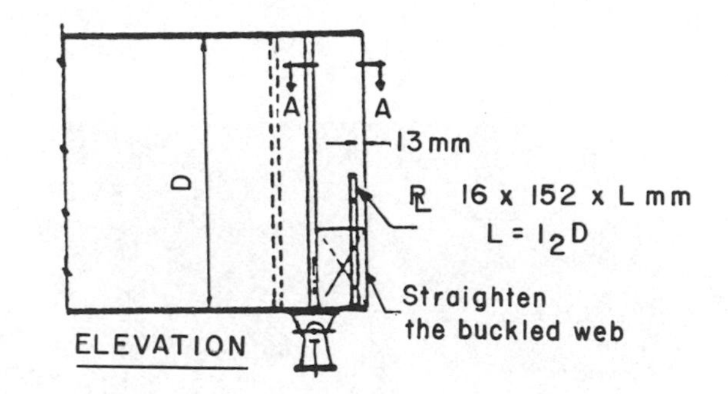

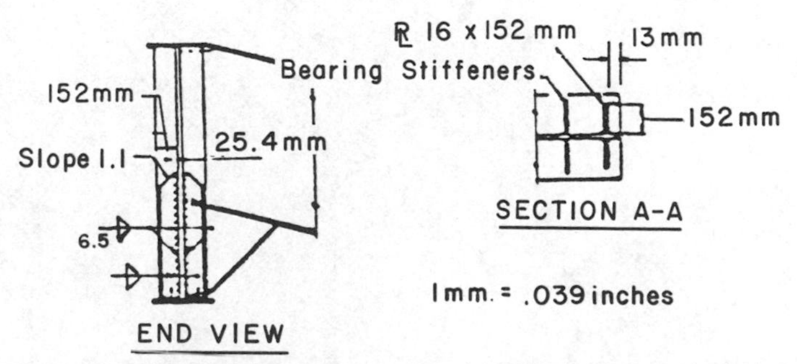

Figure 12.11 Retrofit details for girder ends. (*a*) Retrofit for cracked girder ends; (*b*) retrofit for repair of buckled webs.

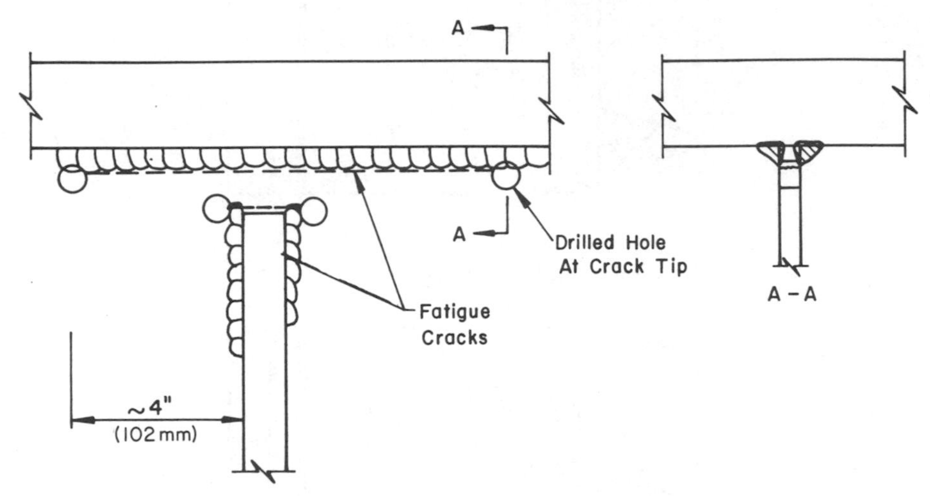

Figure 12.12 Retrofit holes at fatigue crack tips.

flanges, as shown in Figure 12.11. Figure 12.13 shows the retrofitted ends of two girders. Also visible are the neoprene bearings that were installed in place of the rocker supports. The large crack shown in Figure 12.9 was repaired by installing splice plates, as shown in Figure 12.14. Yearly inspections have indicated that no cracks had reinitiated from the drilled holes as of 1982.

Figure 12.13 Retrofitted girder end showing stiffeners and elastomeric bearings (courtesy of Illinois Department of Transportation).

Figure 12.14 Splice plates installed at large web crack (courtesy of Illinois Department of Transportation).

12.2 FATIGUE CRACKING OF THE POLK COUNTY BRIDGE (DES MOINES)

12.2.1 Description and History of the Bridge

Description of Structure

The Des Moines (Polk County) Bridge carries east and westbound traffic over the Chicago Rock Island and Pacific Railroad tracks and the East Four Mile Creek on Route 163. The bridge is located near Des Moines, Iowa. The east- and westbound bridges are two similar but separate structures.

Each five-span skew bridge structure is composed of two continuous welded plate girders with transverse stiffeners 590 ft (179.8 m) long supporting a 30 ft (9.1 m) roadway. The west end span is 91 ft (27.74 m) long, and the east end span is 118 ft (36 m) long. The three interior spans are each 127 ft (38.7 m) long. The slab is $7\frac{1}{4}$ in. (184 mm) thick reinforced concrete supported by two 18 WF 45 stringers and the two main girders, as shown in Figure 12.15. Figure 12.16 shows a view along the inside surface of the longitudinal girder. The steel framing plan is shown in Figure 12.17.

The two main girder webs are 74 in. (1.88 m) deep and $\frac{7}{16}$ in. (11 mm) thick. The floor beams supporting the stringers have $48 \times \frac{5}{16}$ in. (1.22×8 mm) web plates with alternating transverse stiffeners.

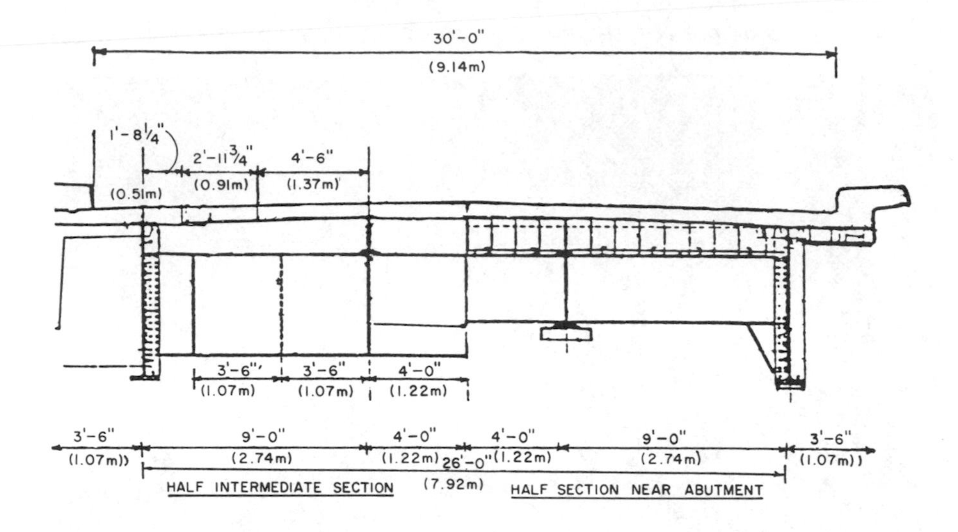

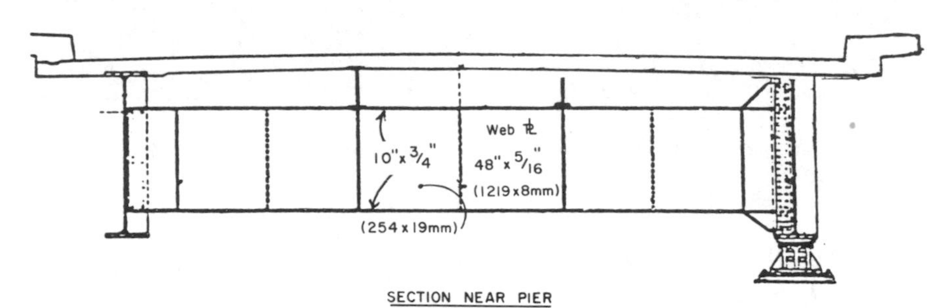

Figure 12.15 Typical cross sections of Des Moines (Polk County) Bridge.

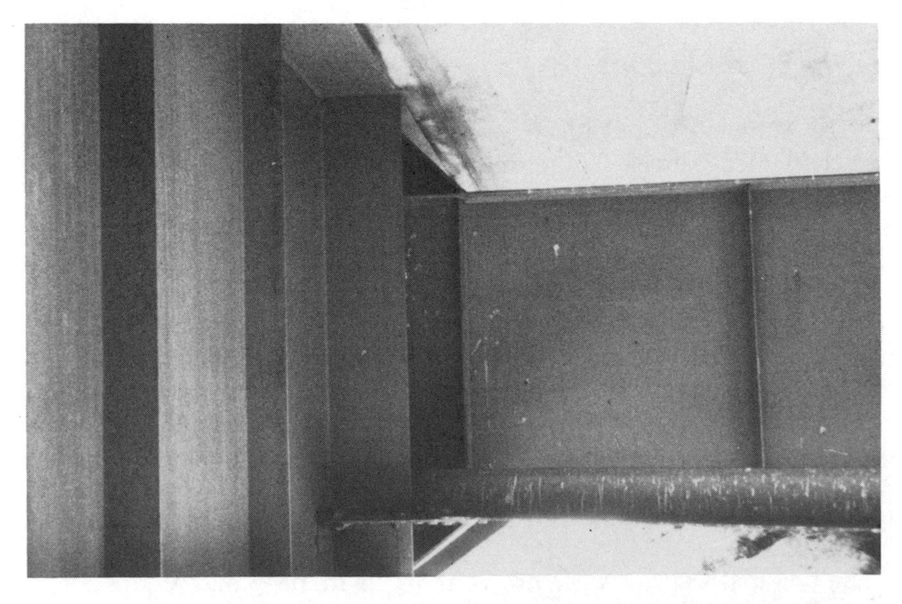

Figure 12.16 Floor-beam framing into a longitudinal girder.

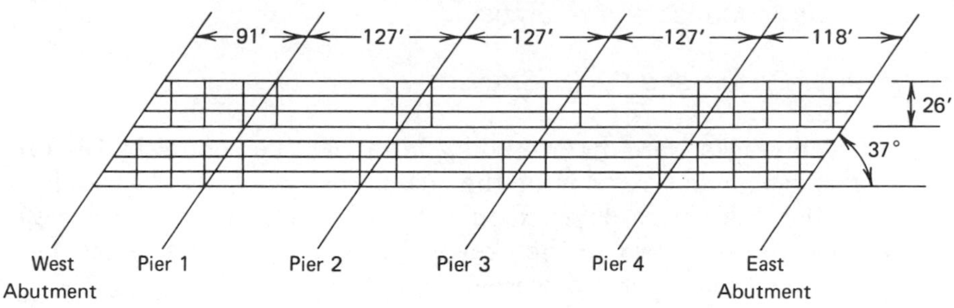

Figure 12.17 Steel framing plan.

The Des Moines Bridge was designed for H20-S16 loading. The design was based on the 1961 AASHO Specifications and 1960 Iowa-DOT Standard Specifications with current Special Provisions and Supplemental Specifications. All members were fabricated from A36 steel.

History of the Structure and Cracking

The Des Moines Bridge was constructed in 1962 and opened to traffic in 1963. The volume of truck traffic crossing the bridge in 1979 was 500 trucks per day each way.

In 1979 several fatigue cracks were discovered in the webs of the main girders of the Des Moines Bridge [12.5]. These cracks were in the web plate in the negative moment regions along the web-to-flange fillet welds above the floor-beam connecting plate. Several vertical cracks were also observed in the fillet welds joining the floor-beam connecting plates to the girder web and in the web several inches below the web flange connection. Figure 12.18 shows the location of the cracks that were initially found in the bridge structures.

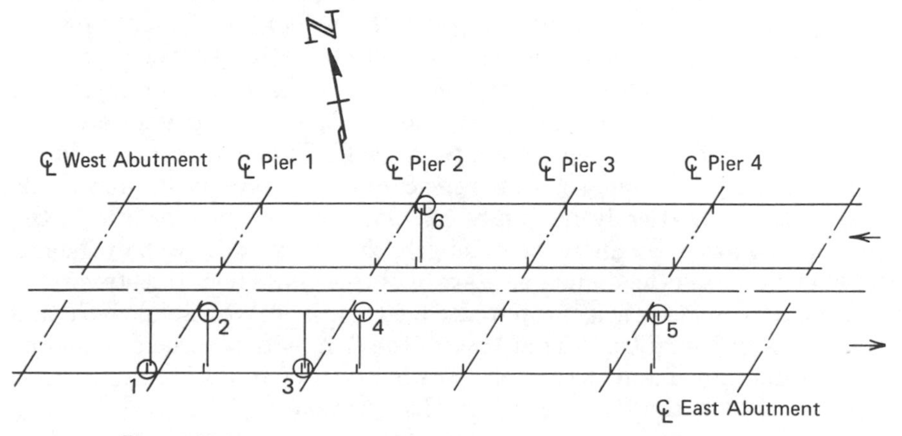

Figure 12.18 Location of crack at floor-beam connection plates.

12.2.2 Failure Modes and Analysis

Location of Cracks and Cyclic Stresses

Several web cracks were discovered by Iowa-DOT personnel in 1979 in the negative moment regions of the main girders of the Des Moines Bridge. All cracks were detected in the girder web at the floor-beam connection plates nearest to the skewed piers in the negative moment regions. Three types of cracks were observed:

1. Horizontal fatigue cracks in the main girder webs along the toe of the web-to-flange fillet welds. These cracks were 3 to 6 in. (76 to 152 mm) long and one such crack is shown in Figure 12.19.
2. Cracking of the inside and outside web surfaces on each side of the web plate about 2 in. (51 mm) below the web-flange weld. The cracks can be seen in the photographs taken of the details that are shown in Figure 12.19.
3. Vertical weld cracks at the top end of the floor-beam connecting plate, as shown in Figure 12.19*b*. At several locations these cracks stopped after propagating downward for an inch or so and intersecting the lower web crack.

On-site examination of the web cracks in the main girders of the Des Moines Bridge indicated that all were fatigue cracks. The principal cause of the cracking was the out-of-plane deformation of the girder webs in the small gaps at the ends of the floor-beam connection plates which is comparable to the distortion in the Poplar Street Bridge. The fatigue crack development in the girder webs in the negative moment regions results as the web is forced out of plane by the floor-beam end rotation in the small gaps between the end of the floor-beam connection plates and web-to-flange welds. Since the slab anchors the girder top flange, and the connection plate has its strong axis perpendicular to the web, end rotation of the floor beam from bending and relative end movement forces the web out of plane. When the gap is very small, the cyclic stresses are high at the upper end of the connection plate weld. This causes the welds to crack from their root and separates the connection plate from the web, as illustrated schematically in Figure 12.20. As the vertical cracks in the connection plate welds increase in length, they permit the web to deform, and this increases the double curvature out-of-plane bending stresses in the web. As a result cracks form in the web along the web flange weld and at a point 2 or 3 in. (50 or 75 mm) below the flange. In skewed bridges the out-of-plane movement is more severe because of the different vertical displacements at the two ends of the floor beams [4.3].

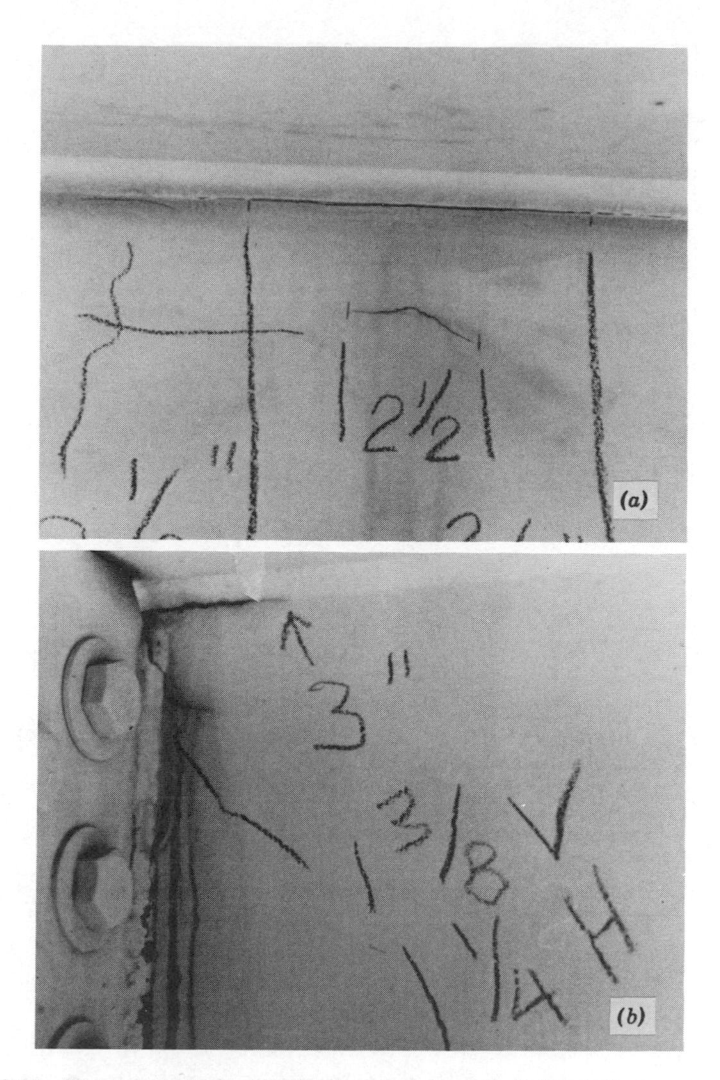

Figure 12.19 Typical cracks at the floor-beam connection plate web gap (courtesy of Iowa Department of Transportation). (*a*) Cracks viewed from outside web surface; (*b*) cracks viewed from inside web surface.

No web gap stresses were measured in the Polk County Bridge near Des Moines. The average daily truck traffic was 500 vehicles in 1979. It seems probable that about 2 million trucks crossed the structure between 1963 and 1979. None of the cracks exhibited a cleavage fracture appearance. Hence the fracture toughness of the material did not play a role in the development of the cracks that formed in the girder webs.

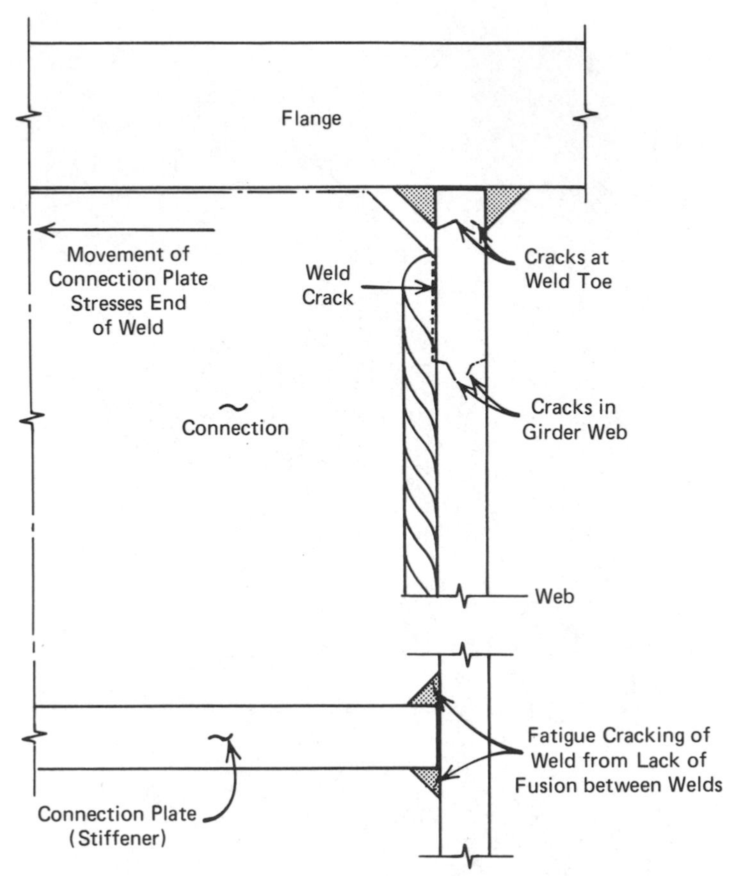

Figure 12.20 Schematic of cracking in connection plate web weld and at toe of flange-to-web weld.

Failure Analysis

No measurements were available to provide an accurate assessment of the cyclic out-of-plane bending stress introduced into the girder web. However, an estimate can be made by evaluating the crack propagation behavior of the detail. Since the variable stress cycles required to propagate the crack are known, the stress intensity factor can be used to estimate the effective stress range. The assumed two million variable stress cycles can be equated to the relationship

$$N = 2 \times 10^6 = \int_{0.015 \text{ in.}}^{0.22 \text{ in.}} \frac{da}{3.6 \times 10^{-10}\, \Delta K^3} \tag{12.5}$$

The stress intensity range can be estimated from Eq. 12.3, with $t_w = \frac{7}{16}$ in.

(11 mm) and $Z = \frac{1}{4}$ in. (6.4 mm). For a 10 ksi (69 MPa) effective stress range, Eq. 12.5 yields the relationship

$$2 \times 10^6 = 2.22 \times 10^6 \left(\frac{10}{S_{r\,\mathrm{Miner}}}\right)^3$$

Hence $S_{r\,\mathrm{Miner}} = 10.4$ ksi (71.4 MPa).

It can be seen that the estimate is in reasonable agreement with the measurements made on the Poplar Street Bridge. Hence the cracking has occurred in a predictable manner and time.

12.2.3 Conclusions

The cracks discovered in the main girders of the Des Moines Bridges were all caused by cyclic stresses. Each of the skew bridges consisted of two main longitudinal girders supporting transverse floor beams and longitudinal stringers. The fatigue cracks developed in the webs of the main girders at the floor beams adjacent to the piers. At the piers the floor beams were bolted to transverse stiffener connection plates that were located 9 in. (229 mm) from the bearing stiffeners. The cracks were caused by out-of-plane displacement of the main girder web at the ends of

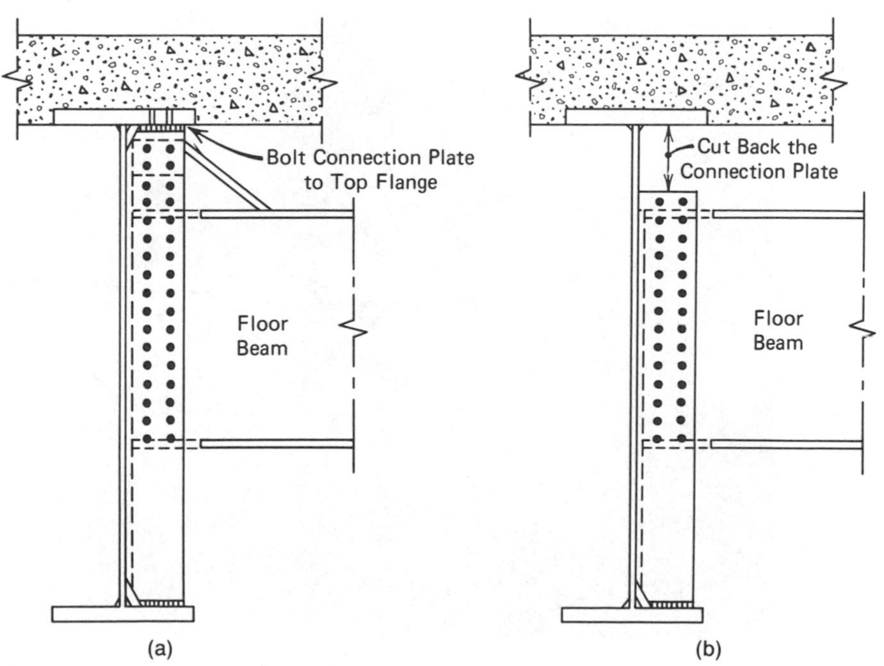

Figure 12.21 Schematic of retrofit details. (*a*) Positive attachment between connection plate and flange; (*b*) increased web gap.

Figure 12.22 Cutting away portion of connection plates (courtesy of Iowa Department of Transportation).

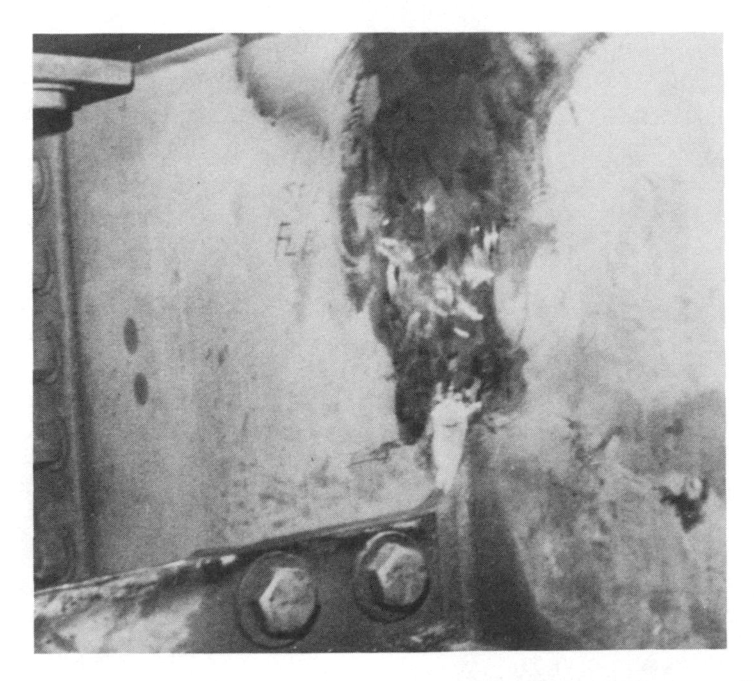

Figure 12.23 Grinding weld from the web surface (courtesy of Iowa Department of Transportation).

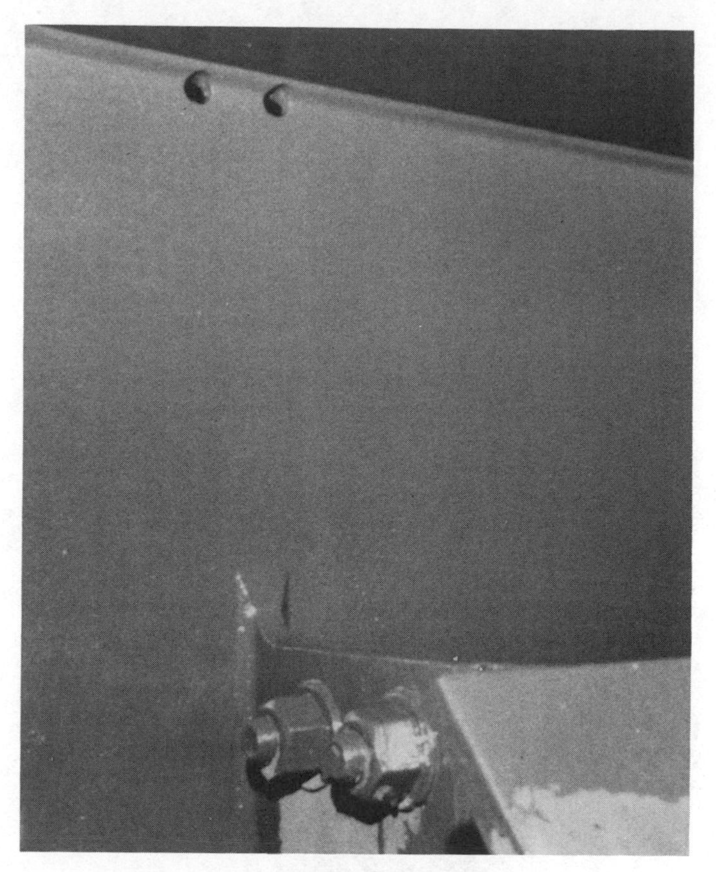

Figure 12.24 Drilled holes at the ends of crack (courtesy of Iowa Department of Transportation).

the connection plates in the negative moment regions. The displacement caused double curvature bending to occur in the web gap, and this resulted in horizontal cracks along the web-flange fillet welds. Vertical cracks also formed in the transverse connection plate welds. This contributed to the formation of additional horizontal web cracks 2 to 3 in. (50 to 75 mm) below the flange.

12.2.4 Repair and Fracture Control

The Des Moines (Polk County) Bridge was retrofitted in 1980. After visual and dye-penetrant inspection for fatigue cracks in the girder web, the retrofit schemes shown in Figure 12.21 were carried out. The following procedures were used:

Figure 12.25 Photographs of retrofit carried out at piers (courtesy of Iowa Department of Transportation). (*a*) Bolting connection plate to top flange; (*b*) view of angle clip between flange and connection plate.

1. In the negative moment regions away from the piers the top portion of the stiffener connection plate for the floor beam was removed by flame cutting, as shown in Figure 12.22, and the remaining weld removed by grinding, as illustrated in Figure 12.23.
2. At the ends of the web cracks $\frac{3}{4}$ in. (19 mm) holes were then drilled in order to isolate the cracks and ensure no further growth. Figure 12.24 shows a drilled section and the final paint coat.
3. In the negative moment regions at the piers the bearing stiffeners were bolted to the girder top flange, using a clip angle on each side of the bearing stiffener. In order to facilitate this repair, a rectangular section of the reinforced concrete deck above the stiffener was cut out, and holes were then drilled in the top flange for the bolted connection, as illustrated in Figure 12.25.

REFERENCES

12.1 Bureau of Design, Bridge and Traffic Structures Section, Investigation Report on Poplar Street Complex Bridge, East St. Louis, St. Clair County, Illinois, Illinois Department of Transportation Report, June 1974 (limited distribution).

12.2 Fisher, J. W., Report on Investigation of the Girder Web Crackings at Floor Beam Connection Plates of Poplar Street Complex, Prepared for H. W. Lochner, Inc., Consulting Engineers, Chicago, Illinois, October 1975 (limited distribution).

12.3 Fisher, J. W., Fatigue Cracking in Bridges from Out-of-Plane Displacements, *Can. J. Civ. Eng.* 5 (1978):542–556.

12.4 Hsiong, W., Repair of Poplar Street Complex Bridges in East St. Louis, Transportation Research Record 664, *Bridge Engineering,* Vol. 1, Proc. St. Louis Conference by Transportation Research Board, September 25–27, 1978, pp. 110–119.

12.5 Letter dated September 28, 1979, from C. A. Pestotnik to J. W. Fisher.

CHAPTER 13

Diaphragm Connection Plates

Where transverse connection plates are welded to longitudinal girders, diaphragms and cross-frames in multiple beam bridges often provide similar conditions to those when crack-initiating conditions are provided by connection plates between floor beams and girders. Generally, the diaphragms and cross-frames are fastened to transverse stiffeners which are welded to the girder web. No connection was usually provided between the stiffener and the girder tension flange. Sometimes these stiffeners are not attached to either flange. Since adjacent beams deflect differing amounts, the differential vertical movement produces an out-of-plane deformation in the web gap at the stiffener ends that are not welded to the beam flange. The magnitude of this out-of-plane movement depends on the girder spacing, skew and type of diaphragm, or cross-frame. As a rule skewed bridges will experience larger relative vertical movement of adjacent girders, which results in larger out-of-plane movement in the web gap. Figure 13.1 shows a schematic drawing of the relative movements, Δ, in the web gap as a result of the differential vertical movement, δ.

At least 10 bridge structures have experienced cracking in the longitudinal girder webs as a result of out-of-plane movement in the web gap. Various types of diaphragms and girder spacing have resulted in this type of cracking. The diaphragms have ranged from simple X-bracing using angles to more rigid rolled sections. Figure 13.2 shows a five girder bridge wi‘h relatively stiff diaphragms composed of a blocked rolled beam with haunched ends. Cracks formed in the web gap along the web-flange weld toe and in the vertical welds, as illustrated in Figure 13.3. In all but one highway bridge these cracks have been confined to the negative mo-

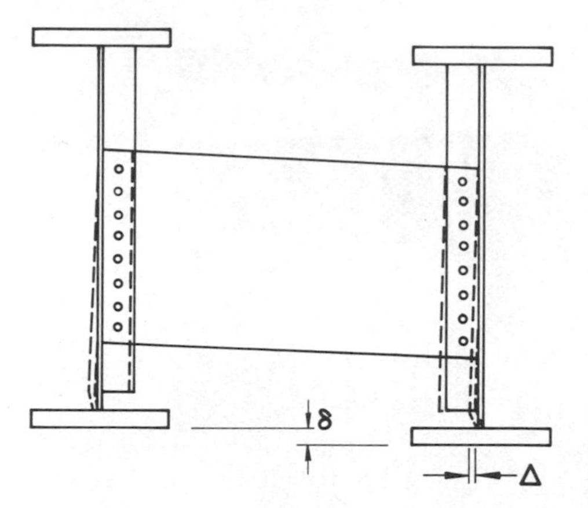

Figure 13.1 Schematic showing web distortion as a result of relative vertical displacement.

ment region because the connection plate was welded to the top flange elsewhere. The slab has provided greater fixity to the top flange, and the adjacent web gap has experienced greater deformation and has cracked first.

Several simple span railroad bridges with relatively rigid diaphragms have developed cracks in the web gap adjacent to the bottom flange, as illustrated in Figure 13.4 [4.3]. These structures were usually skewed, with the diaphragms placed perpendicular to the longitudinal girders.

Figure 13.2 View of diaphragms and girders (courtesy of Iowa Department of Transportation).

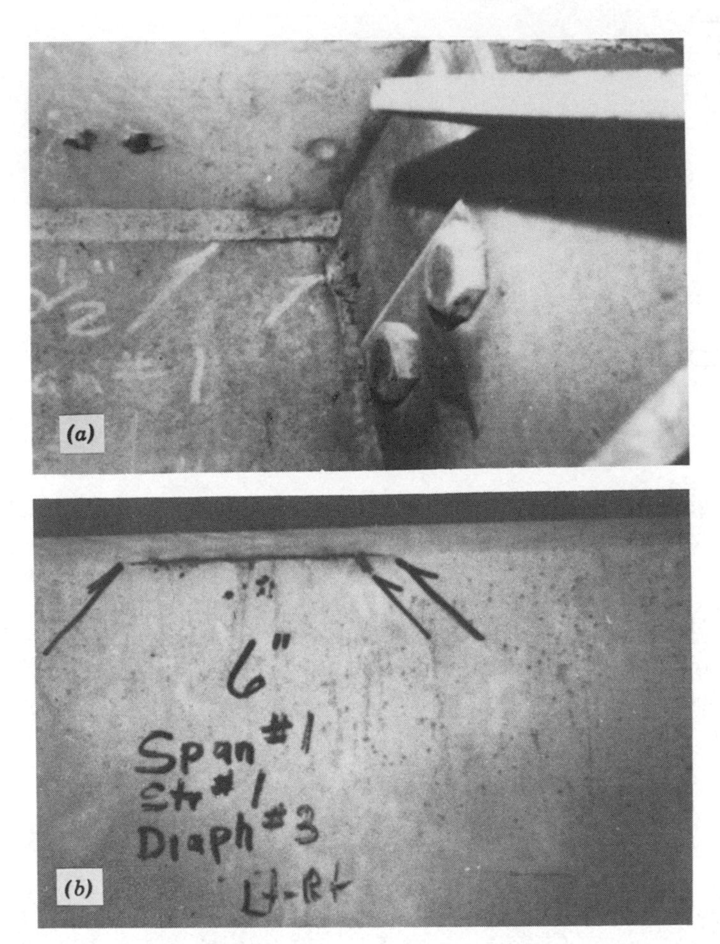

Figure 13.3 Cracks along web-flange weld at an exterior girder diaphragm connection plate (courtesy of Iowa Department of Transportation). (*a*) Inside web surface; (*b*) outside web surface.

Cracking initially was observed at the ends of the transverse connection plate which was cut short of the tension flange by 1 to 2 in. (25 to 50 mm). Cracking has since developed along the web-flange weld after holes were drilled at the crack tips seen in Figure 13.4.

More recently, cracking has been observed in the web gaps of curved box girder bridges with interior diaphragms [13.1]. Most of the cracking was detected in the negative moment region, at the web gap adjacent to the top tension flange. Cracking was also detected in positive moment regions at the web gap adjacent to the bottom flange of the trapezoidal box.

An analogous type of cracking has been observed in skewed riveted structures with transverse diaphragms. Figure 13.5 shows the intermedi-

Figure 13.4 Crack in girder web at transverse diaphragm connection plate (courtesy of Canadian National Railways).

Figure 13.5 Cracking at diaphragms of riveted bridges. (*a*) Cross-frame in a skewed riveted bridge.

Figure 13.5 Cracking at diaphragms of riveted bridges. (*b*) Cracking in cross-frame connection angle.

ate cross-frames of a riveted multiple girder bridge which carries a large number of trucks. Distortion of the transverse angle stiffener which connects the cross-frame to the main girders has resulted in cracking along the fixing line provided by the rivet heads, which can be seen in Figure 13.5*b*. Nearly all cracking was adjacent to the bottom flange in both positive and negative moment regions.

Two structures that experienced cracking in the longitudinal girder webs at X-type cross-frames are reviewed in detail hereafter.

13.1 FATIGUE CRACKING OF THE BELLE FOURCHE RIVER BRIDGE

13.1.1 Description and History of the Bridge

Description of Structure

The Belle Fourche River Bridge carries north- and southbound traffic on U.S. 212 over the Belle Fourche River near the city of Belle Fourche, South Dakota. The structure consists of a 341 ft 9 in. (104.2 m) three-span welded plate girder bridge haunched at the interior supports and four 60 ft (18.3 m) simple span composite beam bridges. Figure 13.6 shows the plan and elevation of the bridge. All spans were supported by seven longitudinal welded built-up girders. Figure 13.7*a* shows a partial cross section of the continuous girder span at an interior support, and Figure 13.7*b* the type of interior diaphragm used in all seven spans. An X-type

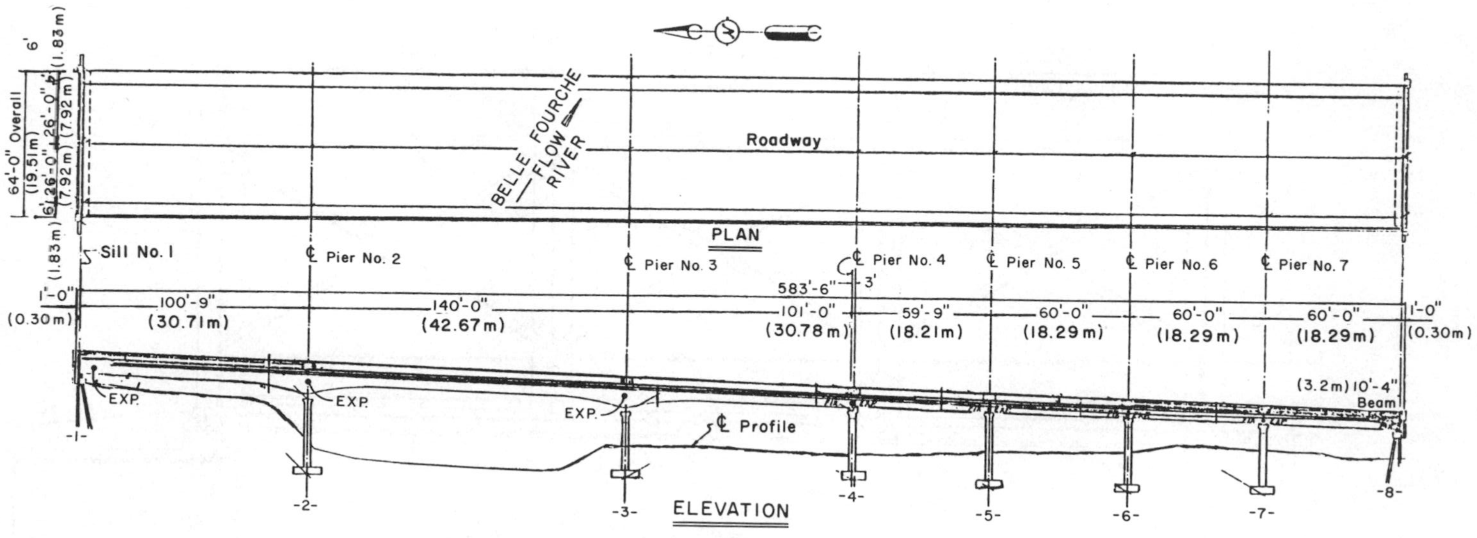

Figure 13.6 Plan and elevation of Belle Fourche Bridge.

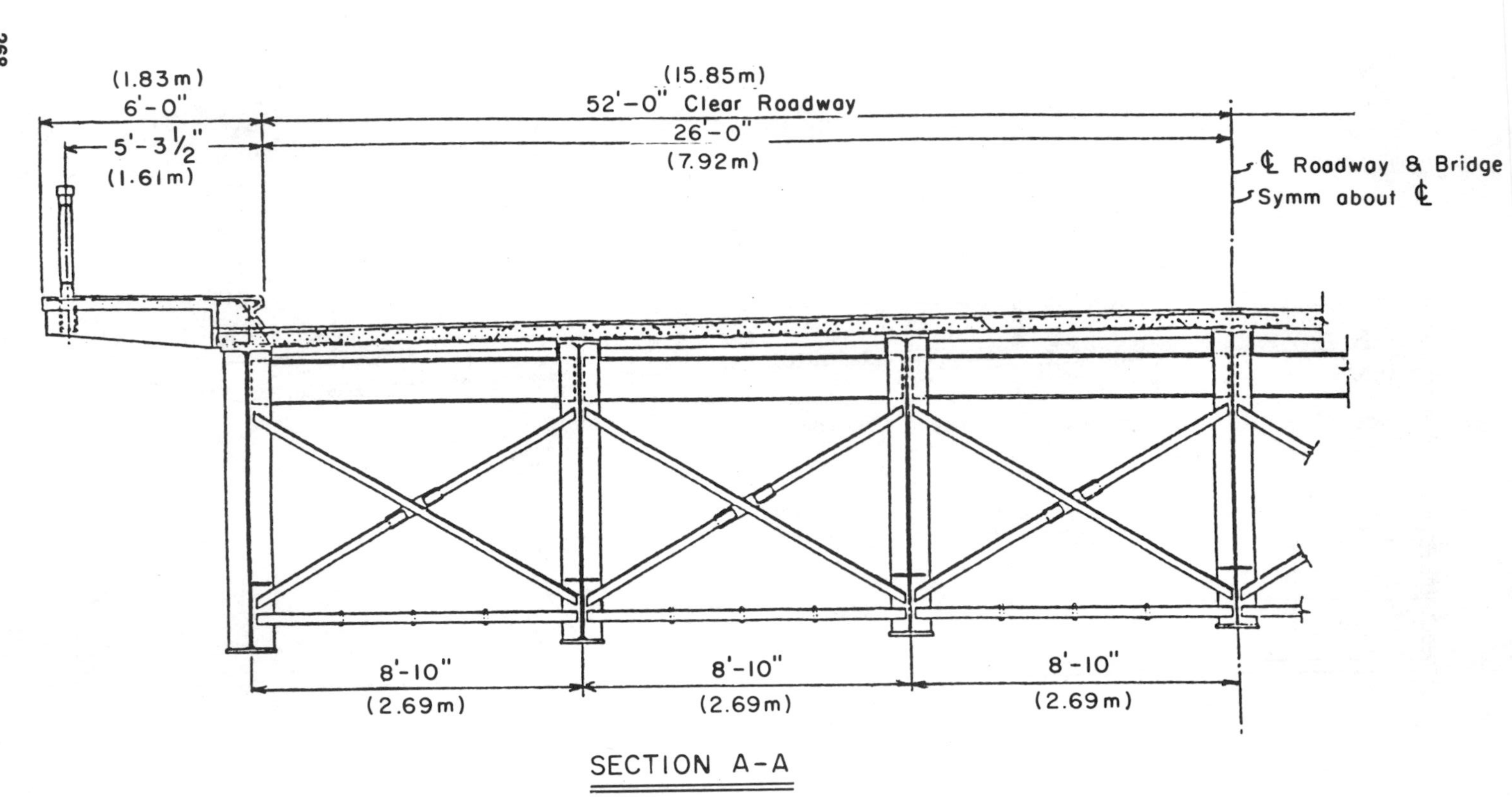

(a)

Figure 13.7 Cross section and interior diaphragms. (*a*) Cross section near support of continuous span.

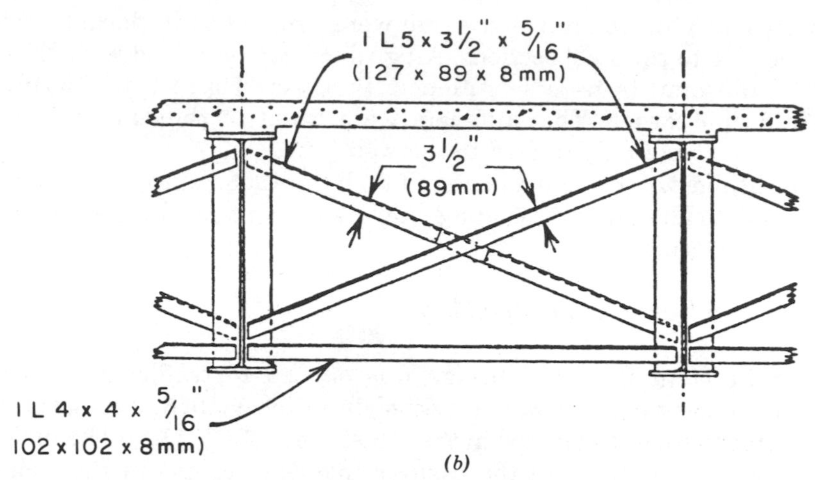

Figure 13.7 Cross section and interior diaphragms. (*b*) Typical interior diaphragm for all spans.

cross-frame was welded to the transverse web stiffeners in the continuous and simple span structures. Figure 13.8 shows several cross-frames in the continuous span structure.

The north and southbound roadways are each 52 ft (15.85 m) wide with two 5 ft (1.5 m) sidewalks. The seven longitudinal girders in each span supported a 6 in. (152 mm) reinforced concrete slab. Composite action was

Figure 13.8 View of diaphragms and longitudinal girders (courtesy of South Dakota Department of Transportation).

provided by 3 in. (75 mm) angles that were 2 in. (50 mm) wide and welded by their toe to the steel sections. All girders were provided with $\frac{3}{8} \times 6$ in. (9.5×150 mm) transverse stiffeners that were coped $\frac{3}{4}$ in. (19 mm) at their inside corner. The stiffeners were fitted to the top and bottom flanges but were only welded to the web.

The bridge was designed in accord with the 1953 South Dakota specifications for H20-44 truck loading. A373 steel was used to fabricate the steel superstructure.

History of Structure and Cracking

The Belle Fourche River Bridge was opened to traffic in 1958. The bridge's superstructure was inspected in October 1976, and several fatigue cracks were discovered in the girder webs [13.2, 13.3]. The majority of the cracks occurred in the positive moment regions in the web gap adjacent to the compression flanges at the transverse stiffener with diaphragms attached. A photograph of a typical fatigue cracked girder web at the end of a transverse stiffener is shown in Figure 13.9. Cracking was observed along the web-flange weld toe and at the end of the transverse stiffener weld.

Most of the cracks were detected in 1976. The largest cracks were found in span 3 of the continuous span structure. These cracks were

Figure 13.9 Fatigue cracks along flange-web weld toe and at end of diaphragm connection plate (courtesy of South Dakota Department of Transportation).

visible on each side of the web surface. One or more cracks were detected in every span by 1978 [13.4].

Three basic types of cracks were observed.

1. Cracks parallel to the web-to-flange weld along the weld toe in the web.
2. Cracks at the top end of the stiffener-to-web fillet weld which extended down on either side of the stiffener at a 45° angle. After a few inches of extension these curved up and became parallel to the tension stress direction in the web.
3. Cracks in the vertical welds starting from the coped end of the stiffener which severed the stiffener-to-weld fillet weld from the web.

Usually a combination of these three types of cracks was observed at the cross-frames.

13.1.2 Failure Analysis

The average daily truck traffic (ADTT) between 1960 and 1978 was estimated to be 280 vehicles. Between 260 and 305 trucks were observed in individual years with no significant trend in growth. Assuming a uniform daily traffic count for the 30-year period of service, this would yield 3.066×10^6 vehicles and variable stress cycles.

No stress measurements or crack surfaces were available for evaluation of the fatigue crack growth behavior. Only the results of a visual surface inspection of the cracked girder webs were available.

None of the cracked girder webs gave any indication of having experienced unstable crack growth (brittle fracture). Hence the temperature and environment did not appear to contribute to the observed behavior.

All cracks gave the appearance of stable fatigue cracks and appear directly comparable to the out-of-plane displacement-induced cracks that have formed at other web gap locations. The crack growth condition at the Belle Fourche Bridge seems comparable to the conditions observed at the Des Moines floor-beam–girder web cracks. Solution of Eq. 12.5 had indicated that an effective stress range at the weld toe of 10 ksi (69 MPa) would yield 2.2×10^6 variable stress cycles. This is reasonably close to the observed behavior of the structure, as it was estimated that about 3 million trucks had crossed the bridge between 1960 and 1978.

13.1.3 Conclusions

The cracks in the girder webs of the Belle Fourche Bridge were fatigue cracks. All cracks originated in the web gap at the top of the stiffener-to-web fillet weld.

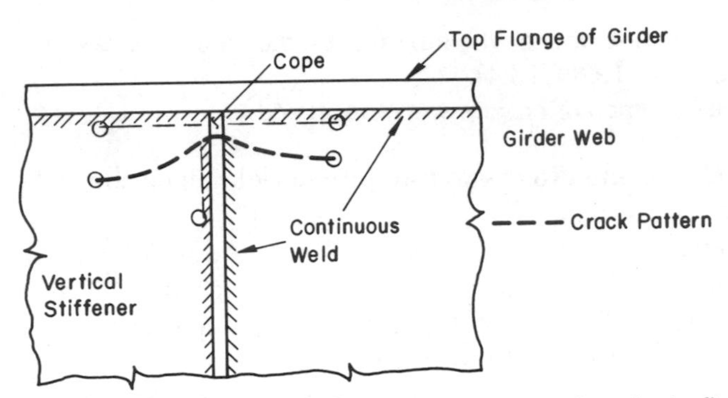

Figure 13.10 Schematic showing web cracks and location of retrofit holes.

The main cause of the cracks in the web was out-of-plane displacement of the girder webs at the transverse stiffener-diaphragm connections which resulted from the relative vertical displacement of the longitudinal girders under cyclic live loads.

The fatigue cracks were oriented more or less parallel to the longitudinal axes of main girders, so that they were parallel to the primary bending stresses in the web. None of the cracks had extended in a brittle manner.

13.1.4 Repair and Fracture Control

In order to arrest fatigue crack growth and prevent subsequent fracture of the cracked girder webs, $\frac{3}{4}$ in. (19 mm) holes were drilled at the tips of the existing cracks, as shown schematically in Figure 13.10. The crack tips were located with dye-penetrant, and the hole was centered on the end of the crack.

The upper end of each cross-frame connection plate was also welded to the top flange with 4 in. (100 mm) long $\frac{5}{16}$ in. (8 mm) fillet welds on each surface of the transverse stiffener. This prevented further out-of-plane movement of the web gap.

13.2 FATIGUE CRACKING OF CHAMBERLAIN BRIDGE OVER MISSOURI RIVER

13.2.1 Description and History of the Bridge

Description of Structure

The Chamberlain Bridge crosses the Missouri River near the town of Chamberlain, South Dakota. The bridge is 2004 ft (610.8 m) long with five-truss spans connected to continuous plate girder structures on each approach. The west approach span is a three-span composite welded plate girder structure with spans of 92 ft 8¾ in., 116 ft, and 93 ft 8½ in. (28.3 m, 35.3 m, and 28.6 m). The east approach span is a two-span composite welded plate girder structure with spans of 91 ft 4½ in. (27.8 m). Figure 13.11 shows plan and elevation of the west approach spans.

The girder spans carry the eastbound and westbound traffic with a concrete median strip separating them. The truss spans are separate superstructures for the eastbound and westbound roadways. They are supported, however, on a single substructure. The first span and the truss spans were from existing bridges. The continuous girders were spaced at 9 ft (2.75 m) and supported a 6 in. (150 mm) reinforced concrete slab. Figure 13.12 shows the cross section of the continuous spans and the interior diaphragms. The X-bracing for the interior diaphragms were $5 \times 5 \times \frac{5}{16}$ in. ($126 \times 126 \times 8$ mm) angles that were welded to $5 \times \frac{5}{16}$ in. (126×8 mm) transverse stiffeners. All transverse stiffeners were only connected to the girder web with intermittent fillet welds.

The bridge structure was designed in 1951 to support H20-44 truck loading. The structure was built in 1952.

History of Cracking

During an inspection of the bridge in September 1973 several cracks were discovered [13.5]. These cracks were detected in the continuous plate girder spans 2, 3, 4, 10, and 11. The stiffener-to-web stitch welds were cracked at the stiffener-to-diaphragm connections, and the girder web was cracked along the web-to-flange fillet weld toe at the ends of the transverse stiffeners used to connect the diaphragm to the girder. Figure 13.13 shows two views of the typical condition that was observed at the diaphragm connection plates.

Most of the cracks occurred in the positive moment region adjacent to the compression flange at the web stiffener-diaphragm connection plate. Most of the cracks developed in the interior girders. Cracks were visible on both sides of the girder web. The largest cracks were located in span 3. In general, cracking was more severe in the interior girders than in the exterior girders.

Additional cracks were detected in 1976 [13.6]. Three basic types of cracking were observed at the diaphragm connection plate web gap. They appeared to occur in a random combination on both sides of the girder web. The conditions observed were as follows:

1. Cracks along the web-to-flange fillet weld toe.
2. Cracks in the girder web at the end of the upper intermittent fillet weld that propagated nearly parallel with the girder axis.
3. Cracks in the intermittent fillet welds. These cracks either propagated through the weld throat or along the leg attached to the girder web.

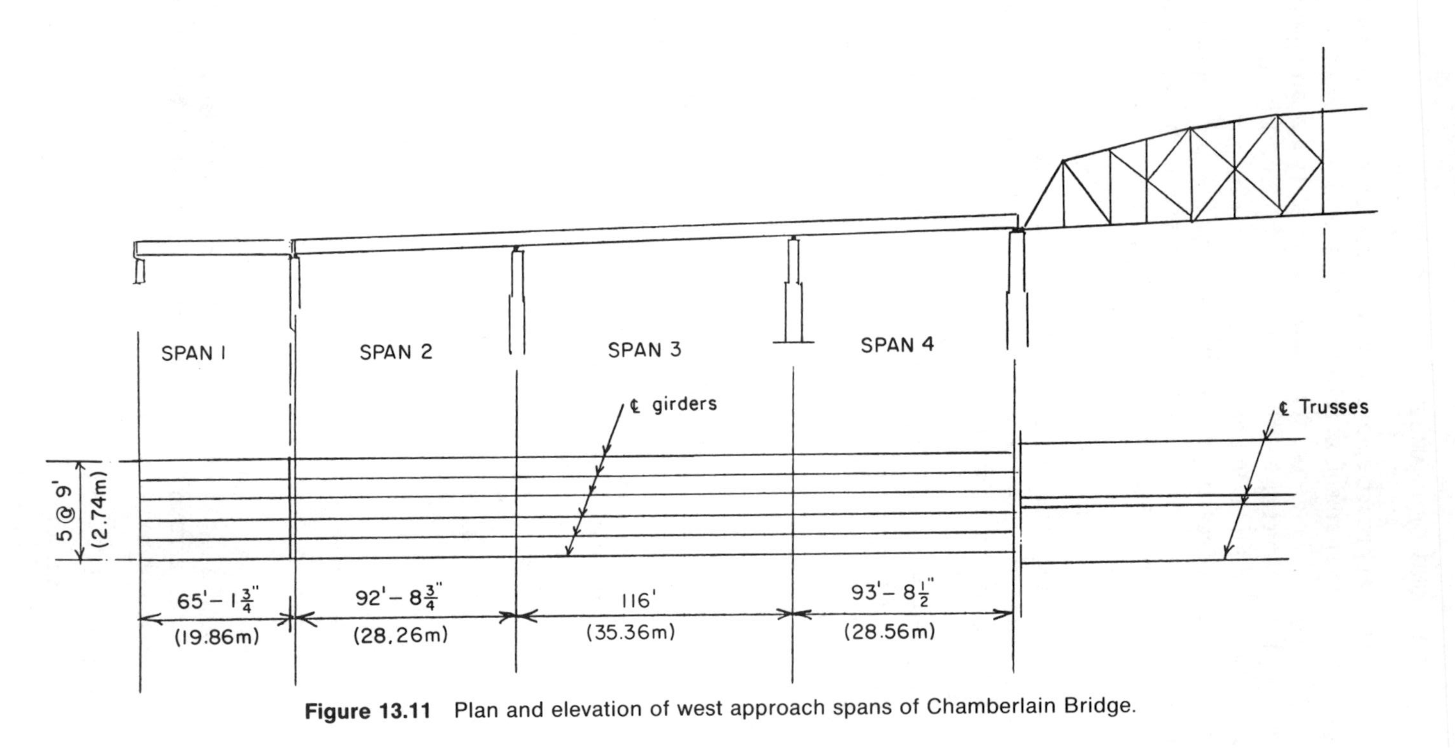

Figure 13.11 Plan and elevation of west approach spans of Chamberlain Bridge.

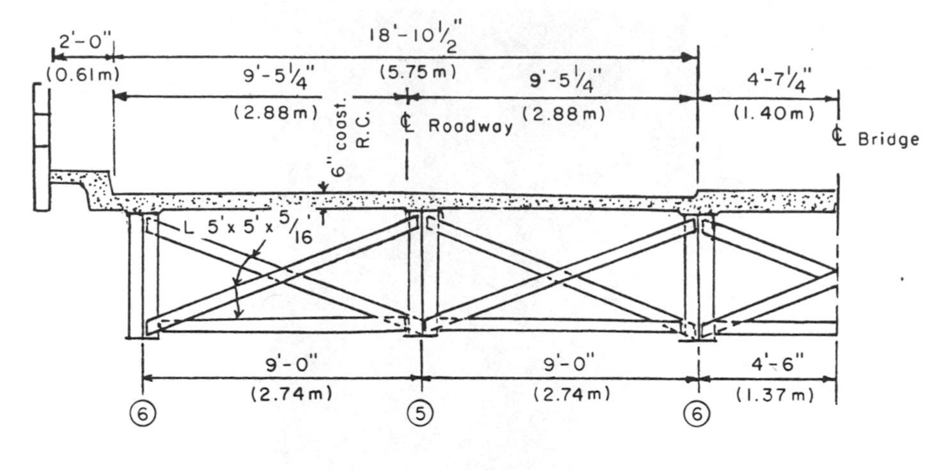

Figure 13.12 Typical cross section of girder spans.

The photographs in Figure 13.13 show the first and third types of cracking. Figure 13.14 shows a schematic of the types of cracking that were observed.

13.2.2 Failure Analysis

The average daily truck traffic using the bridge in 1960, 1965, and 1970 was 388 vehicles with dual tires and two or more axles. Assuming this count applies for the period between 1953 and 1973, 2.8 million variable load cycles result.

Figure 13.13 Typical cracks in the stiffener welds and web gap (courtesy of South Dakota Department of Transportation). (*a*) View of back of angle.

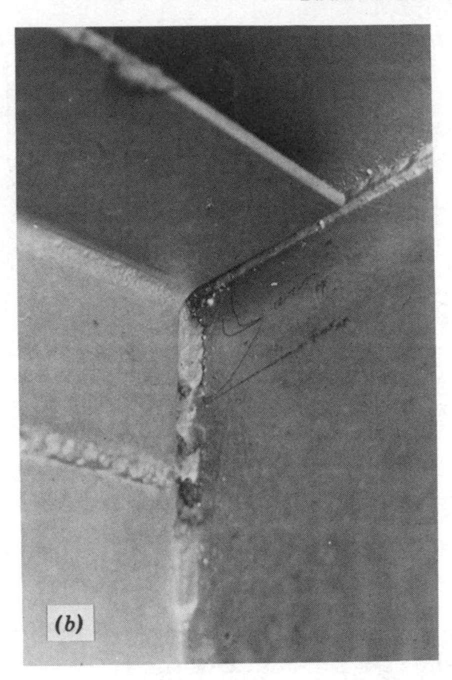

Figure 13.13 Typical cracks in the stiffener welds and web gap (courtesy of South Dakota Department of Transportation). (*b*) View showing angle lap.

No stress measurements or crack surface samples were available for examination, only the surface crack configuration. None of the cracked girder webs gave any indication of unstable rapid crack growth. Hence the material fracture toughness and environment had no significant affect on the observed cracking.

It is apparent that all cracks resulted from the out-of-plane distortion that developed in the web gaps as a result of differential vertical move-

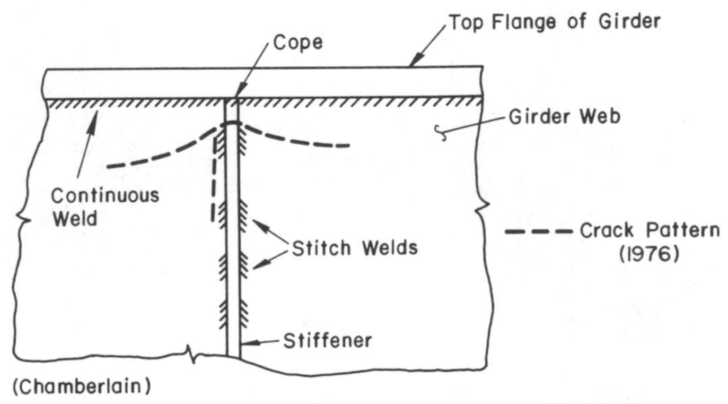

Figure 13.14 Typical crack pattern and weld details.

ment of the adjacent girders. The double curvature in the web gap created high bending stresses in the web along the web-flange weld toe and at the end of the transverse stiffener-web plate intermittent weld. The forces that developed from the deformation in the cross-frame were primarily transmitted from the connection plate into the web by the upper pair of intermittent fillet welds. Lack of fusion at these weld roots causes crack propagation from the root and severs the weld from the web.

Crack propagation at the weld toe is governed by Eq. 12.5, and as noted in the review of the Belle Fourche Bridge, an effective web gap stress range of 10 ksi (69 MPa) would result if 2.2 million stress cycles propagate through the web plate.

The conditions necessary to crack the upper pair of stitch welds from the web are defined by the approximation [6.5]

$$N = \frac{I}{3.6 \times 10^{-10} S_r^3} \tag{13.1}$$

where

$$I \sim \frac{[0.71 - 0.65(2a_i/t_p) + 0.79(H/t_p)]^3}{\sqrt{t_p}}$$

$$H = \text{weld leg} = 0.18 \text{ in. } (4.6 \text{ mm})$$

$$t_p = \text{stiffener thickness} = 0.32 \text{ in. } (8.1 \text{ mm})$$

This yields

$$N = \frac{0.1326}{3.6 \times 10^{-10} S_r^3}$$

If N is taken as 2.5×10^6 cycles, the effective stress range required to achieve that interval of life is 5.3 ksi (36.6 MPa).

13.2.3 Conclusions

All cracks observed in the web gaps between the end of the transverse stiffener-diaphragm connection plates were fatigue cracks caused by cyclic out-of-plane displacement. The cyclic displacement was due to the relative vertical displacement of adjacent girders under traffic. The cross-frame displaced the web gap out-of-plane under these conditions. Most of the cracks were in the interior girders. The girder web cracks were in a plane parallel to the longitudinal axis of the girder. Hence none of the cracks experienced unstable crack extension from the bending moments due to the applied loads.

13.2.4 Repair and Fracture Control

All existing web cracks were arrested by drilling $\frac{3}{4}$ in. (19 mm) holes at the ends of the cracks. Dye-penetrant was used to establish the location of the crack tips along the web-flange weld and at the ends of the transverse welds.

The top and bottom ends of all interior diaphragm connection plates were welded to the top and bottom flanges with 3 in. (75 mm) lengths of $\frac{5}{16}$ in. (8 mm) welds on each surface of the stiffeners. The welded connection prevented subsequent out-of-plane movement in the web gap.

REFERENCES

13.1 Fisher, J. W., and Mertz, D. R., Displacement Induced Fatigue Cracking of a Box Girder Bridge, IABSE, Symposium on Maintenance, Repair and Rehabilitation of Bridges, Washington, D.C., September 1982.

13.2 South Dakota-DOT Letter from A. R. Bachman, Bridge Maintenance Supervisor, to J. F. Wilsey, Engineer of Bridge Maintenance, dated August 10, 1976.

13.3 Tide, R. H. R., South Dakota Inspection Trip, A Brief Report to K. C. Wilson, Bridge Engineer, South Dakota-DOT, from R. H. R. Tide, AISC Regional Engineer, Minneapolis, dated October 5–6, 1976.

13.4 South Dakota-DOT Internal Communication from J. F. Wilsey, Engineer of Bridge Maintenance, to District Bridge Maintenance Supervisors, dated March 17, 1978.

13.5 Young, J. J., Wilsey, J. F., and Fiall, L. L., Report of Inspection of Missouri River Bridge at Chamberlain, South Dakota-DOT Internal Memo, September 27–28, 1973.

13.6 Wilsey, J. F., South Dakota-DOT Internal Memo to the District Bridge Maintenance Supervisors from J. F. Wilsey, Engineer of Bridge Maintenance, dated March 17, 1978.

CHAPTER 14

Tied Arch Floor Beams

A number of tied arch structures have been built with the floor beams framed into the tie girder with web shear connections alone. The tie girders are bending ties that are deeper that the floor beams. No direct connection is provided between the floor-beam flanges and the tie girders. In several older structures the floor-beam end connections are riveted double angles attached to the floor-beam web with the outstanding legs riveted to the tie girder. More recently, constructed arches have utilized transverse welded connection plates on the tie girder webs. The floor beams are bolted to the welded transverse connection plate.

Cracks have formed in the floor beams of structures with either riveted end connections or welded transverse connection plates. The cracks have formed in the floor-beam web along the web-flange connection at the floor-beam web gap that exists between the end of the connection angles and the end of the welded connection plates. These cracks extend parallel to the floor-beam flange along the length of the web gap and then begin to turn and propagate toward the bottom flange.

Figure 14.1 shows a crack that has formed in the floor-beam web above the riveted connection. Visual observations showed that a horizontal displacement in the longitudinal direction of the bridge developed between the tie girder and the top flange of the floor beam at the end connection. This relative movement produces out-of-plane deformation in the small web gap between the end of the floor-beam angle end connection and the top flange of the floor beam.

At least eight tied arch structures have experienced cracking in the floor beams along the web-flange connection of the top flange. In some of these structures the stringers rested on the top of the floor-beam flange. In others the stringers framed into the floor beams.

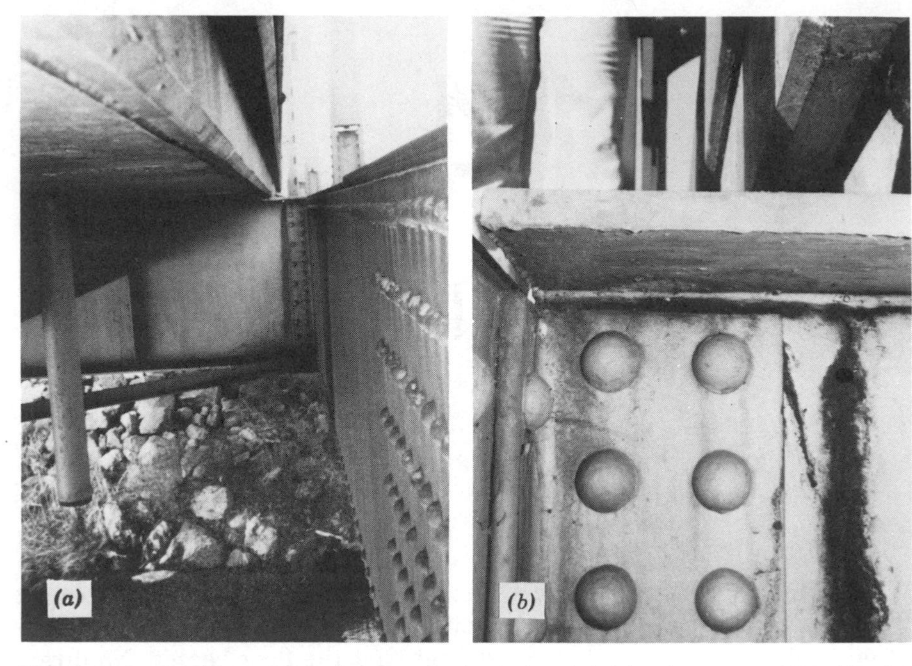

Figure 14.1 Cracking of floor-beam webs in tied arch bridge (courtesy of Washington Department of Transportation). (*a*) Floor-beam tie–girder connection; (*b*) crack above floor-beam end connection.

14.1 FATIGUE CRACKING OF PRAIRIE DU CHIEN FLOOR BEAMS

14.1.1 Description and History of the Bridge

Description of the Structure

The Prairie Du Chien Bridge carries U.S. 18 over the Mississippi River between Prairie Du Chien, Wisconsin, and Marquette, Iowa. The river separates into two channels at the crossing. The west channel is the navigation channel and is crossed by a 462 ft (140.8 m) tied arch span connected on each side to welded plate girder structures. The east approach spans are composed of two welded plate girders with seven 184 ft (56.1 m) spans and one 152 ft (46.3 m) span. There are two hinges in the eight girder spans—one in span 7 and the other in span 10. These can be seen in Figure 14.2 which shows the plan and elevation of the tied arch and plate girder approach spans. The west approach structures are four-span continuous plate girders with interior spans of 181 ft 6 in. (55.3 m) and side spans of 146 ft (44.5 m).

Typical cross sections of the tied arch span and the welded plate girder spans are shown in Figure 14.3. The roadway of both types of structures

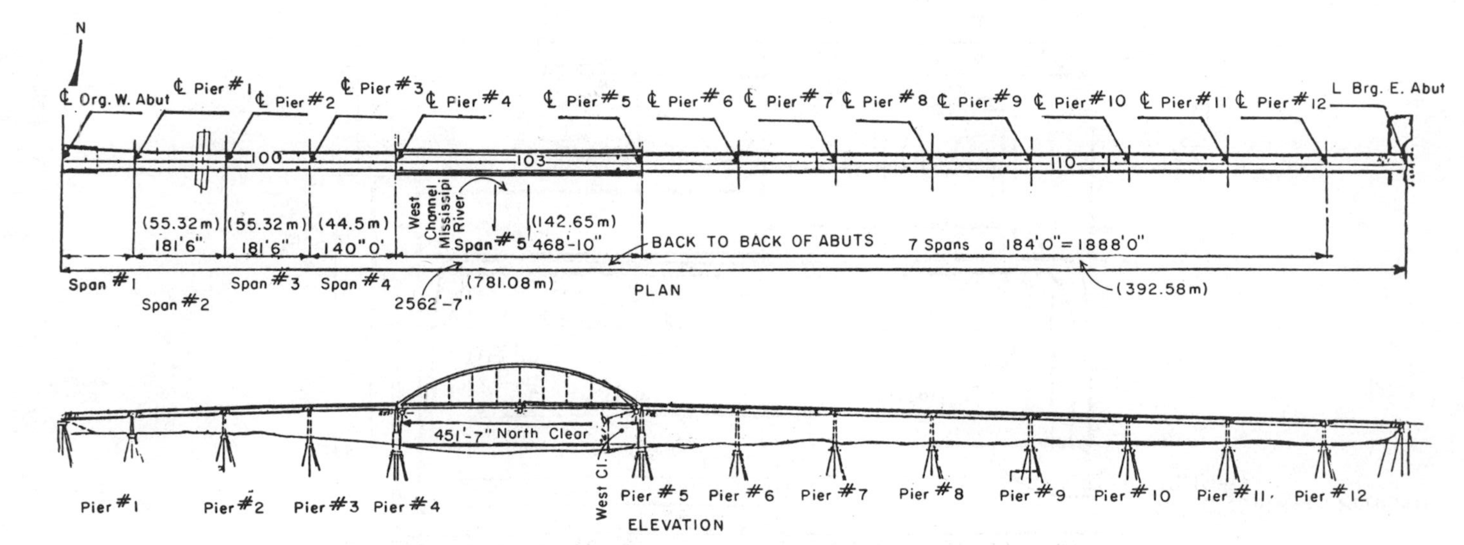

Figure 14.2 Plan and elevation of the tied arch and plate girder approach spans.

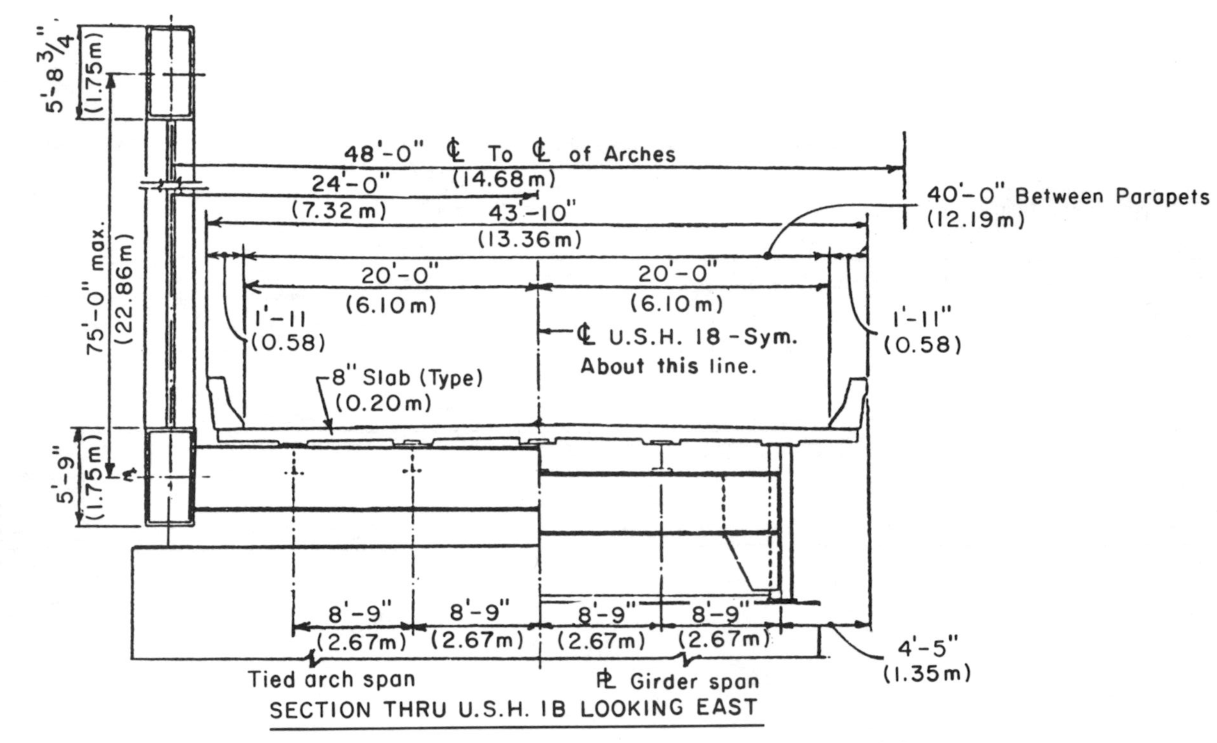

Figure 14.3 Cross sections of the tied arch and plate girder spans.

consists of an 8 in. (203 mm) thick reinforced concrete slab acting compositely with the longitudinal stringers that frame into the floor beams on the tied arch span. The stringers rest on the floor beams in the plate girder spans.

The steel superstructures are fabricated from A36 steel and A441 steel. The bridges were designed in accord with the 1969 AASHO specifications. Construction was started in October 1973 and completed in October 1974.

History of Cracking

During an inspection on August 30, 1979, several cracks were detected in the floor beams of the tied arch structure at the floor-beam–tie girder connection [14.1]. Figure 14.4 shows a schematic of the connection and

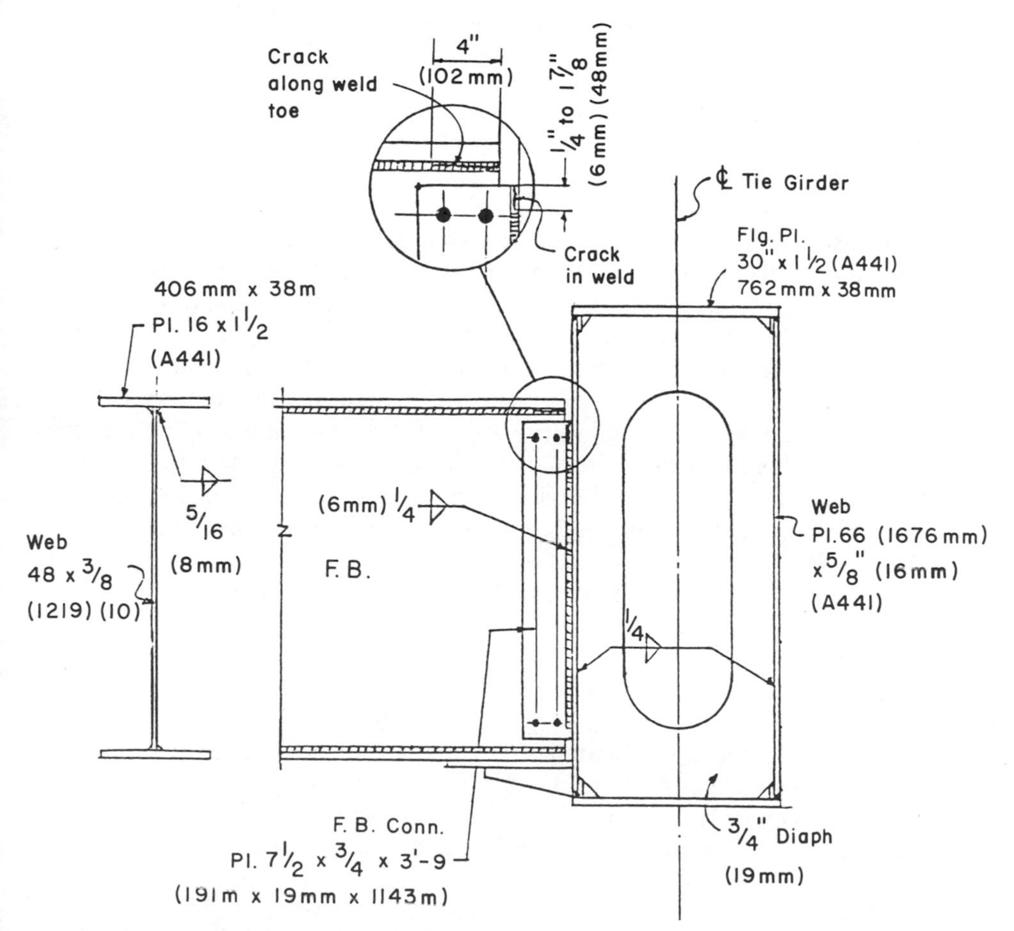

Figure 14.4 Schematic of floor-beam–tie girder connection showing location of cracks.

Figure 14.5 Cracks in floor beam and floor-beam connection plate (drilled holes are placed at crack tip) (courtesy of Wisconsin Department of Transportation).

the location of the cracks that were detected. Cracks were found in the floor-beam web at the web gap between the end of the connection plate and the flange-to-web weld. Cracks were also detected in the upper end of the fillet welds that attached the floor-beam connection plate to the tie girder web. Figure 14.5 shows a photograph of the cracked beam web and connection plate with $\frac{3}{4}$ in. (19 mm) holes drilled at the crack tips.

Cracks were also detected in the webs of the longitudinal plate girders at the floor-beam connection plates. These cracks were identical to the cracks that formed in the structures discussed in Chapter 12. Figures 14.6 and 14.7 show the connection detail and the crack that formed in the girder web above the floor-beam connection plate. All cracks were located in the negative moment regions where the connection plates were not connected to the girder tension flange.

14.1.2 Failure Analysis

A 1972 traffic survey indicated that the average daily traffic count was 3580 vehicles with approximately 400 trucks. By 1979 the ADT was 5500 with approximately 590 trucks. Hence in its five years of service up to the discovery of cracks, about 1,000,000 trucks had crossed the structures.

None of the cracks near the floor-beam connection appear to be influenced by temperature or other environmental conditions. All cracks are fatigue cracks that have formed as a result of high cyclic stress.

Figure 14.6 Floor-beam girder connection near interior pier.

The vertical cracks in the floor-beam connection plate–tie girder welds are due to the restraint at the ends of the floor beams. The torsional resistance of the tie girder and the rigidity of the bolted shear connection between the floor beam and the connection plate subject the upper end of the welded connection to high cyclic stresses. The section modulus of the floor-beam connection plate is only 20% of the section modulus of the floor beam. Since only $\frac{1}{4}$ in. (6 mm) fillet welds are attached to each side of the $\frac{3}{4}$ in. (19 mm) connection plate, a large initial crack exists between the weld roots. The conditions necessary for fillet weld cracking can be evaluated by the approximation [6.5]

$$N = \frac{[0.71 - 0.65(2a_i/t_p) + 0.79(H/t_p)]^3}{3.6 \times 10^{-10} S_r^3 \sqrt{t_p}} \tag{14.1}$$

where $2a_i = t_p = 0.75$ in. (19 mm) and $H = 0.25$ in. (6.4 mm).

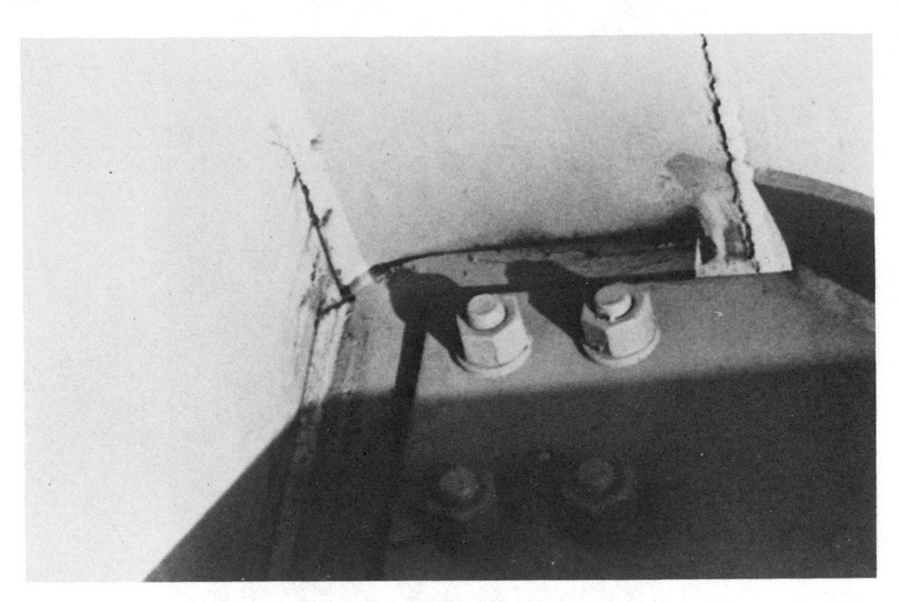

Figure 14.7 Crack in web gap at floor-beam connection (courtesy of Wisconsin Department of Transportation).

Equation 14.1 reduces to the relationship

$$N = 9.39 \times 10^7 S_r^{-3} \tag{14.2}$$

If N is set equal to 1,000,000 cycles, Eq. 14.2 yields an effective stress range of 4.5 ksi (31.4 MPa) for the nominal bending stress at the end of the connection plate. This could correspond to an end restraint bending stress of about 1 ksi (6.9 MPa) in the floor beam. This seems reasonable considering the geometry of the structure.

The horizontal crack along the floor-beam–web-to-flange weld toe is produced by the relative longitudinal movement of the floor system to the tie girder. This results from the general deformation of the structural system and causes an out-of-plane movement in the small web gap at the end shear connection plate, as shown schematically in Figure 14.8. The relative horizontal movement was visible on the Satsop Bridge, as was noted when discussing the crack shown in Figure 14.1. This movement occurred each time a truck crossed the structure.

The cracks that formed in the welded plate girders along the web-flange weld are identical to the cracks discussed in Chapter 12. They occur as a result of the out-of-plane movement that develops in the web gap when the floor beam deforms under load. The slab restrains the top flange, and the web gap is forced into double curvature. An analysis of this type of response was given in Chapter 12.

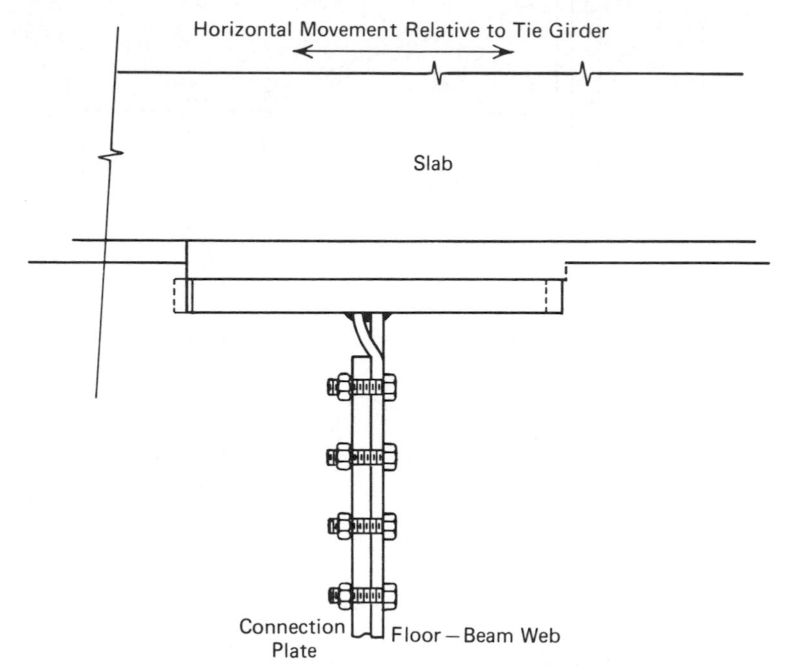

Figure 14.8 Schematic showing distortion of the floor-beam web gap.

14.1.3 Conclusions

The cracks that formed in the tied arch floor beams and the welded plate girder spans were all caused by cyclic out-of-plane movement of a small segment of the girder web. In the floor beams of the tied arch this results because of relative longitudinal displacement between the floor system and the tie girders. In the longitudinal welded plate girders the webs are forced out of plane by the end deformation of the girder span floor beams. The short web gap experienced a high cyclic bending stress that promoted early cracking. Only 1,000,000 vehicle crossings produced significant cracking in the floor beams of the tied arch span and in the plate girders of the approach spans.

14.1.4 Repair and Fracture Control

In order to prevent continued propagation of the cracks in the floor-beam web gap and the floor-beam connection plate–tied arch box girder, $\frac{3}{4}$ in. (19 mm) holes were drilled at the crack tip, as illustrated in Figure 14.5. The position of the crack tip was determined by dye-penetrant and magnetic-particle testing. All holes were positioned at the tip of the crack.

To prevent reinitiation of the cracks, T-sections were used to connect the floor-beam flange to the tie girder web and internal diaphragms, as

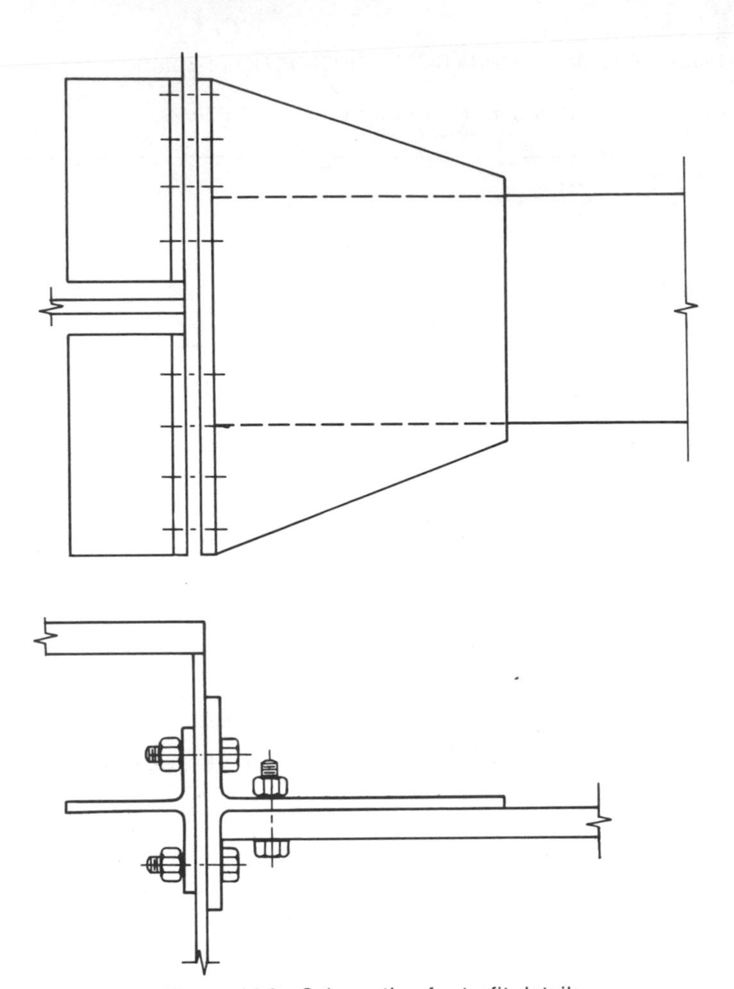

Figure 14.9 Schematic of retrofit details.

Figure 14.10 Retrofit connections between floor-beam flange and tie girder. (*a*) T-section and channel inside tie girder at diaphragm.

Figure 14.10 Retrofit connections between floor-beam flange and tie girder. (*b*) T-connecting floor-beam flange to tie girder.

shown schematically in Figure 14.9. Photos of the T-sections are given in Figure 14.10. Figure 14.10*a* shows the interior diaphragm–box girder web and the T- and channel sections used for a positive attachment. The exterior T-connection between the box girder web and the floor-beam flange is shown in Figure 14.10*b*.

The retrofit procedure adopted for the welded plate girders was identical to the scheme outlined in Section 12.2. Lengths of the floor-beam connection plates in the negative moment regions were removed along with the haunched portion of the floor-beam end bracket. This work was carried out in 1980.

REFERENCE

14.1 Brakke, B. L., Memorandum to D. S. Cook, Assistant Division Administrator, FHWA-Iowa, October 12, 1979.

CHAPTER 15

Stringer-To-Floor Beam (Truss) Brackets

Stringers supported on floor beams of deck truss bridges or on floor-beam trusses of suspension bridges have experienced cracking in the stringer web above web brackets that connected the stringers to the floor beam or floor-beam truss. At least two suspension bridges and three deck truss structures are known to have developed cracks in the stringer webs. The orientation of the cracks and the similarity of the connection details suggest that similar out-of-plane behavior is occurring in both types of structures. All five structures utilized rolled stringers that were connected to the floor beam (truss) with a riveted bracket. The brackets were riveted to the stringer web and to the top flange of the floor beam.

The out-of-plane behavior at the stringer web connection was verified by field measurements on the Walt Whitman Suspension Bridge [15.1]. Those measurements confirmed that out-of-plane bending stresses large enough to initiate fatigue cracks from the surface of the stringer web were developed at the web gap above the bracket attached to the stringer web.

15.1 FATIGUE CRACKING OF THE STRINGER WEBS OF WALT WHITMAN BRIDGE

15.1.1 Description and History of the Bridge

Description of Structure

The Walt Whitman Suspension Bridge spans the Delaware River between Philadelphia and Gloucester, New Jersey. The structure has a 2000 ft (610 m) center span and 769 ft $9\frac{1}{2}$ in. (234.6 m) side spans. The structure provides a 79 ft (24 m) wide roadway. Figure 15.1 shows the plan and elevation of the structure. Floor-beam trusses spaced at 20 ft 2 in. (6.15 m) support 16 longitudinal roadway stringers which are continuous over six spans. Figure 15.2 shows the cross section at a floor-beam truss. The longitudinal stringers support a concrete-filled steel grid deck which is covered with a $2\frac{1}{2}$ in. (64 mm) bituminous concrete overlay.

The structure was constructed between 1954 and 1957.

History of Cracking

In October 1968 during a routine inspection cracks were discovered in several stringer webs and in the attached angles of the stringer–floor-beam brackets, as illustrated in Figures 15.3 and 15.4. The brackets were attached to stringers 2, 5, and 8 on each side of the centerline of the roadway.

As can be seen in Figure 15.3, the cracks were in the stringer web along the web-flange fillet. All cracks occurred at the roadway relief joints where the stringers and the slab were discontinuous, as can be seen in Figure 15.3. No cracks were observed at the brackets attached to the stringers at floor beam trusses where the stringers were continuous over the support. The cracks were also observed to develop at every floor-beam stringer bracket across the roadway width.

The largest cracks were found to occur in the outer lanes of the roadway where gaps were observed between the bottom of the stringer and the floor-beam truss support pads. In beams that were subsequently shimmed, the cracks appeared to arrest after they had propagated away from the web gap, as seen in Figure 15.4.

Cracks were also observed in the connection angles that attached the bracket plates to the stringer web and floor-beam truss. Figure 15.4 shows one such crack in the angle attached to the stringer web. The cracks passed near the edge of a rivet head and extended to the top of the angle. The crack location was comparable when the crack formed in the angle attached to the floor-beam truss.

Most of the cracked angles were located in the brackets under the outside lanes where gaps had developed between the bottom of the string-

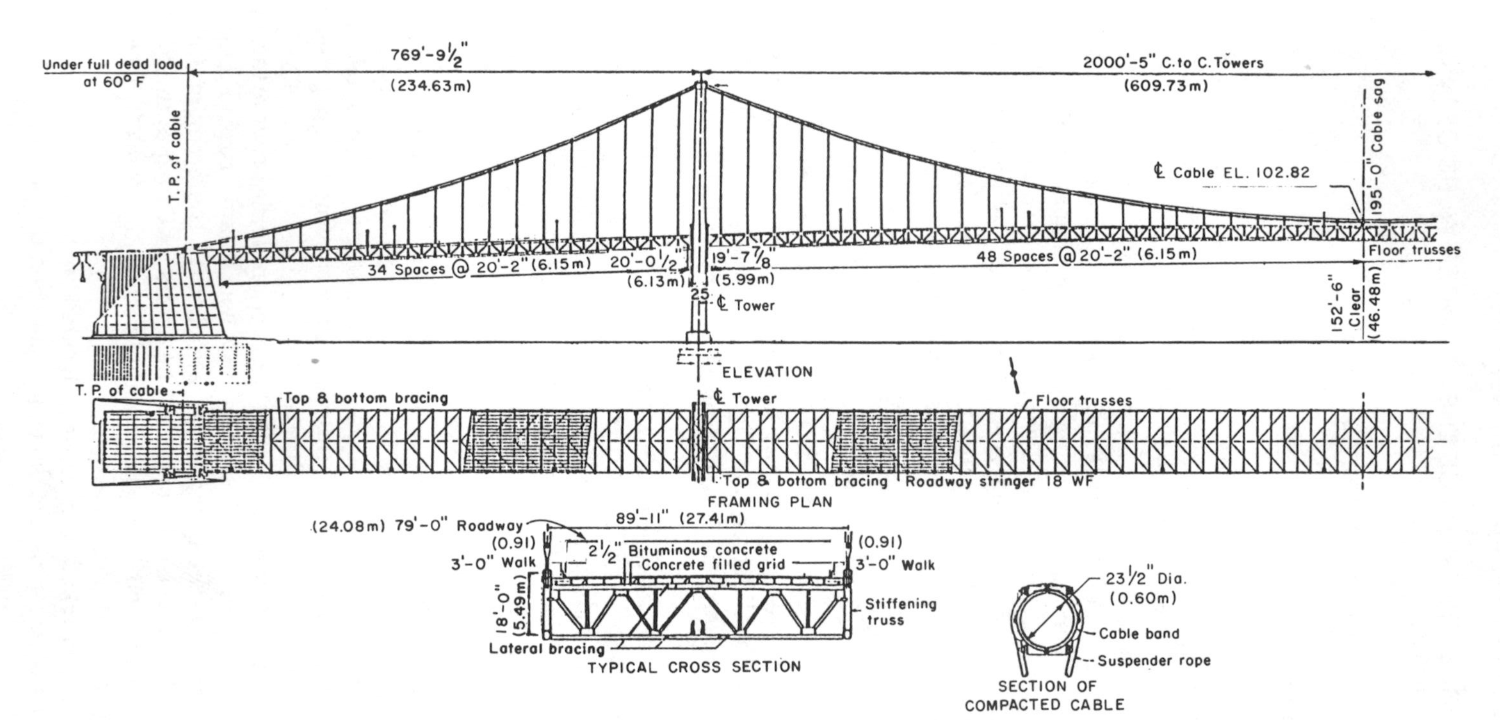

Figure 15.1 Plan and elevation of the Walt Whitman Bridge.

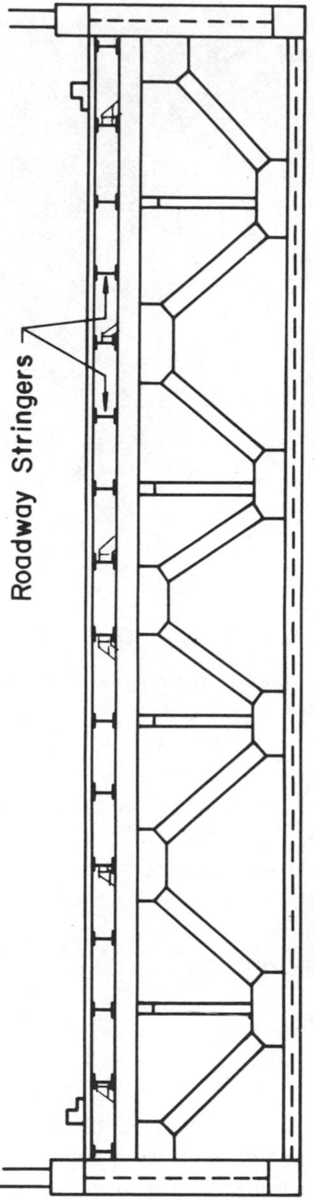

Figure 15.2 Typical cross section showing stringers and floor-beam truss.

Figure 15.3 Cracks in stringer web above bracket at roadway relief joint (courtesy of Delaware River Port Authority).

ers and the floor-beam truss support pads. At the roadway relief joint bolts were used in slotted holes to attach the stringer to the floor-beam truss. At other locations the stringers were riveted directly to the floor-beam truss.

15.1.2 Failure Modes and Analysis

Cyclic Loads and Stresses

A review of toll records indicated that about 3.6 million trucks with three or more axles had crossed the structure between the time it was opened to traffic in 1957 and when the cracks were discovered in 1968. Initial analytical studies did not provide a satisfactory explanation of why the cracks had formed. As a result two sets of experimental studies were carried out to assist with determining the causes of cracking. These measurements acquired information on the strain in the web gap in the constructed condition and after several structural modifications. The modifications included removing rivets and replacing them with loose bolts, removing a portion of the web angle in order to increase the web

Figure 15.4 Close-up view of crack in stringer web and in bracket angle (courtesy of Delaware River Port Authority).

gap, and installation of "rigid" diaphragms between the floor-beam truss and slab.

Figure 15.5 shows the strain-time response of a stringer web gap near the center of the roadway, as several trucks cross the structure in the north and southbound lanes. The measured stress range in gauge *A* can be seen to be 900 microinches (27 ksi, or 186 MPa), and it develops over a significant period of time. About 40 sec elapse between the beginning and end of the traces shown in Figure 15.5. The strain gradient at a point in time is also shown in Figure 15.5 and demonstrates that out-of-plane web bending is developing in the stringer web gap. When extrapolated to the stringer fillet, the web bending strain can be seen to exceed the yield point of the A36 steel stringers. The long time for the stress cycle to develop indicates that one of the primary reasons for the out-of-plane bending stress cycle is the torsional response of the overall structural system.

When modifications were made to the stringer floor-beam bracket, significant reductions in the out-of-plane web bending strain were achieved. This is illustrated in Figure 15.6 where the original design response is compared with the results of two modifications. Removing the rivets and

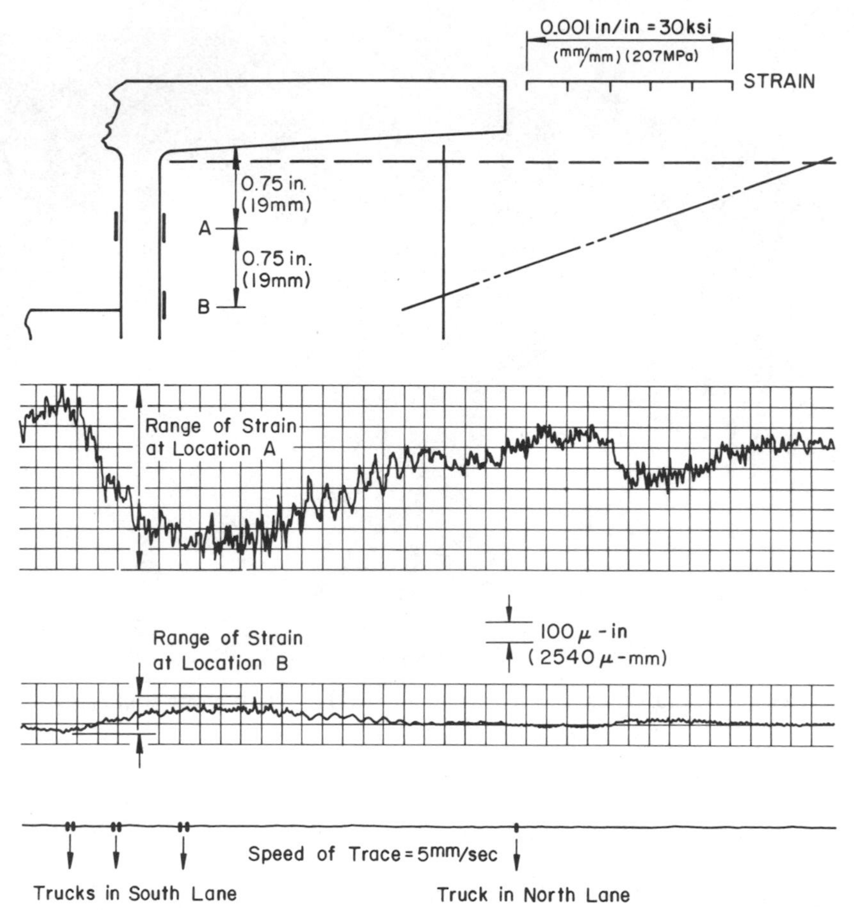

Figure 15.5 Typical strain response in the stringer web gap of the as-built structure.

replacing them with loose bolts significantly reduced the magnitude of strain range. Cutting the bracket in order to increase the length of web gap also reduced the strain range and changed the strain gradient.

Installation of a "rigid" diaphragm connection of the type shown in Figure 15.7 did not significantly reduce the magnitude of the strain range in the stringer web gap. Although some reduction was observed, it was not enough to warrant further consideration.

Failure Analysis

Visual observation of the cracks in the stringer webs indicated that all cracks initiated from the web surface in the gap between the stringer fillet and the lateral support bracket. None of the crack surfaces were removed for examination. However, the crack shape and location demon-

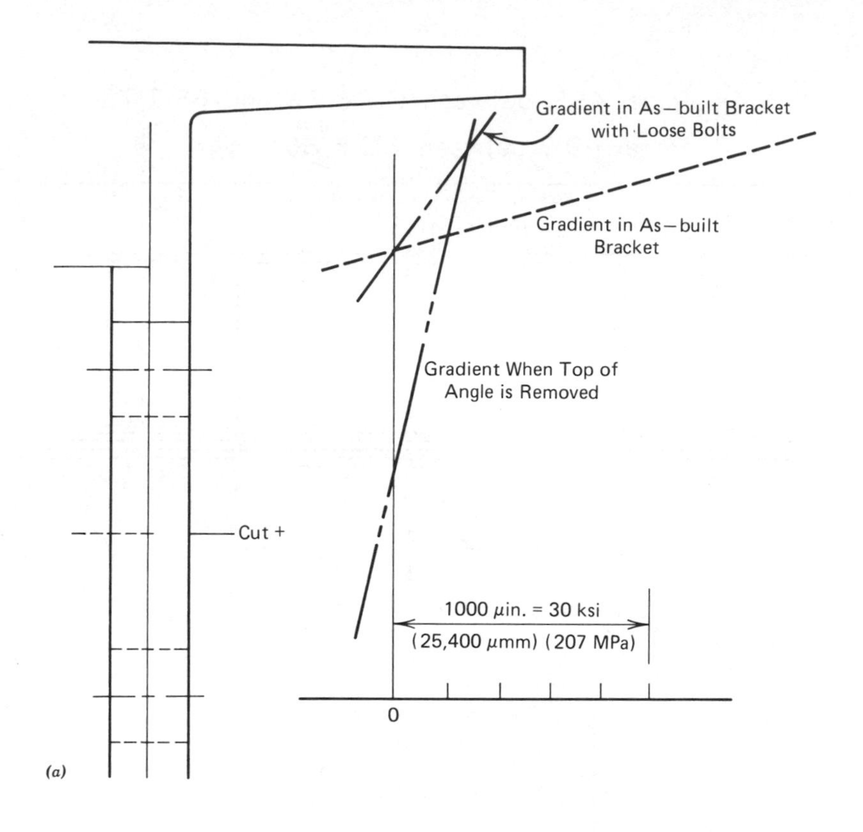

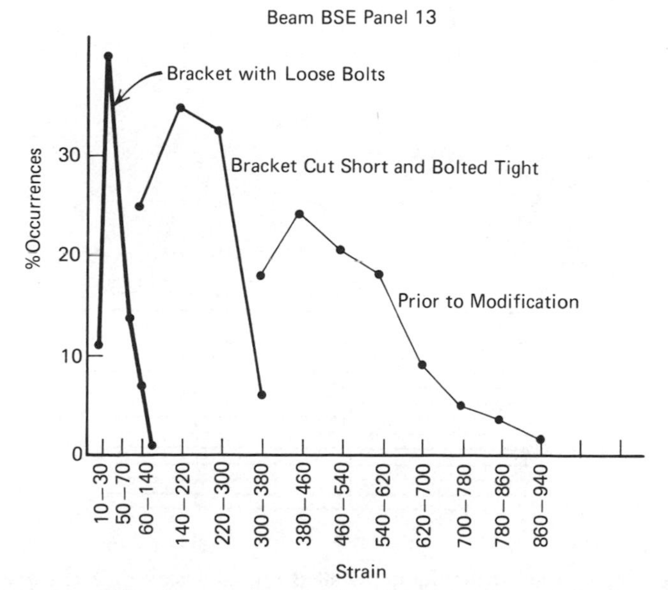

Figure 15.6 Effect of modifications to stringer–floor-beam bracket. (*a*) Strain gradients for as-built and modified connections; (*b*) strain range frequency of occurrence for as-built and modified connections.

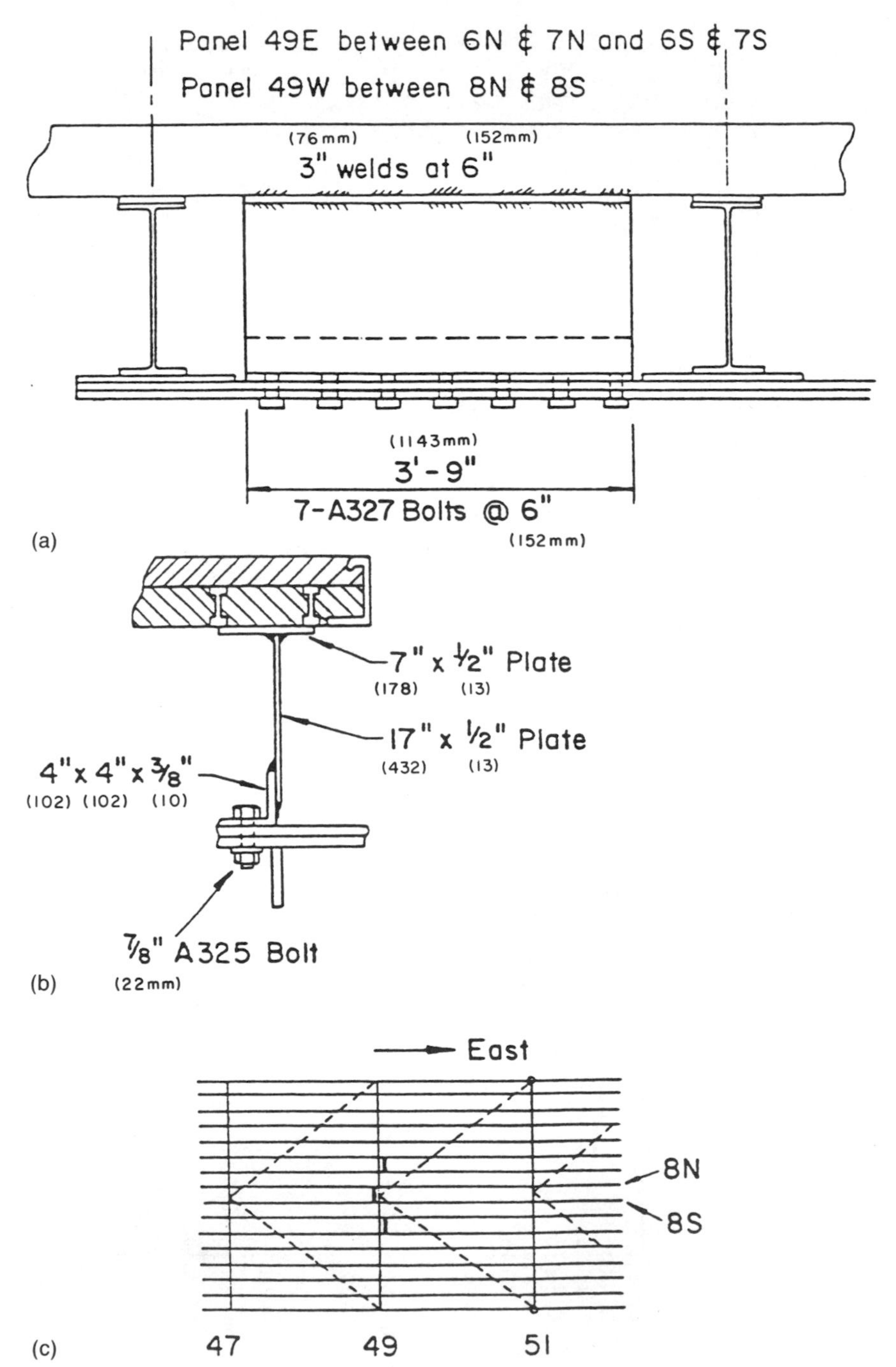

Figure 15.7 "Rigid" diaphragms installed on each side of a roadway relief joint. (*a*) Elevation of diaphragm connection; (*b*) cross section of diaphragm; (*c*) plan showing diaphragms installed at Panel 49.

strated that all of the cracks were created by cyclic loading and were fatigue cracks.

The field measurements indicated that there were three major causes of the out-of-plane distortion in the web gap above the stringer–floor-beam lateral support bracket. The major cause of the distortion and high cyclic stress was found to be the torsional response of the suspended floor system. As was indicated in Figure 15.5, twisting of the torsion box composed of the floor beam and longitudinal trusses and the lateral bracing systems relative to the slab and stringer system introduced a relative movement at the roadway relief joints, which is shown schematically in Figure 15.8.

The lateral connection bracket between the stringer and floor-beam truss was designed to accommodate longitudinal expansion of the stringer. The $\frac{3}{8}$ in. (9.5 mm) plate connected to the angle attachments provides very little restraint to longitudinal movement. However, it acts as a rigid link for vertical or lateral movement between the floor-beam truss and the stringer web. Hence the web gap is forced out of plane when such movement occurs.

The other major cause of out-of-plane movement was the gap that develops at the roadway relief joints between the bottom stringer flange and the floor-beam truss from wear on the bearing pad. Vehicles crossing the roadway relief joint force the gap closed and introduce an out-of-plane displacement into the stringer web. These displacements occur a large number of times after the gap develops, as even the lighter trucks force the gap closed. This type of condition was mainly confined to the outer roadway lanes where most of the truck traffic crossed the structure.

The final source of movement was due to the shortening of the floor-beam truss compression chord, as it deformed somewhat differently than the slab and stringers. This contribution was small and not as significant as the other two sources of distortion.

The strain history measurements acquired in 1969 and 1970 were used to estimate the effective stress range at the web-flange fillet. An effective stress range was estimated as:

$$S_{r\mathrm{Miner}} = (\Sigma \alpha_i S_{ri}^3)^{1/3} \tag{15.1}$$

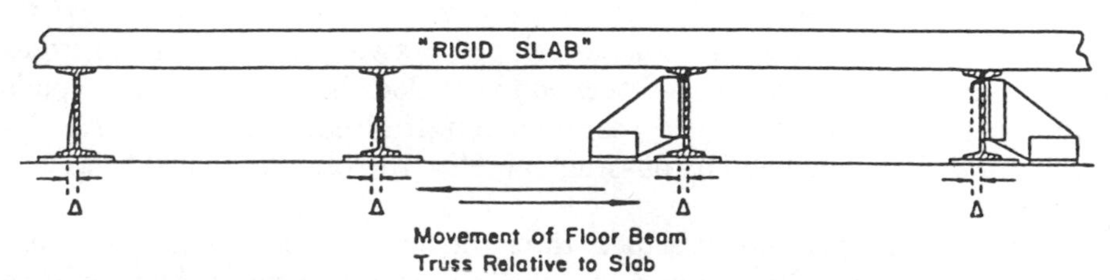

Figure 15.8 Relative deformation between slab and floor-beam truss at roadway relief joints.

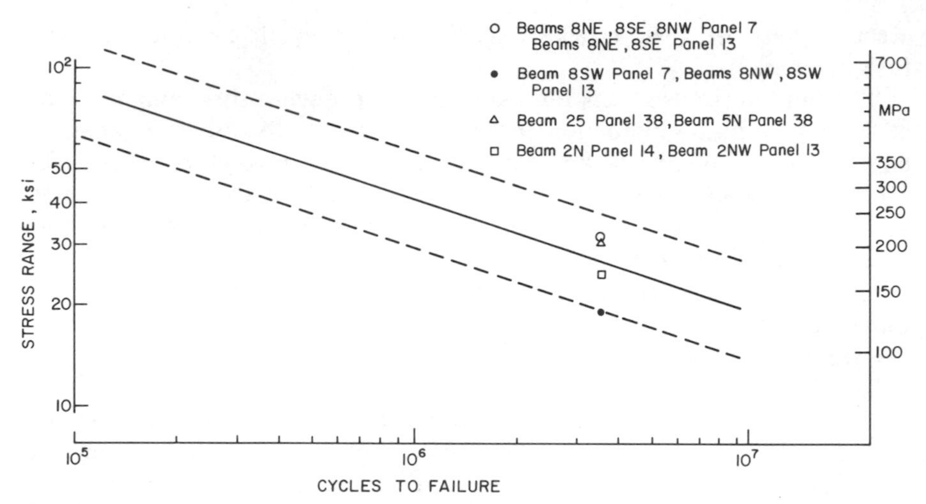

Figure 15.9 Comparison of the effective stress range with the fatigue resistance of rolled beams.

This yielded values between 19 and 35 ksi (131 and 241 MPa) for the web gap at the roadway relief joints. The values of the effective stress range are plotted in Figure 15.9 and compared with the mean and confidence limits developed from fatigue tests on rolled sections. The comparison confirms that cracks would be expected to develop as a result of the high out-of-plane web bending stresses under traffic.

15.1.3 Conclusions

All cracks that formed in the stringer webs or in the angles of the lateral support brackets were caused by out-of-plane distortion between the stringer and the floor-beam truss. The stringer web cracks resulted from the double curvature deformation introduced into the web gap between the lateral support bracket and the top flange of the stringer. The out-of-plane deformation was primarily caused by the torsional distortion of the suspension bridge structure relative to the stringer-slab system. The relative deformation that occurred at the roadway relief joints where the slab and stringers were no longer continuous produced the most severe amount of movement. Out-of-plane movement of the web gap also resulted from the clearances that developed between the bottom stringer flanges and the bearing pads attached to the floor-beam truss. These pads were observed to wear as a result of longitudinal movement. As the stringer was forced down by passing vehicles, the bracket pushed the web out of plane.

Strain measurements verified that the out-of-plane bending stress could be substantially reduced by releasing the connection between the floor beam and the stringer web. The measurements also demonstrated

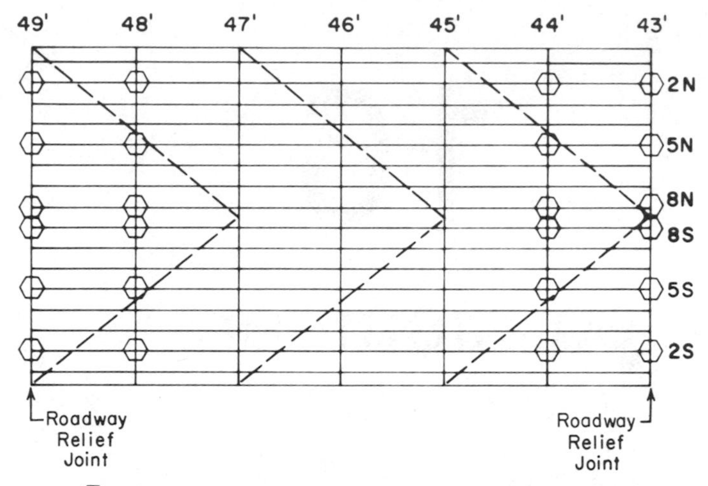

Figure 15.10 Stringer–floor-beam truss lateral support brackets which were eliminated between roadway relief joints.

that it was not possible to prevent the lateral displacement by the installation of rigid diaphragms at the roadway relief joints.

All cracks were fatigue cracks which were initiated on the surface of the stringer web as a result of the very high stress range produced in the gap by truck traffic on the structure.

None of the cracks reduced the load-carrying capability of the stringer and floor system.

15.1.4 Repair and Fracture Control

Since none of the cracks seriously affected the load-carrying capability of the structure, the initial retrofit consisted of removing the gaps between the stringer flange and floor beam by providing shims. Yearly inspections were also carried out in order to monitor the crack growth.

After modifications to the lateral support system were studied and the experimental measurements verified that releasing the lateral support brackets would decrease the cyclic stress and prevent further crack growth, several bays had their lateral support brackets modified. Figure 15.10 shows the locations where the connection between the stringer webs and the floor-beam truss was eliminated by cutting through the $\frac{3}{8}$ in. (9.5 mm) lateral support plate. All connections at the roadway relief joints and the adjacent floor beam truss were released.

REFERENCE

15.1 Fisher, J. W., and Mertz, D. R., Fatigue Cracking in Longer Span Bridges, *Ann. New York Acad. Sci.* 352 (December 1980):193–218.

CHAPTER 16

Coped Members

Coped stringers, floor beams, and diaphragms have been commonly used in bridges to frame one member into another. The coped member may have one or both of the flanges removed, depending on the particular application.

In railroad bridges the stringers were sometimes coped at the bottom flange to permit the stringer to frame into the floor beam. A number of structures of this type have developed cracks because of the fixity of the end connection. Removing the flange from the beam reduces the cross-sectional bending resistance by 80 to 90%. As a result large cyclic stresses can be developed in the coped section web plate. Since the cope is usually made by flame cutting, a high residual tensile stress exists along the burned edge. Hence cracks can be initiated under cyclic compression and can continue to propagate due to the principal stress that results from the end shear.

It has been reported that about 30 Japanese railroad bridges on the high speed line developed cracks at coped flanges of stringers and floor beams [2.2]. These structures were all built between 1965 and 1968.

Similar behavior has occurred in a number of highway bridges. Usually it has involved coped stringers that frame into floor beams or an end diaphragm. Negative bending moments produced by a cantilever arrangement or by continuity across the floor beam have resulted in cracks originating from the cope. At least 10 highway bridges have developed cracks in the stringer webs that originate at these copes [1.1]. Cracks have also occurred at coped ends of spans where hangers were used to support suspended structures or where expansion joints were framed onto the girders. These cracks result from the tension bending stress introduced at the cope. At least three bridges in the United States have developed cracks at this type of detail. Several bridges in Japan have also cracked at this type of detail [16.1].

16.1 FATIGUE CRACKING AT STRINGER COPES OF BRIDGE 51.5 WINDERMERE SUBDIVISION

16.1.1 Description and History of the Bridge

Description of Structure

Bridge 51.5 of the Windermere Division of the Canadian Pacific Railroad consists of two simple span structures crossing Dutch Creek near Ottawa, Ontario. The structures are single track through girder structures 104 ft 5 in. (31.8 m) long. The structures were fabricated from CSA G40.11 Grade B steel (A588).

Figure 16.1 shows the plan and elevation of one of the spans. Welded built-up plate girders 9 ft 4 in. (2.8 m) deep supported the floor-beam–stringer system. The bridge cross section is shown in Figure 16.2. The floor beams were fabricated from W33 × 130 rolled sections, and the stringers from W24 × 76 rolled sections. Figure 16.3 shows the details of the floor beam-stringer connections and geometry of the floor system. The structure was built in 1970.

History of Cracking

In November 1975 seventeen of the coped stringers were found to be cracked at the bottom flange where the stringers were coped to clear the floor beam. Figure 16.4 shows photographs of the coped flange and typical cracks that have formed from the flame-cut edge.

16.1.2 Failure Modes and Analysis

Cyclic Loads and Stresses

Between 1970 and 1975 it was estimated that about 1.3 million significant stress cycles had occurred at the stringer copes from trains crossing the structure.

The bridge was instrumented and tested during 1976. Strain gauge rosettes were installed on several stringers as indicated in Figure 16.5. Since the structure primarily carries unit coal trains, measurements were obtained from several unit trains that crossed the structure at varying speeds [16.2].

The measurements showed that as the stringer was being loaded, one major stress cycle occurred from the trucks of adjacent cars. This can be seen in Figure 16.6 which shows the principal stress-time response of a typical stress cycle at a point $\frac{3}{8}$ in. (9.4 mm) from the edge of the cope, as shown in Figure 16.5. The principal stress vector varied between +4° and −15° during the stress cycle shown in Figure 16.6. The corresponding

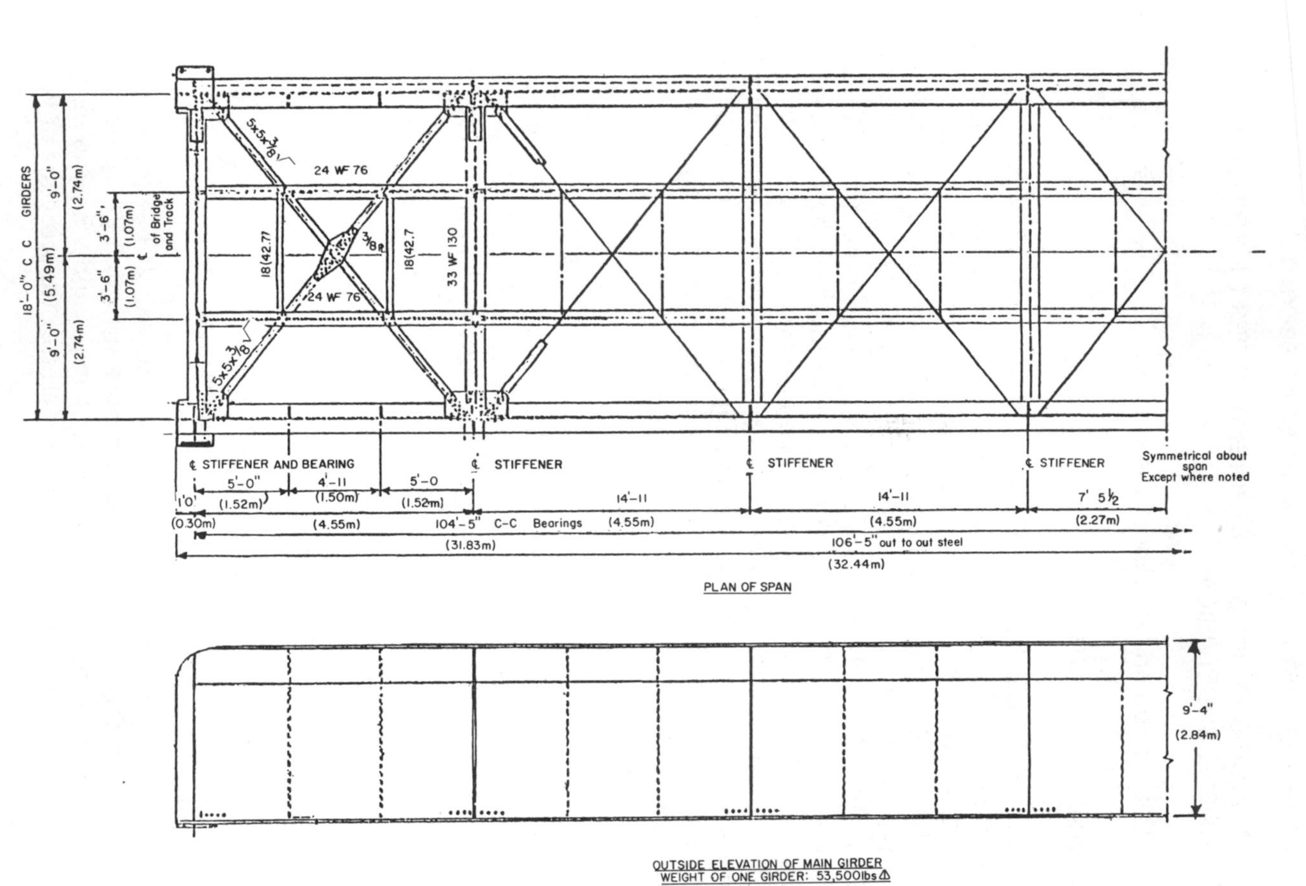

Figure 16.1 Plan and elevation of a girder span.

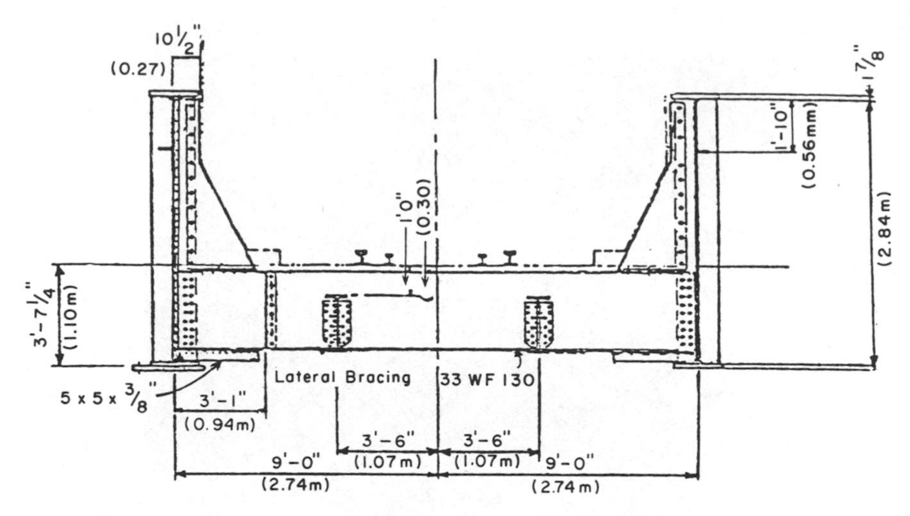

Figure 16.2 Cross sections of the structure.

stress range at the upper two gauges adjacent to the top uncoped flange of the stringer was about −7.5 ksi (52 MPa). This included the effect of bending from deformation of the ties.

The field measurements indicated that the speed of the train had only a minor effect on the magnitude of the cyclic principal stress range.

Of the estimated 1.3 million stress cycles, about 600,000 were produced by loaded 100 ton cars and the locomotives. An equal number of unloaded cars crossed the structures. An additional 100,000 cycles were estimated to have occurred as a result of mixed traffic.

Unloaded cars have been observed to produce about 30% of the stress range caused by a loaded car. Hence the estimated effective stress range at the gauge location for all traffic was estimated as

$$\begin{aligned} S_{r\,\mathrm{Miner}} &= \left[\frac{7}{13}\,(22)^3 + \frac{6}{13}\,0.3 \times 22)^3\right]^{1/3} \\ &= 18 \text{ ksi} \quad (124 \text{ MPa}) \end{aligned} \tag{16.1}$$

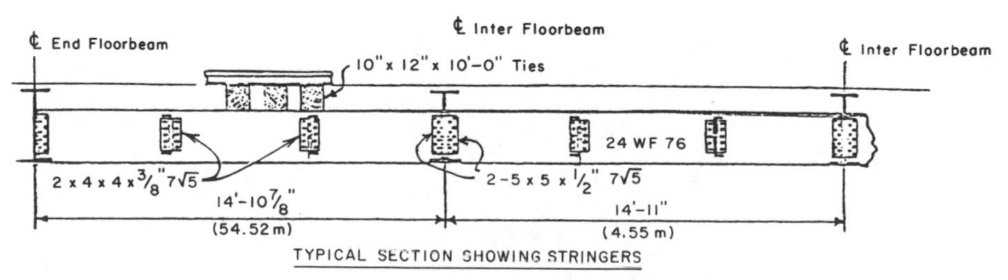

Figure 16.3 Floor-beam–stringer geometry.

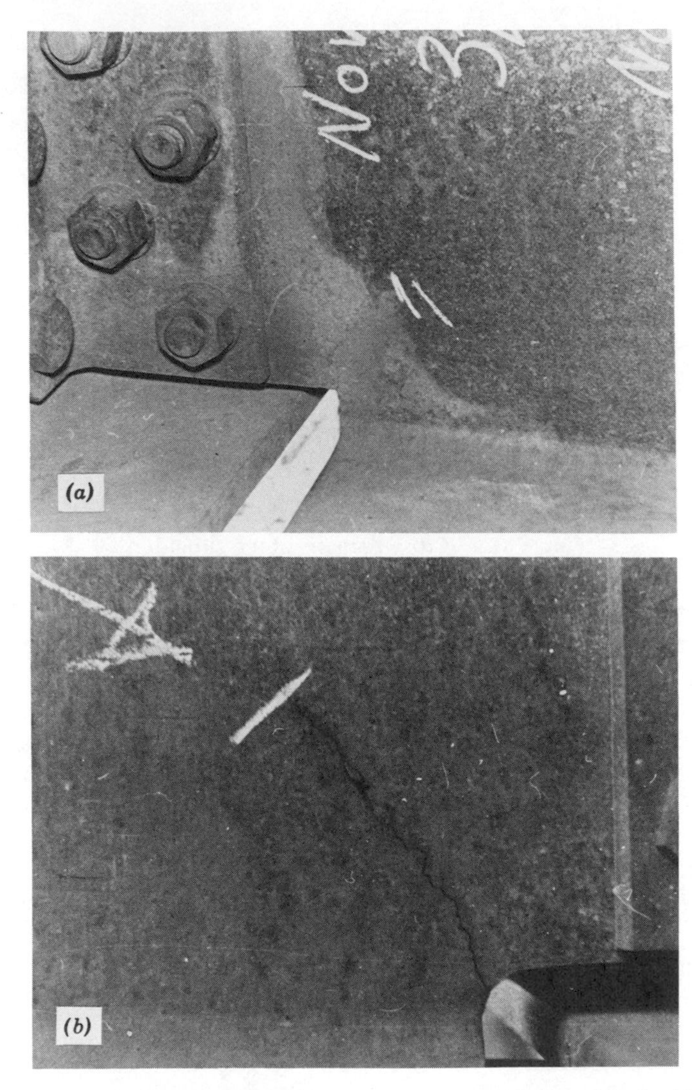

Figure 16.4 Cracks at the coped stringer detail (courtesy of Canadian Pacific Railroad). (*a*) View of the coped bottom stringer flange; (*b*) view of crack originating out of cope.

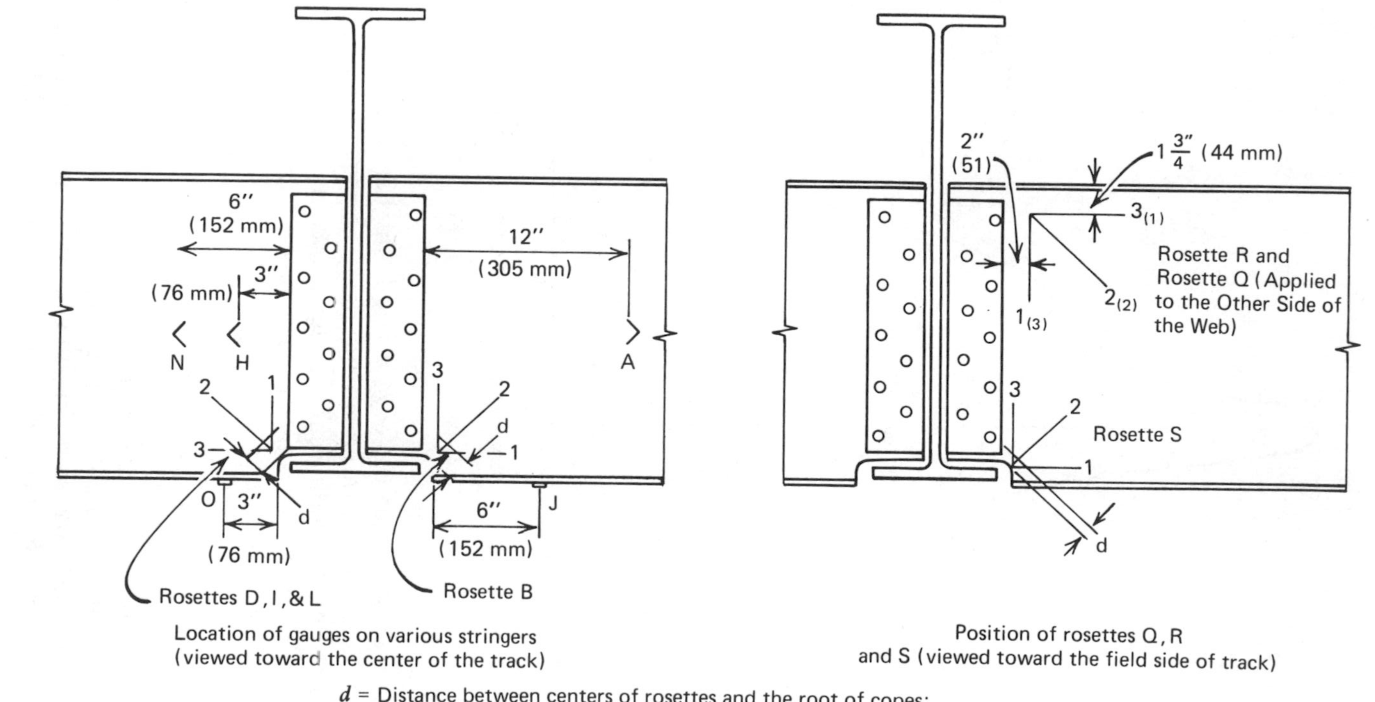

d = Distance between centers of rosettes and the root of copes:

For rosettes B, D, I, and L: $d = 1\frac{3''}{8}$ (41 mm), for rosette S: $d = \frac{3''}{8}$ (10 mm)

Figure 16.5 Details of the strain gauge placement. (*a*) Stringers in north span; (*b*) stringers in south span.

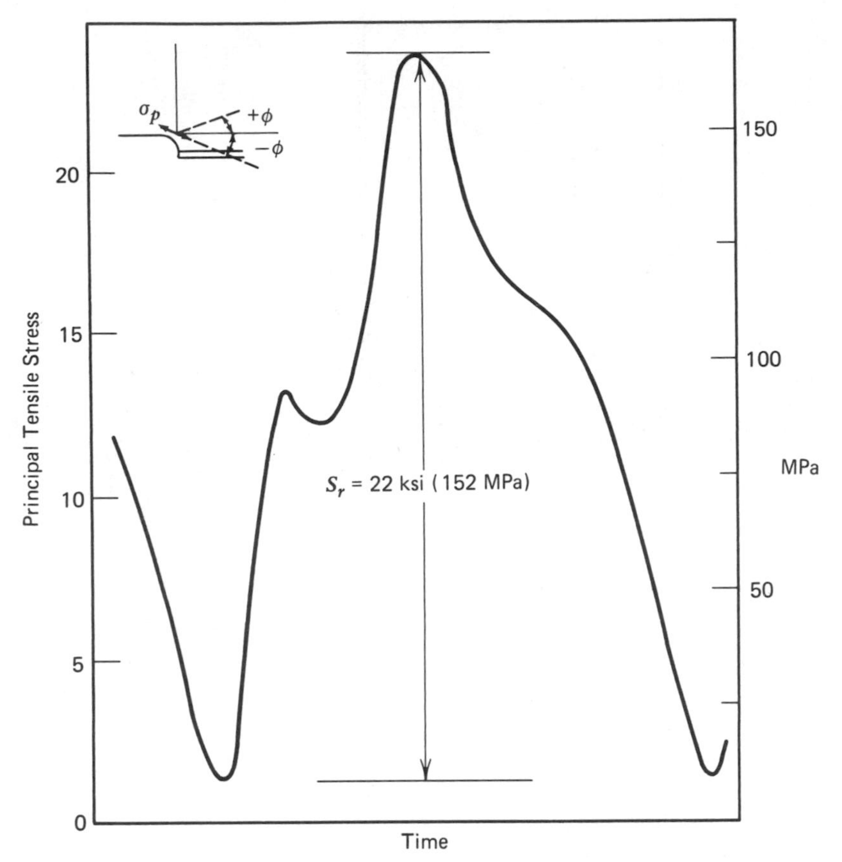

Figure 16.6 Typical principal tensile stress versus time at gauge S, located $\frac{3}{8}$ in. from cope.

The stress range at the flame-cut edge would be higher because of the increase in the stress concentration at the edge. The stress gradient reduction at a point $\frac{3}{8}$ in. (9.4 mm) from the edge was estimated to be about 50%. Hence the effective stress range at the flame-cut edge was estimated to be $S_{r\,\mathrm{Miner}} = 36$ ksi (255 MPa).

Failure Analysis

The nominal fatigue strength of flame-cut copes was estimated to be compatible with category C in [4.3]. Crack propagation from the flame-cut edge can be modeled by considering an edge crack in the martensite surface from flame cutting. Such defects are generally very shallow and wide.

The stem of the T-shaped section created when the flange is coped is subjected to increased bending stress as the section modulus is decreased.

For the W24 × 76 section the T-section provides a section modulus of 53 in.^{3}, whereas the full beam section provides 176 in.^{3}. Hence the bending resistance has been decreased by 70%. This is the reason for the large principal stress range at the coped section. Since bending is the primary reason for the high stress range at the flame-cut edge, the stress intensity factor was defined as

$$K = \left[\frac{0.923 + 0.199(1 - \sin \pi a/2b)^4}{\cos \pi a/2b}\right] \left(\frac{2b}{\pi a} \tan \frac{\pi a}{2b}\right)^{1/2} \sigma\sqrt{\pi a} \quad (16.2)$$

where the width b was taken as 21.5 in. (546 mm), the depth of the coped stringer.

The number of cycles required to propagate an edge crack from the flame-cut martensite surface to a depth of 4 in. (100 mm) was estimated as

$$N = \int_{a_i}^{a_f} \frac{da}{3.6 \times 10^{-10}\, \Delta K^3}$$

where a_i was estimated to be 0.015 in. (0.38 mm), the depth of the martensite, and a_f = 4 in. (100 mm). If the effective stress range is 18 ksi (127 MPa), Eq. 16.3 yields 955,000 variable stress cycles to propagate a 0.015 in. (0.38 mm) edge crack to 4 in. (100 mm). This is in reasonable agreement with the field observations.

The estimated effective stress range of 18 ksi (124 MPa) needs only to decrease by a small amount to correspond to the estimated 1.3 million variable stress cycles.

16.1.3 Conclusions

The primary cause of fatigue cracking at the stringer cope was the high cyclic stress that developed from continuity in the stringer at the bolted connection to the floor beam and the large reduction in bending resistance with the flange removed. The bending resistance of the coped section was only 30% of the bending resistance of the full cross section. Strain measurements demonstrated that a large principal tensile stress range occurred during the passage of each car or locomotive. The flame-cut edge provides a stress concentration and an initial crack condition because of the hard martensite layer produced by the cutting.

Cracking was predicted to develop at the flame-cut edge as a result of the estimated 1.3 million stress cycles and the high level of tension stress range that developed from the stringer continuity and the differential floor-beam deflection.

Cracking could also develop under a compression stress cycle because the flame-cut edge would create a region of yield point tensile residual

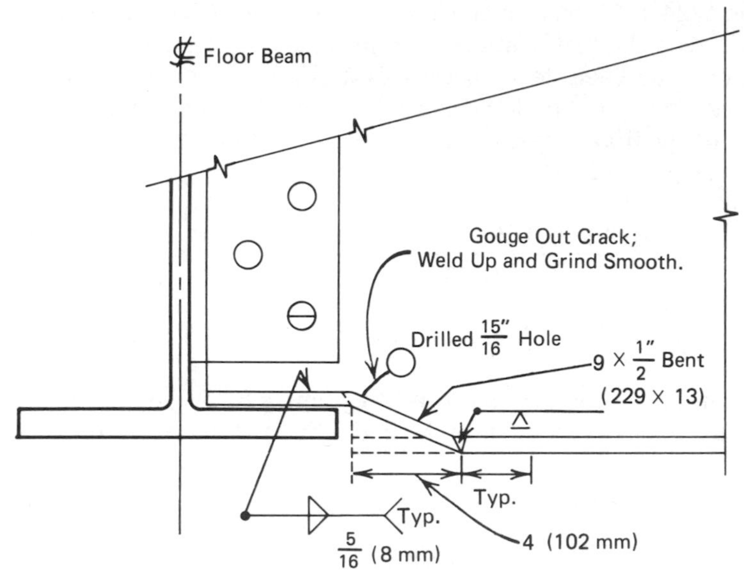

Figure 16.7 Retrofit detail showing reinforcement of cross section.

stress. Eventually, as a result of the high shear and tensile principal stress, crack growth would place the crack tip in a region of cyclic tensile stress.

16.1.4 Repair and Fracture Control

Initially $\frac{15}{16}$ in. (24 mm) holes were drilled at the crack tips to arrest propagation. The final retrofit is shown in Figure 16.7. The crack between the cope and hole was removed by air arc and then repaired by welding. The flange cope was cut back about 4 in. (100 mm), and a 9 × $\frac{1}{2}$ in. (229 × 12.7 mm) plate was bent to shape and groove welded to the bottom flange of the stringer, and then fillet welded to the web of the stringer. This restored a significant portion of the section's bending resistance and reduced the cyclic stress below the fatigue limit.

REFERENCES

16.1 Nishimura, T., Fracture in Welded Bridges, *J. Japan Welding Soc.* 10 (1978, in Japanese).

16.2 Kalousek, J., and Bethune, A. E., Stringer Stress Measurements on Bridge Mile 51.5, Windermere Subdivision, Department of Research, Canadian Pacific Railroad, Report 5538-76, August 1976 (limited distribution).

Author Index

Subject Index